Springer
Berlin
Heidelberg
New York
Barcelona
Hong Kong
London
Milan
Paris
Singapore
Tokyo

PLEASE ASK AT LENDING DESK FOR
ACCOMPANYING CD-ROM
UNIVERSITY OF NOTTINGHAM
10 0418961 6
WITHDRAWN
FROM THE LIBRARY

H. J. Korsch H.-J. Jodl

CHAOS

A Program Collection for the PC

Second Edition
With 250 Figures,
Many Numerical Experiments,
and CD-ROM for Windows 95 and NT

Professor Dr. H. J. Korsch
Professor Dr. H.-J. Jodl
Fachbereich Physik, Universität Kaiserslautern
Erwin-Schrödinger-Strasse
D-67663 Kaiserslautern, Germany
e-mail: korsch@physik.uni-kl.de
jodl@physik.uni-kl.de

The **cover picture** shows a Lorenz attractor generated with the program ODE, ordinary differential equations, included on the CD-ROM

ISBN 3-540-63893-8 2nd ed. Springer-Verlag Berlin Heidelberg New York

ISBN 3-540-57457-3 1st ed. Springer-Verlag Berlin Heidelberg New York

Library of Congress Cataloging-in-Publication Data applied for.

Die Deutsche Bibliothek - CIP-Einheitsaufnahme

CHAOS: a program collection for the PC; with many numerical experiments/
H. J. Korsch; H.-J. Jodl. - Berlin; Heidelberg; New York; London; Paris; Tokyo;
Hong Kong; Barcelona: Springer
Literaturangaben
ISBN 3-540-57457-3 (1. ed., Berlin ...)
ISBN 3-540-63893-8 (ed.)
Buch. - 2. ed. - 1998
CD-ROM. -2. ed. - 1998

Cover Design: E. Kirchner, Heidelberg
Typesetting: Camera-ready copies from the authors using a Springer LaTeX macro package
SPIN: 10661573 56/3144 - 5 4 3 2 1 0 – Printed on acid-free paper

Preface to the Second Edition

The still growing interest in chaotic dynamics in physics and the friendly receipt given to the first edition encouraged us to prepare a second edition of this book.

During the last years, we observed an increasing introduction to chaotic (or nonlinear) dynamics already in basic courses in physics. Here the computer is often used as an ideal tool for the demonstration of chaotic phenomena in computer "experiments" during lectures. More and more students realize that they can benefit from the simultaneous interaction with computer programs and reading of texts, provided that specialized and easy to use programs with many suggestions for such "experiments" exist. The resonance from many students and colleagues gave us the impression that our collection of programs helps the students to explore the highly non-trivial behavior of dynamical systems.

We have taken the opportunity to correct some minor errors and to clarify a few points in the text of the book, but most of the program codes of the first edition remained unchanged. However, the rapid development of computer operating systems made it necessary to modify some of the computer codes and to change the installation routine. The programs in this collection were originally written to run within the operating system DOS. Now they have been tested to run under Windows 95 and NT as well. Because of the increase of speed, the computing times are noticeably reduced and much more elaborate numerical experiments may be performed in acceptable times.

In addition to the students and research associates who contributed considerably to the first edition, we would like to thank Dr. Martin Menzel, Dr. Leo Schoendorff and Bernd Schellhaaß for their assistance in preparing this second edition.

Kaiserslautern, October 1998

H. J. Korsch and H.-J. Jodl

Preface to the First Edition

The problem, expressed in its general form, is an old one and appears under many guises. Why are the clouds the way they are? Is the solar system stable? What determines the structure of turbulence in liquids, the noise in electronic circuits, the stability of a plasma? What is new, that is to say with respect to Newton's Principia published three hundred years ago, and which has emerged over the last few decades, is the heuristic use of computers to enhance our understanding of the mathematics of nonlinear dynamical processes and to explore the complex behavior that even simple systems often exhibit.

It is the **purpose of this book** to teach chaos through a simultaneous reading of the text and interaction with our selected computer programs.

The use of computers is not only essential for studying nonlinear phenomena, but also enables the intuitive geometric and heuristic approach to be developed, taught to students, and integrated into the scientist's skills, techniques, and methods.

- Computers allow us to penetrate unexplored regions of mathematics and discover foreseen links between ideas.
- Numerical solutions of complex nonlinear problems – displayed by graphs or videos – as opposed to analytical solutions, which are often limited due to the approximations made, become possible.
- The visualization of mathematics will also be a focus of this book: one good graph or simulation video that highlights the evolution of a coherent complex pattern can be worth more than a hundred equations.
- Nonlinear problems are almost always difficult, often having unexpected solutions.
- In attempting to understand the details of the computer solution one may uncover a new kind of problem, or a new aspect that leads to deeper understanding.
- Appropriate graphical displays, especially those that are constructed and composed on the screen as one interacts with the computer, will improve the ability to choose from among promising paths. This procedure naturally complements the usual approaches of experiment, theoretical formulation and asymptotic approximation.

- The benefits of the computational approach clearly depend on the availability of various graphical displays: small effects that may signal new phenomena (zoom); proper mapping of data in place of a search for structures in voluminous printouts of columns of numbers; pictures which clearly produce an insight into the physics; color or real-time videos which enhance perception, enabling one to correlate old and new results and recognize unexpected phenomena.

- Interactive software in general, or educational software in physics, must provide the ability to display in one, two or three dimensions, to show spatial and temporal correlations, to rotate, displace or stretch objects, to acquire diagnostic variables and essential summaries or to optimize comparisons.

Are we providing this kind of training in our universities, so that our students may learn this method of working at the 'nonlinear' frontier? We have to find new methods to teach students to experiment with computers in the way that we now teach them to experiment with lasers, or deepen their knowledge of theory. Therefore, the philosophy of our approach in this book is to practice the use of a computer in computational physics directed at a convincing topic, i.e. nonlinear physics and chaos. Our programs are written in such a style that physical problems can quickly be tackled, and time is not wasted to program details such as the use of algorithms to solve an equation, or the input and output of data. Of course, the student will eventually need to master the elements of programming himself as he comes closer to independent research. Therefore, this book is aimed at those who have completed a course of study in physics and are on the threshold of research. Another advantage is that 'mini–research' can be carried out, thanks to the nature of the topic, chaos, and to the tool, the computer. These allow one to discuss physical problems which are only mentioned in textbooks nowadays as a potential topic of study, e.g. the double pendulum.

From the point of view of the complexity of the mathematics and physics, **this book is designed** mainly for students in the third or fourth year in a science or engineering faculty. In a limited way, it might also be useful to those working at the frontiers of nonlinear physics, since this topic is relatively new and far from having well-established solutions or wide applications.

This book is organized in the following way: in Chap. 1 *'Overview and Basic Concepts'*, we attempt to introduce typical features of chaotic behavior and to point out the broad applicability of chaos in science as well as to make the reader familiar with the terminology and theoretical concepts. In Chap. 2 *'Nonlinear Dynamics and Deterministic Chaos'*, we will develop the necessary basis, which will be deepened and applied in subsequent chapters. Chapter 3 *'Billiard Systems'* and Chap. 4 *'Gravitational Billiards'* will treat two of the 'classical examples' of simple conservative mechanical systems. In Chap. 5, the class of different pendula, such as kicked, inverted, coupled, oscillatory or rotating, is representatively discussed through the double pendulum. Phenomena appearing in chaotic scattering systems are represented by the three disk scattering in Chap. 6. The subsequent chapters treat systems explicitly dependent on time: namely, in Chap. 7 a periodically kicked particle in a box *'Fermi Acceleration'*, and in Chap. 8 a driven anharmonic oscillator *'Duffing Oscillator'*. The celebrated one-dimensional iterated maps are the topic of Chap. 9, and the observed period-doubling scenario can be studied via the physical example of nonlinear electronic circuits in Chap. 10. Numerical experiments with two-dimensional maps are considered in Chap. 11 *'Mandelbrot and Julia Sets'*, while Chap. 12 *'Ordinary Differential equations'* provides a quite general platform from which to investigate systems governed by coupled first order differential equations. Finally, further technical questions of hardware requirements, program installation, and the use of the programs are addressed in the appendices. Most chapters follow the same substructure:

- Theoretical Background
- Numerical Techniques
- Interaction with the Program
- Computer Experiments
- Real Experiments and Empirical Evidence
- References

Many books and articles have been written on chaotic experiments, and some of the 'classical' experiments are mentioned in Chap. 1. Therefore, the last subsection in each chapter is intended to give the reader confidence to progress from his studies on the computer to real experiments and empirical evidence; e.g. comparing the trajectory of a double pendulum in reality and on the screen. Of course, some aspects of the system are better studied in the computer experiment, others in the real one; in addition, they complement each other, e.g. looking for bifurcations in a nonlinear electronic circuit on the oscilloscope and on the screen (Chap. 10). One is at first impressed by the apparently chaotic motion of a real double pendulum, but deeper insight into the structure of this chaotic behavior is gained by looking at Poincaré maps in phase space. The experiments chosen here and briefly reported (for details see the cited literature) are mainly for educational purposes, to be reconstructed and used in student laboratories or in lectures. Therefore, they are not meant to represent those experiments investigated in current nonlinear research.

The most effective way of **using this book** may be to read a chapter while working simultaneously on the computer using the appropriate program. As already mentioned, the reader should use the programs — rather than program major parts himself — in much the same way as he would use standard service software in combination with commercial research apparatus. On the other hand, the use of our programs should not be a simple push button procedure, but involve serious interaction with the software. For example, some parameters, initial values, boundary conditions are already pre-set to execute numerical experiments discussed in the book, while other numerical experiments described in detail require changes in the pre-set parameters. Further studies are suggested and the reader may proceed independently, guided by some hints and the cited literature. The programs are flexible and organized in such a way that he can set up his own computer experiments, e.g. define his own boundary in a billiard problem or explore his favorite system of nonlinear differential equations.

Reading this book and working with the programs requires a knowledge of classical mechanics and a basic understanding of chaotic phenomena. The short overview on chaotic dynamics and chaos theory in Chaps. 1 and 2 cannot serve as a substitute for a textbook. Within the last decade a number of such books have been published reflecting the rapid growth of the field. Some focus on experiments, some concentrate on theory, some deal with basic concepts, while others are simply a selection of original articles. The reader should consult some of these texts while exploring the nonlinear world by means of the computer programs in this book.

We hope that the selected examples of chaotic systems will help the reader to gain a basic understanding of nonlinear dynamics, and will also demonstrate the usefulness of computers for teaching physics on the PC.

Most of the programs, at least in their preliminary version, were developed by students during a seminar *'Computer Assisted Physics'* (1985–1990). With the aid of two grants (PPP 1987–1989, PPPP 1991–1994), we were able to improve, test, standardize and update those programs. Chapter 1 contains a list of all coauthors for every program. We wish to acknowledge funding from the Bundesministerium für Bildung und Wissenschaft (BMBW), from the University of Kaiserslautern via the Kultusministerium of Rheinland Pfalz and the Deutsche Forschungsgemeinschaft (DFG) for the hardware. We are indebted to the Volkswagenstiftung (VW) for supporting one of us (H.-J. J.) to finish this book during a sabbatical.

Finally, we wish to recognize the contribution of undergraduate and graduate students and research associates who worked so enthusiastically with us on problems associated with chaos and on the use of computers in physics teaching. We would particularly like to thank the graduate students Björn Baser and Andreas Schuch for their considerable assistance in developing and debugging the computer codes, the instructions for interacting with the programs as well as the large number of computer experiments. Finally, we would like to thank Frank Bensch and Bruno Mirbach, who read parts of the manuscript and made many useful suggestions.

Kaiserslautern, May 1994 — H. J. Korsch and H.-J. Jodl

Table of Contents

1. Overview and Basic Concepts

1.1 Introduction

Unexpected and unpredictable behavior. In everyday life we feel safer and more comfortable with predictability and determinism: in technically controlled processes, small mechanical forces are expected to cause minor changes; the time-table of trains is hopefully reliable; the motion of the earth and moon around the sun are thought to be regular and stable.

In fact, we are surprised if the opposite happens: a coin placed on its rim may fall to the left or right due to small perturbations; a ball rolling down in a pin-ball machine may be scattered left or right; an amplified signal in an acoustic source–microphone feedback loop may produce irregular noise. Such unpredictable behavior is, however, assumed to be due to random or stochastic processes or forces, which are not under our control. The discovery that fully deterministic systems without random influences show random behavior therefore came as a big surprise. It was even more astonishing that this *'deterministic chaos'* can be found in simple systems with very few degrees of freedom that, moreover, it is quite typical and does not occur only in rare cases. On the contrary, one had to get used to the fact that the well-known paradigmatic systems showing regular behavior treated in lecture courses and textbooks are the exception and not the rule.

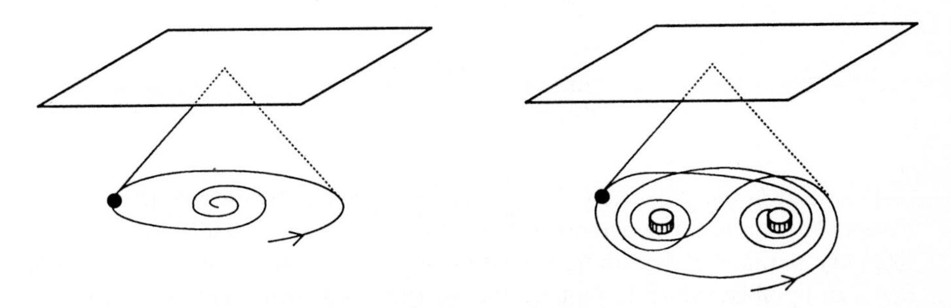

Fig. 1.1. String pendulum.

Imagine, for example, a two-dimensional string pendulum whose trajectory is a circle or, taking friction into consideration, a spiral. A discussion of this system — potential, forces, equations of motion — can be found in almost every introductory textbook on physics. A slight modification of this pendulum (see Fig. 1.1) already exhibits many features of chaotic motion. Let us use a small iron ball and place two or more magnets close to the swinging pendulum. Immediately, the motion becomes very complicated, unexpected, and even unpredictable. Starting the pendulum at a given initial position and 'measuring' the final position of the iron ball at one of the magnets, we can construct a mapping, such as 'number of the target magnet as a function of the initial position x_0, y_0'. It turns out that this function is very complicated. Varying the initial conditions of the pendulum to a high degree of accuracy — a few decimal digits in x_0, y_0 — we may find an extraordinary sensitivity of the subsequent trajectory and the final attracting magnet with respect to these initial conditions.

Applicability of chaotic concepts in science. In the course of fifty years of research into chaos, several classic or prototype models have been successfully established, such as the Lorenz, Duffing, van der Pol, Ueda, Fermi-Ulam, and Hénon-Heiles models. Most of them will be extensively discussed in the following chapters. For example, in 1963 Lorenz derived a set of differential equations to model thermal convection. This phenomenon plays an important role in describing dynamics in gases and liquids or, quite generally, in many systems showing a transition from a laminar flow to turbulence. In a cell containing a fluid, which is heated from below, the onset of convection leads to the formation of organized patterns of stable 'rolls', the so-called *Bénard convection*.

Originally, Lorenz studied heat convection in the atmosphere. He defined a velocity distribution $\mathbf{v}(\mathbf{r}, t)$ and a temperature field $T(\mathbf{r}, t)$ in space; then — on the basis of the Navier-Stokes equation, the equation for thermal conductivity, and the continuity equation — he made some additional simplifying assumptions and modeled the following dimensionless equations for the trajectory in phase space (x_1, x_2, x_3):

$$\begin{aligned}\frac{\mathrm{d}x_1}{\mathrm{d}t} &= \sigma(x_2 - x_1)\\ \frac{\mathrm{d}x_2}{\mathrm{d}t} &= (r - x_3)x_1 - x_2\\ \frac{\mathrm{d}x_3}{\mathrm{d}t} &= x_1x_2 - bx_3\,,\end{aligned}$$

the so-called Lorenz equations. The variables x_1, x_2, and x_3 are related to the stream function and the perturbed temperature, σ is the Prandtl number, b a geometry parameter describing the size of the 'rolls', and r the (normalized) Rayleigh number, which is proportional to the temperature gradient. A characteristic long-time limit of a trajectory in phase space (x_1, x_2, x_3) approaching the so-called 'Lorenz attractor' is shown on the cover of this book.

The power and scope of the Lorenz model — originally valid for heat convection in aero and hydrodynamics — was very well confirmed in 1975 by Haken, who applied it to laser systems by showing the equivalence of the respective differential equations. The (normalized) variables of the electric field E, the polarization P, and the population inversion D satisfy the differential equations

$$\begin{aligned} \frac{\mathrm{d}E}{\mathrm{d}t} &= -\gamma_c E + \gamma_c P \\ \frac{\mathrm{d}P}{\mathrm{d}t} &= \gamma_\perp ED - \gamma_\perp P \\ \frac{\mathrm{d}D}{\mathrm{d}t} &= \gamma_\parallel R - \gamma_\parallel D - \gamma_\parallel EP \,. \end{aligned}$$

Here, R is the pump rate creating the population inversion normalized to unity at the threshold, and γ_c, $\gamma_\perp$, $\gamma_\parallel$ are relaxation rates describing the decay of E, P, and D, respectively.

These equations are isomorphic to the Lorenz equations and the time evolution of a system trajectory in a three-dimensional (E, P, N) space shows the same behavior as the Lorenz system in (x_1, x_2, x_3) space. If the control parameter, e.g. the pump rate R or the temperature gradient, respectively, exceeds a critical value, irregular chaotic behavior of the system is observed. The Lorenz system is studied in more detail in Chap. 12. Here, it should be pointed out, however, that these results are not restricted to a special application, but are *universal* and hold for a large variety of physical, chemical, and biological systems.

Linearity and Nonlinearity. It is a well known method and standard procedure in textbooks to linearize physical laws in order to simplify the analysis. Usually, the first term of a Taylor series is discussed, higher order terms are neglected and this simplified system instead of the real one is discussed. For example, the electric polarization is proportional to the applied electric field

$$\mathbf{P} = \epsilon_0 \, \chi \, \mathbf{E} \,.$$

For strong fields the material constant, the electric susceptibility χ, is itself a function of the external field. Expanding in powers of E, we gain

$$P_i = \epsilon_0 \left[\sum_{j=1}^{3} \chi_{ij}^{(1)} E_j + \sum_{j,k=1}^{3} \chi_{ijk}^{(2)} E_j E_k + \ldots \right] ,$$

where the tensors $\chi_{ij}^{(1)}$, $\chi_{ijk}^{(2)}$, ... are material constants. The nonlinearity is obvious and must be taken into account for large values of the field strength.

On the other hand, as will be shown later, nonlinearity is a necessary, but not a sufficient condition for generating chaotic motion. Thus, in performing experiments or modeling a theory in dynamical systems, one should understand the nature of the nonlinearity in these systems. The origin of nonlinearity can be many-fold. Some examples are:

Nonlinear matter: stress versus strain in elasticity; voltage versus current, e.g. for non-metallic conductors; ferroelectric and ferromagnetic materials.

Nonlinear variables: frictional force versus velocity; convective acceleration in fluid mechanics; acceleration in non-inertial systems.

Nonlinear forces: all elementary forces are nonlinear and realistic potentials are anharmonic.

Nonlinear geometry: force versus displacement in mechanics; magnetization versus magnetic field (hysteresis); voltage/current characteristics in some semiconductors in electrodynamics.

The necessity for, and capability of, computers to solve these nonlinear equations numerically is obvious and well demonstrated in the subsequent chapters.

Chaos as a powerful concept in physics. The more powerful a physical model or concept turned out to be, the broader its use: e.g. oscillations, waves, and fields. During the last thirty years of research into chaos, a similar trend was to be expected in chaotic dynamics. For example, period doubling bifurcations are essential qualities describing an important route for a system to become chaotic; i.e. if a control parameter is varied, the system may react such that the period is 2, 4, ..., 2^n, and chaotic if a critical value is reached. Nowadays, we can demonstrate this behavior in any area of physics.

An example in mechanics is a bouncing ball on a vibrating membrane; if the exciting frequency of the membrane is tuned, at a critical frequency the ball may jump with two different amplitudes.

In electronics, the frequency of an oscillating circuit as a function of the driving frequency may show a bifurcation pattern if the oscillator is driven by a nonlinear feedback element.

In magnetism, the motion of a spinning magnet in an external time-periodic magnetic field may be one-, two-, four-, ... periodic or chaotic depending on the excitation frequency or the strength of the magnetic field.

In optics, a laser may show instability with respect to the frequency or the intensity of the light emitted, due to resonator conditions and excitation mechanisms.

How to demonstrate chaotic motion. During the last decade, many qualitative and quantitative demonstrative experiments have been developed to teach chaos in regular lecture courses or in special lectures. The following collection covers almost all areas in physics:

Mechanics:
any kind of coupled nonlinear pendula — periodically kicked rotators — magneto-elastic ribbons — dripping faucets — three body problems (e.g. in celestial mechanics) — Kundt tubes (nonlinear acoustics)

Thermodynamics:
crystal growth (dendrites) — diffusion — Ising model — spin glasses — cellular automata

Fluid mechanics:
Couette-Taylor flow — Rayleigh-Bénard convection — turbulent flow, vortices, Reynolds number

Optics:
multistability of lasers — nonlinear optical devices — nonlinear feedback systems — chaotic behavior of electric discharges in gas tubes

Electrodynamics:
RLC oscillators with nonlinear elements — chaotic behavior of electric conductivities — bistability in semiconductor devices — Josephson junctions — deterministic noise — plasma waves — particle accelerators (e.g. motion in storage rings or detuned cyclotrons) — spinning magnets in time-periodic magnetic fields — coupled dynamos modeling fluctuations of the earth's magnetic field ('Rikitake')

Others:
Belousov-Zhabotinskii reactions (chemical clocks and chemical waves) — population dynamics (prey–predator model) — self-organization of biological systems

We found many of the demonstration experiments to be well described in the books listed at the end of this chapter, or in many articles appearing in scientific journals with a pedagogical contents, such as the American Journal of Physics. Most of those experiments, which are closely related to the simulation programs discussed in this book, are described and cited at the end of each chapter in sections entitled *'Real experiments and empirical evidence'*.

1.2 The Programs

The collection of computer programs presented in this book can be used without any prior knowledge of computing. The interaction with the program is described in detail in each chapter; some general remarks can be found in Appendix B. The programs are easy to use and allow extensive computational experiments.

All programs have their origin in a course 'Computer Assisted Physics' taught by the authors of this book to undergraduate and graduate physics students at the University of Kaiserslautern during the period 1985 to 1993. The course aimed to teach physics using simulations on microcomputers. This makes it possible for the students to investigate problems, which are otherwise almost inaccessible at such an early stage. Various areas of physics, e.g. electrodynamics, quantum mechanics, relativity, statistical mechanics, have been explored during the course. The programs presented here are devoted exclusively to nonlinear dynamics and can be used to study deterministic chaos in simple dynamical systems. The authors of the computer programs are students at the Fachbereich Physik of the University of Kaiserslautern.

The following programs are included in the present collection:

- **Billiard** — *'Point mass on a billiard table'* written by Thomas Kettenring, Achim Schramm, Andreas Schuch, and Matthias Urban:
  Simulates frictionless motion on a plane with a hard convex boundary. Coordinate space and phase space dynamics is displayed. Arbitrary boundary curves can be specified.

- **Wedge** — *'A particle jumping in a wedge under gravitational force'* written by Stefan Steuerwald:
  Simulates the motion of a point mass in a homogeneous gravitational field inside a hard 'wedge'. The dynamics shows a very complex dependence on the wedge angle ('breathing chaos').

- **Dpend** — *'The double pendulum'* written by Björn Baser, Ralf Getto, Christian Laue, and Boris Ruffing:
  Allows the study of the large amplitude oscillation of a double pendulum in coordinate and phase space.

- **3Disk** — *'Scattering off three disks'* written by Dieter Eubell and Andreas Schuch:
  Classical planar scattering of a point mass off three fixed hard disks. A model for chaotic scattering systems. Deflection angles and collision numbers are computed as a function of the impact parameter.

- **Fermi** — *'The Fermi-acceleration'* written by Thomas Kettenring, Thomas Pütz, Andreas Schuch, and Frank Werner:
  A model for chaotic dynamics in periodically driven systems. Simulates the motion of a point mass between a fixed and an oscillating hard wall. Various wall oscillations can be chosen.

- **Duffing** — *'The Duffing-oscillator'* written by Jorg Imhoff, Hans Jürgen Roth, Andreas Schuch, and Franz Speckert:
  Solves the equations of motion for the Duffing oscillator, i.e. a forced, damped, and harmonically driven oscillator with a cubic nonlinearity in the restoring force, for arbitrary values of the parameters. The solution is presented as a phase space curve or a Poincaré section. The dynamical behavior is extremely rich. Strange attractors can be studied.

- **Feigbaum** — *'One-dimensional iterative maps of an interval'* written by Alexander Keller and Andreas Schuch:
  Allows the numerical iteration of mappings of the unit interval (unit circle)

as a model for nonlinear dynamics, in particular the logistic map and circle map. The behavior of the iterations, fixed points, period doubling, chaotic dynamics can be studied.

- **Chaosgen** — *'An electronic chaos-generator'* written by Jürgen Grohs, Thomas Kettenring, and Andreas Schuch:
  Simulates a nonlinear electronic circuit. When a resistance R_m is decreased, the amplitude and hence the nonlinearity effects increase. The oscillation shows a typical sequence of period doublings and chaotic dynamics.

- **Mandelbr** — *'Mandelbrot and Julia sets'* written by Jorg Imhoff, Thomas Kettenring, Andreas Schuch, Michael Schwarz, and Franz Speckert:
  Allows the numerical iteration of second order polynomial mappings of the plane, in particular quadratic complex valued maps, e.g. the Mandelbrot map. Julia and Mandelbrot sets can be studied. The results are displayed graphically. Various algorithms for exploring and coloring the plane can be selected.

- **ODE** — *'Ordinary differential equations'* written by Wolfgang Langbein and Andreas Schuch:
  Provides a numerical solution of a system of first order ordinary differential equations (ODE) and allows various graphical presentations of the solutions.

The first four programs simulate the dynamics of time-independent Hamiltonian systems, i.e. systems without dissipation, for bounded and open (scattering) systems. The programs BILLIARD, WEDGE, and 3DISK simulate systems where the dynamics is (mainly) determined by collisions with hard walls. Such systems can be reduced to discrete mappings between the impacts and allow fast computation of the trajectories. These programs are recommended as a first introduction to chaotic dynamics. The program DPEND simulating a double pendulum, which is the most widely used example for demonstrating chaotic motion in a classroom experiment, requires a solution of differential equations. Computation times are longer here, and it is therefore necessary to plan the numerical experiment more carefully.

In the programs FERMI and DUFFING, characteristic time-dependent systems are studied. In the first case, a one-dimensional billiard-type model with oscillating hard walls, where the dynamics is again an impact mapping, is treated. The second, an anharmonic forced oscillator with friction, also offers a very simple starting point into the field of chaotic dynamics, since it is a direct extension of the well-known forced harmonic oscillator.

In the study of dynamical systems, one often constructs discrete mappings which simplify the discussion and analysis. In the programs FEIGBAUM and MANDELBR, one can investigate the fascinating properties of such mappings for one-dimensional and two-dimensional cases. One of these properties is a route to chaos via an infinite sequence of period doubling bifurcations. Such a scenario can be observed, e.g., in electronic circuits containing nonlinear elements, which are simulated in the program CHAOSGEN.

The last program, ODE, is a general-purpose program for studying dynamical systems described by differential equations. This program can be used to study numerous phenomena in chaotic dynamics. A selection of such applications is discussed in Chap. 12.

Installation of the Programs. The programs described above are found on the disk. These programs should be installed on the hard disk. It is also possible, however, to run the programs directly fom the disk. The hardware requirements for running the programs, as well as a convenient method to install and start the program package on the PC, are described in Appendix A.

1.3 Literature on Chaotic Dynamics

Working with the programs without some prior knowledge of nonlinear dynamics cannot be recommended. In the present book, we do *not* intend to present a systematic and definitive treatment of chaotic dynamics. The short overview of the theory of nonlinear dynamics given in the following chapter can only serve as an introduction to the basic ideas and results in an effort to simplify the analysis of the computer experiments.

The purpose of the limited number of numerical experiments described in some detail in the following chapters is twofold. First, the reader should become familiar with the various options of the program. Secondly, the reader is introduced into the basic features of the simulated system. In any case, the enormous variety of fascinating features of the systems can only be touched upon. Some additional experiments are suggested and some useful references to such studies found in the literature are given. We expect that the reader will soon begin to explore the nonlinear world himself, and in doing so it is highly recommended to consult from time to time the textbooks available. Some of these books are listed below.

The following books are considered to be helpful when working with the programs in the present collection. It is evident that this list is incomplete, and we apologize if the reader finds that his favorite book has been omitted.

G. L. Baker and J. P. Gollub, *Chaotic Dynamics — An Introduction* (Cambridge Univ. Press, Cambridge 1990)

R. L. Devaney, *An Introduction to Chaotic Dynamical Systems* (Addison–Wesley, New York 1987)

J. Frøyland, *Introduction to Chaos and Coherence* (IOP Publishing, Bristol 1992)

J. Guckenheimer and P. Holmes, *Nonlinear Oscillations, Dynamical Systems, and Bifurcations of Vector Fields* (Springer, New York 1983)

A. J. Lichtenberg and M. A. Lieberman, *Regular and Stochastic Motion* (Springer, New York 1983)

F. C. Moon, *Chaotic Vibrations* (J. Wiley, New York 1987)

R. Z. Sagdeev, D. A. Usikov, and G. M. Zaslavsky, *Nonlinear Physics – From the Pendulum to Turbulence and Chaos* (Harwood Acad. Publ., Chur 1988)

H. G. Schuster, *Deterministic Chaos* (VCH, Weinheim 1988)

M. Tabor, *Chaos and Integrability in Nonlinear Dynamics* (John Wiley, New York 1989)

J. M. T. Thompson and H. B. Stewart, *Nonlinear Dynamics and Chaos* (John Wiley, Chichester 1986)

The following books also consider, in particular, numerical and computational aspects of nonlinear dynamics:

H. Koçak, *Differential and Difference Equations through Computer Experiments* (Springer, New York 1986)

T. S. Parker and L. O. Chua, *Practical Numerical Algorithms for Chaotic Systems* (Springer, New York 1989)

W.-H. Steeb, *Chaos and Fractals* (BI Wissenschaftsverlag, Mannheim 1992)

N. B. Tufillaro, T. Abbott, and J. Reilly, *An Experimental Approach to Nonlinear Dynamics and Chaos* (Addison–Wesley, New York 1991)

Experimental aspects — in particular, demonstrative experiments also suitable for lecture courses — are discussed in:

F. C. Moon, *Chaotic Vibrations* (J. Wiley, New York 1987)

N. B. Tufillaro, T. Abbott, and J. Reilly, *An Experimental Approach to Nonlinear Dynamics and Chaos* (Addison–Wesley, New York 1991)

S. Vohra, Ed., *Experimental Chaos Conference, Arlington 1991* (World Scientific, Singapore 1992)

A selection of reprints of important original articles is found in the following books:

B.-L. Hao, *Chaos* (World Scientific, Singapore 1984)

P. Cvitanović, *Universality in Chaos* (Adam Hilger, Bristol 1984)

R. S. MacKay and J. D. Meiss, *Hamiltonian Dynamical Systems* (Adam Hilger, Bristol 1987)

For the convenience of those readers having little experience with chaotic dynamical systems a glossary of the important general terms has been included at the end of this book.

2. Nonlinear Dynamics and Deterministic Chaos

The aim of this chapter is to provide an introduction to the theory of nonlinear systems. We assume that the reader has a background in classical dynamics and a basic knowledge of differential equations, but most readers of this book will only have a vague notion of chaotic dynamics. The computer experiments in the following chapters will (hopefully) lead to a better understanding of this new and exciting field. These chapters form the core of the book and are written at a level suitable for advanced undergraduate students. An understanding and interpretation of the numerical results is, however, impossible without a knowledge of the relevant theory.

The theory of dynamical systems is well-developed and a number of excellent textbooks [2.1]–[2.8], as well as collections of important original articles [2.9]–[2.11], are available. The reader is invited to consult these references while exploring nonlinear dynamics on the computer. The overview of the theory of nonlinear dynamics presented in the following serves as a short survey, describing the basic phenomena and clarifying the notation used in the setup and analysis of the computer experiments.

The first three sections introduce *Deterministic Chaos* and the special cases of chaotic dynamics in *Hamiltonian Systems* and *Dissipative Dynamical Systems*. It is hoped that the reader will gain a basic understanding on going through these sections. Many features, however, will become clearer only later on in the context of the computational studies, which are linked to this introductory chapter by ample cross-references.

The last section of this chapter, *Special Topics*, contains additional material, which is useful for a more detailed understanding of certain aspects of chaotic dynamics introduced in the subsequent chapters. This section can (and should) be omitted in a first reading of the text.

2.1 Deterministic Chaos

'Deterministic chaos' is a term used to denote the irregular behavior of dynamical systems arising from a strictly deterministic time evolution without any source of noise or external stochasticity. This irregularity manifests itself in an extremely sensitive dependence on the initial conditions, which precludes any

long-term prediction of the dynamics. Most surprisingly, it turned out that such chaotic behavior can already be found for systems with a very small degree of freedom and it is, moreover, *typical* for most systems.

A *dynamical system* can be described simply as a system of N first order differential equations

$$\frac{\mathrm{d}x_i}{\mathrm{d}t} = f_i(x_1, \ldots, x_N, r) \quad , \quad i = 1, \ldots, N \, , \tag{2.1}$$

where the independent variable t can be read as time and the $x_i(t)$ are dynamical quantities whose time dependence is generated by (2.1), starting from specified initial conditions $x_i(0)\,,\, i = 1, \ldots, N$. It should be noted that the system (2.1) is autonomous because it is not explicitly t-dependent. The f_i are nonlinear functions of the x_i and are characterized by the parameter(s) r. The equations lead to chaotic motion, which develops and changes its characteristics with varying control parameter(s) r. The assumption of an autonomous system is not essential, because otherwise it can be converted into an autonomous one by introducing the time t as an additional variable x_{N+1}. Examples of dynamical systems are the Hamiltonian equations of motion in classical mechanics, the rate equations for chemical reactions or the evolution equations in population dynamics.

A *discrete dynamical system* is an iterated mapping

$$x_i(n+1) = f_i(x_1(n), \ldots, x_N(n), r) \quad , \quad i = 1, \ldots, N \tag{2.2}$$

starting from an initial point $x_i(0)\,,\, i = 1, \ldots, N$. Such a discrete system may appear quite naturally from the setup of the problem under consideration, or it may be a reduction of the continuous system (2.1) in order to simplify the analysis, as for example the Poincaré map described below.

Basically, one can make a distinction between *conservative* (e.g. mechanical systems governed by Hamilton's equations of motion) and *dissipative* systems with 'friction'. In the first case, volume elements in phase space are conserved, whereas dissipative systems contract phase space elements. This results in markedly different behavior.

2.2 Hamiltonian Systems

In a so-called Hamiltonian system with N degrees of freedom, the dynamics is derived from a Hamiltonian $H(\mathbf{p}, \mathbf{q}, t)$, where $\mathbf{q} = (q_1, \ldots, q_N)$ and $\mathbf{p} = (p_1, \ldots, p_N)$ are the canonical coordinates and momenta. The Hamiltonian equations of motion

$$\dot{p}_i = -\frac{\partial H}{\partial q_i} \, , \qquad \dot{q}_i = \frac{\partial H}{\partial p_i} \, , \qquad i = 1, \ldots, N \tag{2.3}$$

generate trajectories $\mathbf{p}\,(t)\,,\, \mathbf{q}\,(t)$ in $2N$-dimensional phase space. In more global terms, it is said that the Hamiltonian produces a flow in phase space. This flow conserves the phase space volume

$$\mathrm{d}p_1 \ldots \mathrm{d}p_N \, \mathrm{d}q_1 \ldots \mathrm{d}q_N \tag{2.4}$$

as well as the phase space 'area'

$$\mathrm{d}\mathbf{p} \cdot \mathrm{d}\mathbf{q} = \sum_i \mathrm{d}p_i \mathrm{d}q_i \,. \tag{2.5}$$

2.2.1 Integrable and Ergodic Systems

In many cases, the Hamiltonian does not depend explicitly on time and the energy $E = H(\mathbf{p}, \mathbf{q})$ is conserved along the trajectory, i.e. it is a constant of motion. A simple example of such a conservative system is the motion of a particle with mass m in a potential V with Hamiltonian

$$H(\mathbf{p}, \mathbf{q}) = \frac{\mathbf{p}^2}{2m} + V(\mathbf{q}) \,. \tag{2.6}$$

There may be more constants of motion, which can be a consequence of obvious symmetries of the system. Well-known examples are translational symmetry leading to conservation of the momentum or rotational symmetry resulting in angular momentum conservation. In other cases, the (possibly existing) constants of motion are far less obvious and it is a non-trivial problem to find them or to prove their existence. It is, however, simple to show that a given phase space function $F(\mathbf{p}, \mathbf{q})$ is a constant of motion. This can be most elegantly done by writing the equation of motion for F in terms of the Poisson bracket

$$\{A, B\} = \sum_i \left(\frac{\partial A}{\partial p_i} \frac{\partial B}{\partial q_i} - \frac{\partial B}{\partial p_i} \frac{\partial A}{\partial q_i} \right) \tag{2.7}$$

between two arbitrary phase space functions $A(\mathbf{p}, \mathbf{q})$ and $B(\mathbf{p}, \mathbf{q})$ as

$$\frac{\mathrm{d}F}{\mathrm{d}t} = \{H, F\} \tag{2.8}$$

(a simple example of these generalized equations comprises the canonical equations of motion (2.3)). Therefore, F is a constant of motion if the Poisson bracket with H vanishes identically

$$\{H, F\} \equiv 0 \,, \tag{2.9}$$

which can easily be checked for a given F.

The existence of a constant of motion severely restricts the dynamical possibilities of a trajectory $(\mathbf{p}(t), \mathbf{q}(t))$, because it must follow the (hyper)surface $F(\mathbf{p}(t), \mathbf{q}(t)) = \text{constant}$. The motion is even more restricted if additional constants of motion exist. In the extreme case, the dynamical equations can be integrated in closed form. A system is said to be *'integrable'* if N independent constants of motion F_j exist:

$$F_j(\mathbf{p}, \mathbf{q}) = c_j = \text{constant} \,, \quad j = 1, \ldots, N \,. \tag{2.10}$$

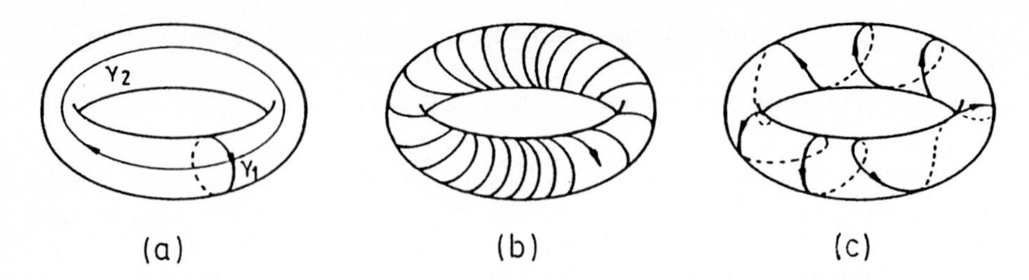

Fig. 2.1. Two-dimensional torus in phase space: (a) Topologically different paths γ_i. (b) A typical quasiperiodic trajectory. (c) Torus supporting periodic orbits.

One of the F_j can be the energy. In addition, the F_j must be 'in involution', i.e. their pairwise Poisson brackets must vanish:

$$\{F_j, F_k\} = \sum_i \left(\frac{\partial F_j}{\partial p_i} \frac{\partial F_k}{\partial q_i} - \frac{\partial F_k}{\partial p_i} \frac{\partial F_j}{\partial q_i} \right) = 0\,. \tag{2.11}$$

In this case, it can be shown that the phase space trajectory is confined to the surface of an N-dimensional manifold in $2N$-dimensional phase space. Furthermore, it can be shown that this surface has the topology of a torus. The whole phase space is filled with such tori and the phase space flow is therefore highly organized and regular.

For systems with two degrees of freedom as, for example, the motion of a mass m in a two-dimensional potential $V(x,y)$ with Hamiltonian

$$H(p_x, p_y, x, y) = \frac{1}{2m}\left(p_x^2 + p_y^2\right) + V(x,y)\,, \tag{2.12}$$

the solution of the equations of motion

$$\dot{p}_x = -\frac{\partial V}{\partial x} \quad , \quad \dot{x} = \frac{p_x}{m} \tag{2.13}$$

$$\dot{p}_y = -\frac{\partial V}{\partial y} \quad , \quad \dot{y} = \frac{p_y}{m} \tag{2.14}$$

describes the time evolution of the system in four-dimensional phase space (x, y, p_x, p_y). Because of energy conservation, the trajectory is restricted to the three-dimensional manifold $H(p_x, p_y, x, y) = E$, the 'energy surface'. For an integrable system, because of (2.11), there exists a second integral of motion with vanishing Poisson bracket with the Hamiltonian H. To take a specific example, we can assume that the potential is rotationally symmetric, $V(x,y) = V(x^2 + y^2)$, with constant angular momentum $L = xp_y - yp_x$.

This restricts the trajectory to a two-dimensional submanifold of the energy surface, which has the topology of a torus, as illustrated in Fig. 2.1. Such a torus is invariant under the equations of motion and is, therefore, referred to as an *'invariant torus'*. It should be stressed, however, that a two-dimensional torus embedded in four-dimensional phase space can be bent and twisted in a complicated manner, and that Fig. 2.1 is only a schematic illustration.

The torus topology can now be utilized in a theoretical analysis to introduce new canonical coordinates and momenta, namely the so-called 'action-angle variables' with actions $\mathbf{I} = (I_1, I_2)$ and angles $\boldsymbol{\varphi} = (\varphi_1, \varphi_2)$, allowing a solution — an integration — of the equation of motion in closed form. The actions are given by the phase integrals

$$I_i = \frac{1}{2\pi} \oint_{\gamma_i} \mathbf{p} d\mathbf{q}\,, \ i = 1, 2\,, \tag{2.15}$$

where γ_1 and γ_2 are two topologically different closed paths on the torus (see Fig. 2.1(a)), as discussed in any textbook of analytical mechanics. Written in action-angle variables, the Hamiltonian is simply given by $H = H(I_1, I_2)$ and — introducing the frequencies $\omega_i = \partial H/\partial I_i\,, i = 1, 2$ — the equations of motion simplify to

$$\dot{\mathbf{I}} = 0\,, \quad \dot{\boldsymbol{\varphi}} = \boldsymbol{\omega}\,, \tag{2.16}$$

with the solution

$$\mathbf{I} = \text{constant} \quad \boldsymbol{\varphi}(t) = \boldsymbol{\varphi}(0) + \boldsymbol{\omega} t\,. \tag{2.17}$$

The trajectory is quasiperiodic, characterized by the two frequencies ω_1 and ω_2, and, typically, it covers the entire torus in the long-time limit. For the case of commensurable frequencies, i.e. frequencies whose ratio is rational, the trajectory returns precisely to its starting point and the identical orbit is traced out again: the trajectory is periodic (Fig. 2.1(c)). Moreover, the rationality condition is valid for the entire torus and all trajectories on this torus are periodic with the same period. The frequencies vary, of course, in phase space and therefore, typically, a dense, countable subset of the tori filling the phase space carry periodic trajectories. This picture can easily be extended to the case of more than two degrees of freedom.

Integrable systems are rare. Until recently, however, such systems were almost exclusively treated in textbooks on classical dynamics, as for instance a forced or parametrically excited harmonic oscillator, a point mass in a three-dimensional spherically symmetric potential, N-dimensional coupled harmonic oscillators, etc. This led to the impression that the well-organized behavior of integrable cases is typical for systems with few degrees of freedom.

As another extreme example we have *ergodic* systems. Here, almost every trajectory approaches arbitrarily close to every energetically allowed point in phase space, i.e. it fills the phase space. Ergodic systems have been discussed in physics mainly in the context of a statistical description of systems with many degrees of freedom, e.g. the N-particle gas in thermodynamics ($N \sim 10^{23}$). Only in the last decades has it been realized that low dimensional physical systems can be ergodic.

Ergodicity, however, does not guarantee the decay of correlations in the long time limit, which is demanded for the so-called *mixing systems*. These systems

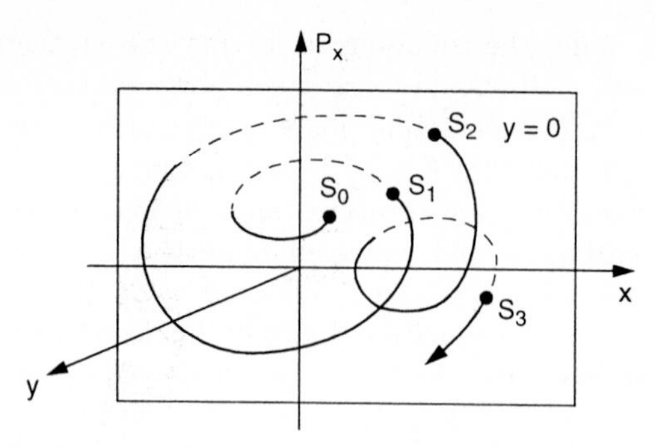

Fig. 2.2. A Poincaré section reduces the dynamics to a discrete mapping.

lose the memory of their history and show irregular (or 'chaotic') behavior. A characteristic of such chaotic dynamics is an extreme sensitivity to initial conditions (exponential separation of neighboring trajectories), which puts severe limitations on any forecast of the future fate of a particular trajectory. This sensitivity is known as the 'butterfly effect': the state of the system at time t can be entirely different even if the initial conditions are only slightly changed, i.e. by a butterfly flapping its wings.

A quantitative measure of this exponential growth of deviations is given by the *Lyapunov exponent* (see Sect. 2.4.3). The most chaotic systems are the so-called *K-systems*. These are systems where nearby orbits separate exponentially. Very few systems have been proven to be mixing, as are, for example, the hard-sphere gas and the stadium billiard (see Chap. 3). A typical system (for two and more degrees of freedom) shows an intricate mixture of regular and irregular motion.

2.2.2 Poincaré Sections

In order to analyze complicated dynamics, one introduces a surface of section in phase space and, instead of studying a complete trajectory, one monitors only the points of its intersection with this surface. In this manner, we obtain a discrete mapping — the *Poincaré map* — which maps an intersection point onto the next one. As an illustrative example we again consider the Hamiltonian (2.12) with two degrees of freedom and take a section through phase space at, e.g., $y = 0$ for a given value of the energy E. Any point on this surface of section with coordinates (x, p_x) uniquely determines an initial point for a trajectory. Solving for p_y yields $p_y = \pm\{2m(E - p_x^2 - V(x,0)\}^{1/2}$. The value of p_y is uniquely determined by the coordinates (x, p_x) in the Poincaré section if we fix the sign of p_y, following the convention of considering only those trajectories which intersect the surface in the positive y-direction $p_y > 0$.

The dynamics then reduces to the discrete two-dimensional Poincaré mapping

$$S_n = (p_{xn}, x_n) \xrightarrow{T} (p_{xn+1}, x_{n+1}) = S_{n+1}\,, \tag{2.18}$$

as illustrated in Fig. 2.2 . A simple proof shows that this map is area-preserving. The set of all such intersection points

$$T^n(p_{x0}, x_0) = \underbrace{T \circ T \circ \cdots \circ T}_{n\,-\text{times}}(p_{x0}, x_0) \tag{2.19}$$

allows a much better insight into the dynamical behavior of the orbit as, for example, the trajectory in coordinate space, and a synoptic picture of different orbits quite easily provides an overview of the global properties of the phase space flow.

Orbits with

$$T^k(p_{x0}, x_0) = (p_{x0}, x_0) \tag{2.20}$$

are called k-periodic. These fixed points of the mapping T^k and their stability properties (see Sects. 2.4.4 and 2.4.5) can play an important role in the organization of the phase space flow and the Poincaré map. It should be noted that k-periodic orbits appear as a set of k discrete points in such a map. A quasiperiodic trajectory fills a closed curve in the Poincaré section, which is the intersection of the invariant torus covered by the trajectory and the Poincaré section. For an integrable system at fixed energy E, the surface of section is filled with a family of such *'invariant curves'*, i.e. curves invariant under the Poincaré map (2.18), as shown in Fig. 2.3. A countable subset of these invariant curves is filled with orbits, which are periodic under the Poincaré map.

Poincaré sections and mappings are used in most of the computer programs described in the following chapters. In some cases, they arise directly from the physics of the dynamical system, as for example the mapping between subsequent impacts in billiard systems (Chap. 3 and 4). Another kind of Poincaré mapping appears for time-periodically driven systems (period T), where the state of the system is only observed at discrete times $t_n = nT$, generating a so-called stroboscopic mapping. Such a mapping is used in the studies of the Duffing oscillator (Chap. 8) and the Chaos generator (Chap. 10).

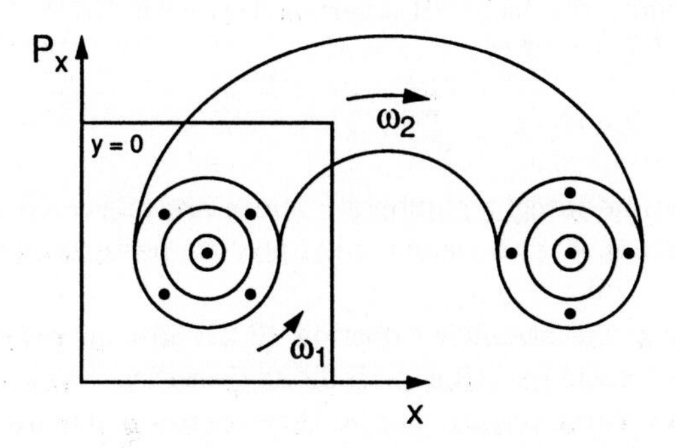

Fig. 2.3. A family of invariant tori in phase space appears as invariant curves in a Poincaré section. A subset of these invariant curves is filled with periodic orbits.

2.2.3 The KAM Theorem

For an integrable system, the entire phase space is filled with invariant tori and any trajectory will remain on the particular torus selected by the initial conditions. The proof of the existence of these tori is, however, based on the integrability of the system, i.e. the existence of N integrals of motion. It is now of considerable interest to investigate the behavior of the system when the integrals of motion are destroyed. This can be achieved by a small perturbation. To take an example, in a two-dimensional potential with rotational symmetry, the symmetry can be disturbed by a superimposed field and the angular momentum is no longer conserved. In this case, there is a priori no confinement to a two-dimensional submanifold of phase space, and the trajectory may explore all those parts of the phase space which are energetically accessible. This is, however, not the case.

The transition from an integrable to a nonintegrable system is most clearly analyzed in one of the most fundamental results in the theory of Hamiltonian (conservative) systems: the celebrated theorem of Kolmogoroff, Arnold, and Moser, which describes the influence of perturbations on an integrable system. Here, we formulate this so-called *'KAM theorem'* for two degrees of freedom.

The KAM-theorem assumes a perturbation of an integrable system H_0 by a term H_1:

$$H(\mathbf{p},\mathbf{q}) = H_0(\mathbf{p},\mathbf{q}) + \varepsilon H_1(\mathbf{p},\mathbf{q}) \tag{2.21}$$

(the mathematical proof requires certain differentiability properties of H_1). The parameter ε measures the strength of the perturbation. The essence of the KAM theorem is the surprising result that most of the invariant tori survive, i.e. the tori still exist in the perturbed system, although slightly deformed, and the trajectory still covers a two-dimensional subset of the phase space. The stability of an invariant torus against perturbation depends on the degree of non-periodicity of the motion on the torus, i.e. the irrationality of the ratio of the two frequencies ω_1 and ω_2 (we will assume that ω_1 is the smaller of the frequencies).

Formulated more precisely, all invariant tori with

$$\left|\frac{\omega_1}{\omega_2} - \frac{r}{s}\right| > \frac{K(\varepsilon)}{s^{5/2}} \tag{2.22}$$

for arbitrary coprime integer numbers r and s are preserved in the perturbed system. The constant $K(\varepsilon)$ depends only upon the perturbation strength ε, and goes to zero for $\varepsilon \to 0$.

Excluded from the stability criterion (2.22) are, in particular, tori with rational frequency ratio carrying periodic trajectories. Moreover, around each rational frequency ratio $\omega_1/\omega_2 = r/s$ there exists a narrow region of width $K(\varepsilon)s^{-5/2}$, in which (2.22) is not satisfied. For increasing 'irrationality' of the frequency ratio, i.e. larger denominator s in r/s, the region (2.22) guaranteeing preservation of the invariant tori decreases. Since the set of rational values of

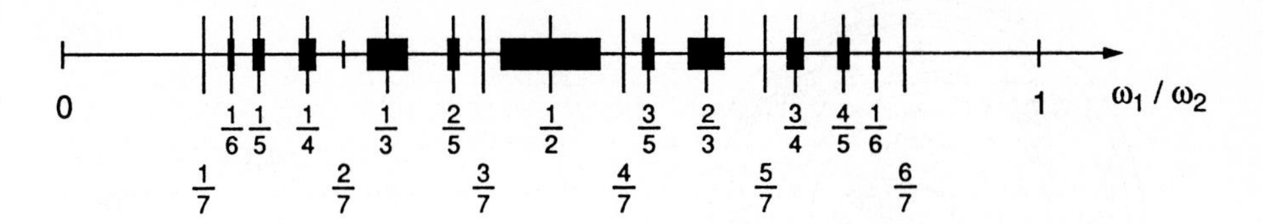

Fig. 2.4. Destroyed zones of length $K/s^{2.5}$ at rational frequency ratio r/s.

ω_1/ω_2 in the interval $[0,1]$ is dense and for every rational frequency ratio r/s an entire interval

$$\left|\frac{\omega_1}{\omega_2} - \frac{r}{s}\right| \le \frac{K(\varepsilon)}{s^{5/2}} \tag{2.23}$$

is excluded, one may conclude that (2.22) is practically never satisfied. This is, however, not the case. Figure (2.4) illustrates the destroyed zones for the most important r/s resonances, i.e. those with a small denominator ($s = 2, \ldots, 7$). Here, the value of the constant $K(\varepsilon)$ is chosen as 0.3. One observes that the width of the destroyed zones decreases rapidly with increasing s. But, nevertheless, there are infinitely many of them. A simple estimate yields for the union of all these intervals (2.23)

$$\sum_{r<s}\left|\frac{\omega_1}{\omega_2} - \frac{r}{s}\right| \le \sum_s \sum_{r<s} K(\varepsilon)s^{-5/2} \le \sum_s sK(\varepsilon)s^{-5/2} = K(\varepsilon)\sum_s s^{-3/2}. \tag{2.24}$$

The last sum converges and, therefore, the sum over all intervals (2.23) goes to zero for $\varepsilon \to 0$, so that for a sufficiently weak perturbation the phase space volume not filled with invariant curves can be made arbitrarily small.

The invariant tori in the zones excluded by the KAM-condition are in most cases destroyed. In the centers of the destroyed zone we have a torus with rational frequency ratio and, hence, periodic motion. In a simplified picture, one can imagine that, for such a periodic motion, the perturbation is also felt periodically, so that initially small changes induced in the trajectories may in time blow up, giving rise to large scale deviations. In addition to the rational torus, an interval given approximately by (2.23) is also destroyed.

The characteristic scenario for the subsequent destruction of invariant tori with increasing perturbation strength is discussed in more detail in Sect. 2.4.1. Here, we only note that a torus filled with periodic orbits with frequency ratio $\omega_1/\omega_2 = r/s$ quite typically decays. A number of $\nu\, s$ stable and $\nu\, s$ ($\nu \in \mathbb{N}$) unstable periodic orbits are, however, still existent, where the integer ν is often equal to one (a periodic trajectory is called *'stable'* if nearby trajectories remain close to it for all times).

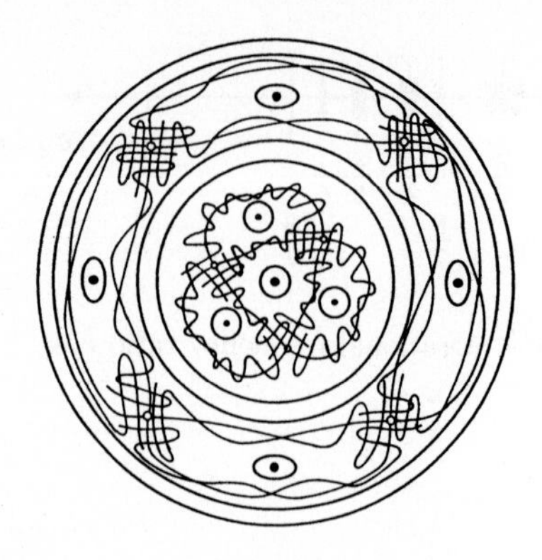

Fig. 2.5. Intact invariant curves and destroyed zones with chains of alternating elliptic (•) and hyperbolic (○) fixed points.

In a Poincaré section, these s-periodic trajectories which persist in the perturbed system appear as stable or unstable fixed points of the iterated Poincaré map T^S. The stability properties of such fixed points are studied in Sect. 2.4.5, where the use of the terms *'elliptic'* and *'hyperbolic'* for stable and unstable fixed points, respectively, is explained. The stable fixed points appear as centers of elliptic stability islands, as shown schematically in Fig. 2.5. A magnification of the neighborhood of such a fixed point appears to be almost self-similar to the original Poincaré section. The fixed point is surrounded by invariant curves, and between these invariant curves we find destroyed zones with alternating elliptic and hyperbolic fixed points. Magnifying the elliptic points again yields a similar picture. This process can be continued down to arbitrarily small scales (see Sect. 2.4.1 for more details).

2.2.4 Homoclinic Points

The neighborhood of the hyperbolic (i.e. unstable) fixed points looks very different and it is in this region that chaotic dynamics first develops. We define the *'stable manifold'* H^+ of the hyperbolic fixed point (p_h, x_h) as the set of points in the Poincaré section approaching the hyperbolic fixed point after infinitely many iterations:

$$H^+ = \{(p,x)| \lim_{n\to\infty} T^n(p,x) = (p_h, x_h)\}\,. \tag{2.25}$$

The *'unstable manifold'* H^- is the set of all points emanating from the fixed point after an infinite number of iterations:

$$H^- = \{(p,x)| \lim_{n\to\infty} T^{-n}(p,x) = (p_h, x_h)\}\,. \tag{2.26}$$

This is illustrated in Fig. 2.6.

One can convince oneself that the stable manifold H^+ has no self-crossings, because of the continuity of the map. The same is true for H^-. There are, however, crossings of H^+ and H^- and, moreover, these crossings occur generically. Such crossing points are called *'homoclinic points'* if the two manifolds belong

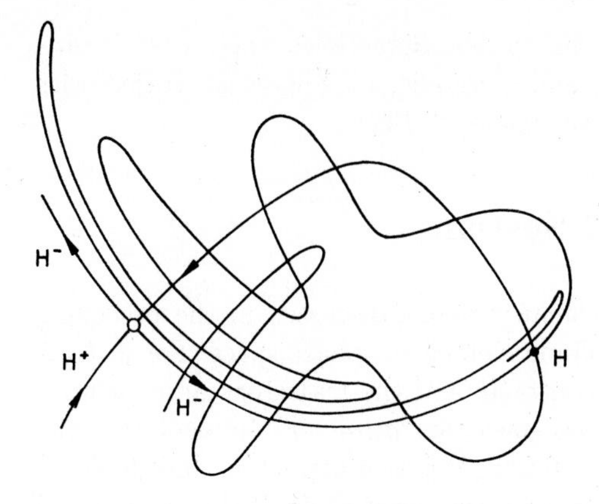

Fig. 2.6. The stable and unstable manifold H^+ and H^- of a hyperbolic fixed point (∘) intersect in a homoclinic point H (•) generating a complex network of loops.

to the same fixed point, or *'heteroclinic points'* for a crossing of stable and unstable manifolds of different fixed points.

The future fate and past history of such a homoclinic point is quite interesting. The continuity of the map implies that the iterates $T^n(X)$ of a neighborhood X of a homoclinic point must resemble the original set X, which contains an interval of the manifolds H^+ and H^-. Therefore H^+ and H^- must also cross in each of the iterates $T^n(X)$, $n = 1, 2, \ldots$, and we have a series of infinitely many homoclinic points, which are mapped onto each other and converge to the hyperbolic fixed point. Moreover, the map is area-preserving and the area enclosed by the branches of H^+ and H^- between two such homoclinic crossings is equal to the area enclosed by the branches between the subsequent crossings. Unavoidably, this leads to the formation of infinitely many thin loops as a consequence of the forbidden self-crossing of the manifolds. As already stated by Henri Poincaré in 1892,

> "The intersections form a kind of lattice, web or network with infinitely tight loops; neither of the two curves (H^+ and H^-) must ever intersect itself but it must bend in such a complex fashion that it intersects all the loops of the network infinitely many times. One is struck by the complexity of this figure which I am not even attempting to draw. Nothing can give us a better idea of the complexity of the three body problem and of all the problems in dynamics..." (cited after Tabor [2.6, p.144]).

It is in the neighborhood of the hyperbolic fixed points that chaotic dynamics can first be observed. With increasing perturbation ε, the destroyed zones grow, and the chaotic area-filling orbits in phase space increase. The 'chaotic sea' is, initially, still enveloped by intact invariant tori. With increasing ε, a growing number of invariant tori is destroyed, depending on the irrationality of the frequency ratio. The last surviving invariant tori are, in most cases, the so-called *'noble tori'*, whose frequency ratio is, in some sense, the most irrational (see Sect. 2.4.2 for more details). Finally, for large perturbation we find wide, extended chaotic regions. Only small islands surrounded by invariant curves remain, which are embedded in the chaotic sea. Sometimes, these islands ultimately disappear and the system becomes ergodic, i.e. almost all trajectories get arbitrarily close to any point on the energy surface.

2.3 Dissipative Dynamical Systems

An important feature of the Hamiltonian systems discussed in the preceding section is the conservation of the volume element in phase space. For a *dissipative system*, this volume element contracts and the trajectory approaches a (lower-dimensional) subset of the phase space, an *'attractor'*. Immediately, the problem arises of characterizing the different types of attractors and determining all coexisting attractors dependent on the parameters of the system and the organization of their *'basins of attraction'* in phase space, i.e. the set of all initial conditions of trajectories converging in the long time limit to a particular attractor, and the structural changes in the attractors when a parameter of the system is varied, i.e. their bifurcation properties.

Let us discuss a few general features of dynamical systems formulated as a system of autonomous differential equations

$$\frac{\mathrm{d}\mathbf{x}}{\mathrm{d}t} = \mathbf{v}(\mathbf{x}) \,, \tag{2.27}$$

where $\mathbf{x} = (x_1, \ldots, x_N)$ is a vector in N-dimensional phase space, and $\mathbf{v} = (v_1, \ldots, v_N)$ a vector field in phase space. For a given initial condition $\mathbf{x}_0 = \mathbf{x}(t_0)$, the differential equation (2.27) generates a flow $\mathbf{x}(t) = \mathbf{x}(t, \mathbf{x}_0)$ in phase space. A Poincaré map can be constructed on a surface of section, as illustrated in Fig. 2.2 above, which reduces the dynamics to the study of a discrete dissipative map.

The flow generated by (2.27) contracts the volume element

$$\Delta\tau = \Delta x_1 \, \Delta x_2 \cdots \Delta x_N \tag{2.28}$$

in phase space at a rate

$$\frac{1}{\Delta\tau}\frac{\mathrm{d}\Delta\tau}{\mathrm{d}t} = \operatorname{div}\mathbf{v} \,. \tag{2.29}$$

For some important model systems, this rate is equal to a (negative) constant, $-\gamma$, and the phase space volume elements shrink uniformly in phase space according to $\Delta\tau \sim \mathrm{e}^{-\gamma t}$. A well-known example of such a system is the forced and damped harmonic oscillator or, more generally, the Duffing oscillator

$$\ddot{x} + r\dot{x} + {\omega_0}^2 x + \beta x^3 = f \cos\omega t \,, \tag{2.30}$$

which is studied in detail in Chap. 8. For $\beta = 0$, we recover the driven harmonic oscillator. This explicitly time-dependent second order differential equation can be rewritten as

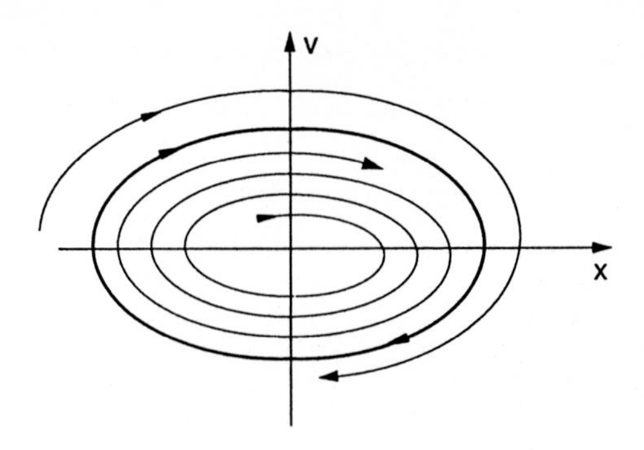

Fig. 2.7. Attracting limit cycle for the damped and driven harmonic oscillator.

$$\begin{aligned}\dot{x} &= v \\ \dot{v} &= -rv - \omega_0^2 x - \beta x^3 + \cos \omega s \\ \dot{s} &= 1\,,\end{aligned} \tag{2.31}$$

where the introduction of the auxiliary variable s removes the explicit time dependence (note that integration of $\dot{s} = 1$ with $s(0) = 0$ yields $s(t) = t$). The phase space volume contracts at the constant rate

$$\frac{1}{\Delta_\mathcal{T}} \frac{\mathrm{d}\Delta_\mathcal{T}}{\mathrm{d}t} = \frac{\partial}{\partial x} v + \frac{\partial}{\partial v}(-rv) + \frac{\partial}{\partial s} 1 = -r\,. \tag{2.32}$$

There are other systems, where the contraction rate varies in phase space, and one can introduce the long time average of the contraction rate along a trajectory. This is required to be negative for all initial conditions $\mathbf{x}_0$ for a dissipative system.

For the case of a discrete map

$$\mathbf{x}_{n+1} = \mathbf{f}(\mathbf{x}_n) \tag{2.33}$$

the N-dimensional phase space volume $\Delta_\mathcal{T}$ contracts per iteration by a factor

$$\left| \det \frac{\partial(f_1 \ldots f_N)}{\partial(x_1 \ldots x_N)} \right| , \tag{2.34}$$

i.e. the Jacobian determinant of the mapping function. For the most famous dissipative map, the one-dimensional logistic map $x_{n+1} = 4r\,x_n(1 - x_n)$ with $0 \le x_n \le 1,\ 0 < r < 1$ (see Chap. 9), the contraction per iteration is equal to $4r\,(1 - 2x_n)$, which is x-dependent and smaller than unity.

2.3.1 Attractors

The attractors in dissipative systems are of special interest. Basically, one can distinguish simple and strange attractors. Two types of simple attractors are already familiar from the driven harmonic oscillator

$$\ddot{x} + r\dot{x} + {\omega_0}^2 x = f \cos \omega t \tag{2.35}$$

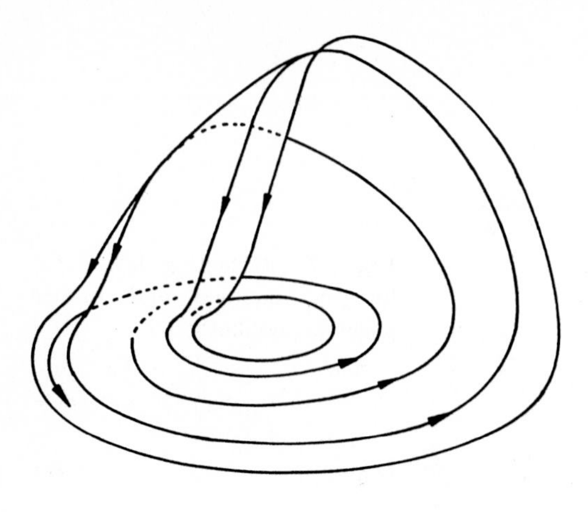

Fig. 2.8. Dissipative flow in three-dimensional phase space converging to a strange attractor.

with $r > 0$. For the unforced case $f = 0$, any trajectory converges to the stationary solution $x = 0$, a *'limit point'* or *'point attractor'*. For the driven case, $f \neq 0$, the function

$$x(t) = A\cos(\omega t - \phi) \tag{2.36}$$

— with an amplitude $A = f\{({\omega_0}^2 - \omega^2)^2 + (r\omega)^2\}^{-1/2}$ and a phase shift given by $\phi = r\omega/({\omega_0}^2 - \omega^2)$ — is a periodic solution oscillating with the same period as the external excitation. Any solution converges in the long time limit to (2.36) as illustrated in Fig. 2.7. Such an attracting periodic solution is called a *'limit cycle'*.

For the harmonic oscillator there is only one limit cycle, but for nonlinear systems as, for example, the Duffing oscillator (2.30), various limit cycles may coexist. This is explored in more detail in Chap. 8. In addition, the period of the limit cycle may be entirely determined by intrinsic properties of the system, and not by any external driving function. Other simple attractors are two-dimensional tori embedded in phase space and supporting a limiting quasiperiodic oscillation characterized by two frequencies.

When a parameter of the systems is varied slowly, the attracting limit points and limit cycles also change, while remaining structurally similar. At certain critical values, however, they can undergo sudden changes, e.g. change their character, split into pairs, disappear, etc. Such phenomena are called *'bifurcations'*. A characteristic example is the so-called *'Hopf bifurcation'*, where a limit point changes into a limit cycle (see Sect. 2.4.6 for more details).

A *'strange attractor'* is a limit set, which is much more complicated [2.12]. First, it is a fractal object, characterized by a noninteger, fractal dimension. A strange attractor has self-similar properties, i.e. a magnification of a part of it is similar to the whole set. Secondly, the dynamics *on* the strange attractor is chaotic, and characterized by an extreme sensitivity to small changes in the initial conditions. Nearby trajectories diverge exponentially, which can be quantitatively measured by the *'Lyapunov exponent'*, as described in more detail in

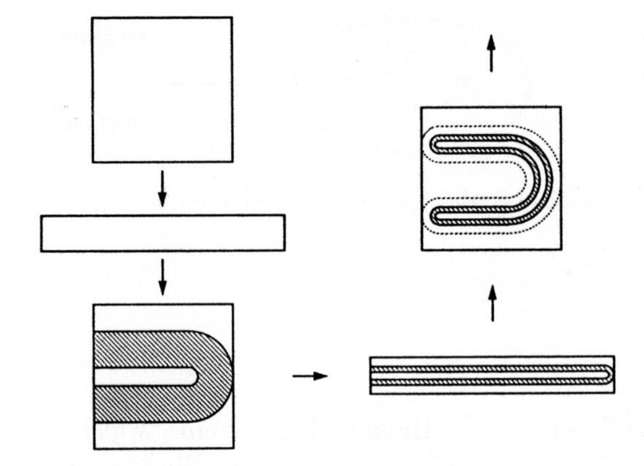

Fig. 2.9. Illustration of the Smale horseshoe map of a square, which consists of a stretching and a compression of the area followed by a folding over. This process is iterated.

Sect. 2.4.3. Both the non-trivial geometry and the irregular dynamics will be explored by means of computer experiments in the following chapters, in Chap. 8 and Chap. 12 in particular.

The two characteristic features, namely 'exponential divergence of neighboring trajectories' and 'contraction of phase space', seem to be incompatible at first sight and, in fact, they cannot simultaneously occur in two-dimensional phase space. For $N \geq 3$, however, such chaotic dissipative dynamics is possible. This can be understood using a simple model: a thin band of trajectories in three-dimensional phase space is first stretched (exponential divergence) and compressed (phase space contraction). It is then folded over and re-injected into itself, as illustrated in Fig. 2.8. This process is then iterated. It should be noted that phase space trajectories cannot cross and that the limiting set of this process will be a very complicated geometrical object.

The process of stretching and folding can be illustrated by the *'Smale horseshoe map'* [2.13]. The map consists of a *stretching* of a two-dimensional square followed by a folding over into the shape of a horseshoe, as illustrated in Fig. 2.9. This defines a map of the square onto itself, which contracts the area and separates nearby points. When this process is iterated, a very complex set of points is generated. A vertical section through the square reveals that the set consists of 2^n disjoint segments after n iterations. Mathematically, such a limiting set is called a *'Cantor set'*, i.e. a compact, uncountable, and totally disconnected set, a *'fractal'* object, which can be characterized by a non-integer dimension.

2.3.2 Routes to Chaos

When the external parameters of a system are varied, the dynamical behavior can change in character. In particular, it is of interest to study the transition from a parameter region with regular dynamics to a chaotic regime. Several characteristic routes of a system from regularity to chaos have been observed. Such a route is called a *'scenario'*.

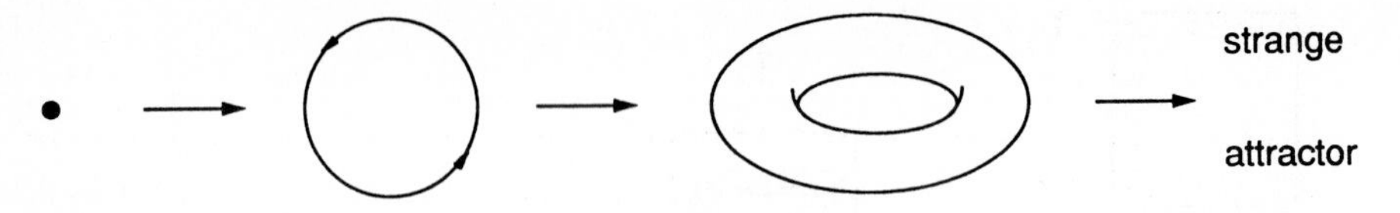

Fig. 2.10. The Ruelle-Takens-Newhouse scenario: bifurcation of attractors in phase space.

An example is the *'Poincarè scenario'* in Hamiltonian systems, which is characterized by the consecutive destruction of invariant tori according to their degree of 'rationality', as described in the preceding section. For dissipative systems other scenarios are known and, to some extent, understood, but the full theory of the transition to chaos has still to be developed.

The *'period-doubling scenario'* consists of a sequence of period-doubling bifurcations (see Sect. 2.4.6), where a stable periodic orbit becomes unstable, while a stable orbit of period two is born, which bifurcates again into a four-periodic orbit and so on. Finally, after an infinite number of such bifurcations, the system becomes chaotic.

Another important route to chaos is the *'Ruelle-Takens-Newhouse scenario'* [2.14, 2.15], which is characterized by a sequence of three bifurcations, as illustrated in Fig. 2.10. First, we find a point attractor, i.e. the system approaches a stable equilibrium. For a critical parameter value, this point attractor turns into a limit cycle (a Hopf bifurcation, as discussed in Sect. 2.4.6) and the system oscillates periodically. After a further parameter change, this periodic orbit loses its stability and bifurcates into a two-dimensional torus attractor in phase space (see Fig. 2.10). The motion on the attractor is quasiperiodic, characterized by two frequencies related to the two different rotations on the torus. When this torus-attractor is destabilized by a further parameter change, it turns into a strange attractor.

Finally, a system can become chaotic via the *'intermittency'* route, as proposed by Manneville and Pomeau [2.16, 2.17]. Such intermittent behavior can most easily be discussed in terms of one-dimensional maps (see Sect. 2.4.4). A stable fixed point, i.e. a point attractor, and an unstable fixed point approach each other when a parameter is varied. At a critical point they coalesce and disappear. In this region the system shows characteristic intermittent behavior, an almost regular and seemingly organized dynamics in the vicinity of the destroyed fixed point interrupted by long intervals of irregular motion (see Sects. 2.4.4 and 2.4.6 for more details).

2.4 Special Topics

In order to assist the reader in a somewhat deeper analysis of the computer experiments, we present in this section some more specialized topics in nonlinear dynamics, which will be useful in many of the systems studied in the following computer experiments. The material is, however, slightly more technical and the connection and relevance to chaotic dynamics is not directly obvious. This section can be omitted at a first reading.

2.4.1 The Poincaré-Birkhoff Theorem

According to the KAM theorem for Hamiltonian systems with two degrees of freedom (see Sect. 2.2) invariant tori with a sufficiently irrational ratio of the two basic frequencies ω_1/ω_2 remain invariant under a small perturbation of the Hamiltonian. Tori with a rational frequency ratio are excluded by the conditions of the KAM theorem and are, in most cases, destroyed.

Let us consider this rational case in more detail. For convenience, we consider a two-dimensional mapping of the plane, which models, for example, a Poincaré map of a two-dimensional Hamiltonian H_0:

$$\begin{aligned} \rho_{i+1} &= \rho_i \\ \theta_{i+1} &= \theta_i + 2\pi\alpha(\rho_i)\,, \end{aligned} \tag{2.37}$$

or simply

$$\begin{pmatrix} \rho_{i+1} \\ \theta_{i+1} \end{pmatrix} = \mathbf{T} \begin{pmatrix} \rho_i \\ \theta_i \end{pmatrix} . \tag{2.38}$$

The mapping $\mathbf{T}$ is called a *'twist map'*, where the radial coordinate ρ and the angular coordinate θ model action angle variables, and α plays the role of the frequency ratio. The mapping is area-preserving:

$$\left| \frac{\partial(\rho_{i+1}, \theta_{i+1})}{\partial(\rho_i, \theta_i)} \right| = \begin{vmatrix} 1 & 0 \\ 2\pi\,\alpha'(\rho_i) & 1 \end{vmatrix} = 1 \tag{2.39}$$

($\alpha' = \mathrm{d}\alpha/\mathrm{d}\rho$). The dynamics of the twist map is simple. The radial coordinate ρ is conserved and all points move along concentric circles, where the winding number α varies with the radius. For circles with a rational α-value

$$\alpha = r/s\,, \tag{2.40}$$

all points are s-periodic, i.e. fixed points of $\mathbf{T}^S$. If the twist map (2.37) is perturbed as

$$\begin{aligned} \rho_{i+1} &= \rho_i + \epsilon f(\rho_i, \theta_i) \\ \theta_{i+1} &= \theta_i + 2\pi\alpha(\rho_i) + \epsilon g(\rho_i, \theta_i)\,, \end{aligned} \tag{2.41}$$

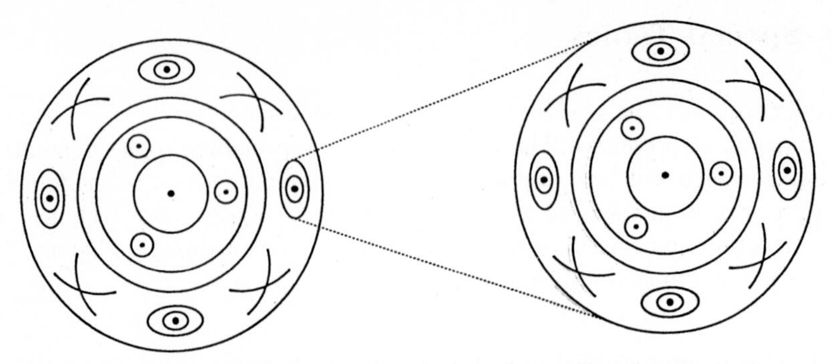

Fig. 2.11. Decay of invariant curves into chains of alternating elliptic and hyperbolic fixed points leading to a self-similar structure of the phase space.

where the (sufficiently well-behaved) functions f and g are chosen so that the mapping

$$\begin{pmatrix} \rho_{i+1} \\ \theta_{i+1} \end{pmatrix} = \mathbf{T}_\epsilon \begin{pmatrix} \rho_i \\ \theta_i \end{pmatrix} . \tag{2.42}$$

is still area-preserving, the perturbed map $\mathbf{T}_\epsilon$ can again be considered as a Poincaré map generated by a Hamiltonian $H_0 + \epsilon H_1$. The KAM theorem then guarantees that the invariant circles of $\mathbf{T}$ having a sufficiently irrational value of α are slightly transformed invariant curves of $\mathbf{T}_\epsilon$ for small values of ϵ.

The *Poincaré-Birkhoff theorem* [2.18] states that invariant circles with *rational* values $\alpha = r/s$ (r and s are coprime integer numbers) consisting entirely of fixed points of $\mathbf{T}$ are not completely destroyed. A number of $2ks$ points, $k \in \mathbb{N}$, transform into fixed points of $\mathbf{T}_\epsilon^S$. Half of these fixed points are elliptic and half are hyperbolic, forming an alternating chain (a short outline of the proof can be found in Ref. [2.4] or [2.19]). The schematic illustration in Fig. 2.11 shows two destroyed invariant curves with $\alpha = 1/3$ and $1/4$ forming chains of three and four elliptic and hyperbolic fixed points, respectively.

The vicinity of the hyperbolic fixed points show the complex irregular dynamics governed by the homoclinic tangle described in Sect. 2.2.4 above. Let us now explore the neighborhood of the elliptic fixed points of the chain $\mathbf{T}_\epsilon^S$ in more detail. At first sight, these 'small' fixed points seem to be surrounded by invariant curves. This is, however, an illusion. As before, the KAM and Poincaré-Birkhoff theorems can now be applied to the 'small' elliptic fixed points, showing that the seemingly invariant curves encircling them are also broken up into alternating elliptic and hyperbolic fixed points for rational rotation numbers. Figure 2.11 shows, for example, a magnification of such a region. Repeating this process by again magnifying the magnified fixed points reveals a fascinating self-similar structure of the phase space down to an arbitrarily small scale, as illustrated in Fig. 2.11.

2.4.2 Continued Fractions

The behavior of dynamical systems is, in many cases, sensitively dependent on the number-theoretic character of the value of certain parameters. To take an example, in the study of billiard systems in Chap. 3, an increasing deformation of the regular circular billiard with quasiperiodic motion destroys an increasing fraction of these regular orbits, depending on the irrationality of the quasi-periodicity. The KAM theorem requires that a frequency ratio be 'sufficiently irrational'. It is therefore necessary to gain some understanding of the relationship between rational and real numbers.

As is well known, a real number x can be arbitrarily closely approximated by rational numbers r/s. Here and in the following, r and s are coprime integers. The most familiar rational approximation is the decimal expansion, e.g.

$$\pi = 3.141592654\ldots \approx \frac{3}{1}, \frac{31}{10}, \frac{314}{100}, \ldots . \tag{2.43}$$

The quality of the decimal approximation is given by

$$\left| x - \frac{r}{s} \right| < \frac{1}{s} \tag{2.44}$$

and the approach is one-sided, i.e. the rational approximations are always smaller or larger than x for $x > 0$ or $x < 0$, respectively.

Let us now discuss the continued fraction expansion of a real number x:

$$x = a_0 + \cfrac{1}{a_1 + \cfrac{1}{a_2 + \cfrac{1}{a_3 + \cdots}}}, \tag{2.45}$$

($a_0 \in \mathbb{Z}$, $a_1, a_2, \ldots \in \mathbb{N}$). This expansion is unique and can easily be constructed for any x by taking a_0 as the integer part of x, defining

$$x_0 = x - a_0 \quad , \quad a_1 = [x_0^{-1}], \tag{2.46}$$

where the Gauss bracket [ ] denotes the largest integer number less than or equal to the real number inside. Iterating the construction (2.46)

$$x_n = x_{n-1}^{-1} - a_n \quad , \quad a_{n+1} = [x_n^{-1}], \quad n = 1, 2, \ldots \tag{2.47}$$

yields the series of the integers a_n.

The series of rational numbers obtained by cutting the continued fraction expansion (2.45) at a_n defines a series of rational approximations

$$x \approx \frac{r_0}{s_0}, \frac{r_1}{s_1}, \ldots, \frac{r_n}{s_n}, \ldots . \tag{2.48}$$

Such a r_n/s_n is the *best* rational approximation of x, i.e. there is no rational number with $s < s_n$ and

$$|x - \frac{r}{s}| < |x - \frac{r_n}{s_n}| . \tag{2.49}$$

A well-known example of such an approximation in terms of continued fractions is the series

$$\pi \approx \frac{3}{1}, \frac{22}{7}, \frac{333}{106}, \frac{355}{113}, \ldots . \tag{2.50}$$

It can be shown that the quality of the continued fraction approximation is given by

$$|x - \frac{r_n}{s_n}| < \frac{1}{s_n s_{n-1}}, \tag{2.51}$$

which is much stronger than, for example, the quality of the simple decimal approximation (2.43). In addition, the convergence is alternating, i.e. x lies between two subsequent approximations.

From (2.45), we see that the convergence of the continued fraction approximation is better for larger values of the natural numbers a_n. The slowest convergence is evidently found for $a_1 = a_2 = \cdots = 1$, corresponding to

$$g^* = \cfrac{1}{1 + \cfrac{1}{1 + \cfrac{1}{1 + \cdots}}}, \tag{2.52}$$

which is least well approximated by the rational numbers. g^* is the so-called *'golden number'*, *'golden mean'*, or *'golden ratio'*. It is closely related to the *'Fibonacci numbers'*

$$\{F_n\} = \{0, 1, 1, 2, 3, 5, 8, 13, \ldots\}, \tag{2.53}$$

which are defined by the recursion $F_{n+1} = F_n + F_{n-1}$ with $F_0 = 0,\ F_1 = 1$. From the ratio

$$u_n = \frac{F_n}{F_{n+1}} = \frac{F_n}{F_n + F_{n-1}} = \frac{1}{1 + u_{n-1}} \tag{2.54}$$

and $u_0 = 0$ we recover by iteration the continued fraction expansion (2.52) and, in addition,

$$g^* = \lim_{n\to\infty} u_n = \lim_{n\to\infty} \frac{1}{1 + u_{n-1}} = \frac{1}{1 + g^*} . \tag{2.55}$$

This yields the quadratic equation $g^*(1 + g^*) = 1$ having the solution

$$g^* = \frac{1}{2}\left(\sqrt{5} - 1\right) = 0.61803\ldots , \tag{2.56}$$

which is known as the *'golden mean'*. This number can be considered as the 'most irrational' number in the interval $[0, 1]$. In fact, there is a whole class of similar numbers, where the continued fraction expansion has the form

$$\{a_0, a_1, \ldots, 1, 1, 1, \ldots\} . \tag{2.57}$$

Such numbers are called *'noble numbers'*. The most important noble numbers in the interval $[0, 1]$ are those of the form

$$g_k^* = \frac{1}{k + g^*} \quad , \quad k = 1, 2, \ldots \, , \tag{2.58}$$

with $g_1^* = g^*$, $g_2 = 1/(2 + g^*) = 1 - g^* \approx 0.38197\ldots$.

As an example, we consider the perturbed two-dimensional Hamiltonian system discussed in Sect. 2.2.3, where the KAM-theorem guarantees the preservation of all invariant tori, whose frequency ratio satisfies

$$\left| \frac{\omega_1}{\omega_2} - \frac{r}{s} \right| > \frac{K(\varepsilon)}{s^{5/2}} \, , \tag{2.59}$$

i.e. whose frequency ratio is sufficiently irrational. With increasing perturbation these zones shrink to zero until finally all invariant tori are destroyed. The last invariant tori are, in many cases, those whose frequency ratio equals a noble number, the so-called *'noble tori'* (compare, e.g., the numerical experiment for the double pendulum in Sect. 5.4.3).

2.4.3 The Lyapunov Exponent

Chaotic dynamics is characterized by an exponential divergence of initially close points. Let us first discuss the case of one-dimensional discrete maps of an interval

$$x_{n+1} = f(x_n) \, , \; x \in [0, 1] \, , \tag{2.60}$$

which are studied numerically in Chap. 9. The so-called *'Lyapunov exponent'* is a measure of the divergence of two orbits starting with slightly different initial conditions x_0 and $x_0 + \Delta x_0$. The distance after n iterations

$$\Delta x_n = |f^n(x_0 + \Delta x_0) - f^n(x_0)| \tag{2.61}$$

increases exponentially for large n for a chaotic orbit according to

$$\Delta x_n \approx \Delta x_0 \, \mathrm{e}^{\lambda_L n} \, . \tag{2.62}$$

One can now relate the Lyapunov exponent analytically to the average stretching along the orbit x_0, $x_1 = f(x_0)$, $x_2 = f(f(x_0)), \ldots , x_n = f^n(x_0) = f(f(f \ldots (x_0) \ldots))$. From (2.61) and the chain rule of differentiation, we have

$$\begin{aligned} \ln \frac{\Delta x_n}{\Delta x_0} &\approx \ln \left| \frac{f^n(x_0 + \Delta x_0) - f^n(x_0)}{\Delta x_0} \right| \\ &\approx \ln \left| \frac{\mathrm{d} f^n(x)}{\mathrm{d} x} \right| = \ln \prod_{j=0}^{n-1} |f'(x_j)| \\ &= \sum_{j=0}^{n-1} \ln |f'(x_j)| \, , \end{aligned} \tag{2.63}$$

and finally

$$\lambda_L = \lim_{n\to\infty} \frac{1}{n} \ln \frac{\Delta x_n}{\Delta x_0} = \lim_{n\to\infty} \frac{1}{n} \sum_{j=0}^{n-1} \ln |f'(x_j)| \,, \tag{2.64}$$

where the logarithm of the linearized map is averaged over the orbit x_0, $x_1, \ldots x_{n-1}$.

Negative values of the Lyapunov exponent indicate stability, and positive values chaotic evolution, where λ_L measures the speed of exponential divergence of neighboring trajectories. At critical bifurcation points the Lyapunov exponent is zero.

For an interpretation of the Lyapunov exponent, it is instructive to note its relationship to the loss of information during the process of iteration. When the interval $[0, 1]$ is partitioned into N equal boxes, one needs $\mathrm{ld}\, N$ bits of information (ld is the logarithm of base two) to localize the particular box containing the point x_j, i.e. one has to ask $\mathrm{ld}\, N$ 'yes' or 'no' questions on average. After an iteration this box is stretched by a factor $|f'(x_j)|$ and we have an information loss

$$\Delta I(x_j) = -\mathrm{ld}\, |f'(x_j)| \tag{2.65}$$

regarding the position of the iterated point. The Lyapunov exponent can therefore be interpreted as the average loss of information:

$$\lambda_L = -\ln 2 \lim_{n\to\infty} \frac{1}{n} \sum_{j=0}^{n-1} \Delta I(x_j) = -\ln 2 \, \overline{\Delta I(x_j)} \,, \tag{2.66}$$

where the factor $\ln 2$ converts binary ('bits') to natural ('nats') units of information.

It is also of interest to express the Lyapunov exponent in terms of the asymptotic density of points covered by the orbit x_0, x_1, x_2, ...

$$\varrho(x) = \lim_{n\to\infty} \frac{1}{n} \sum_{j=0}^{n-1} \delta(x - x_j) \,, \tag{2.67}$$

which is, by virtue of its construction, invariant under the mapping function $f(x)$. This *'invariant density'* satisfies the integral equation

$$\varrho(x) = \int \mathrm{d}x' \, \varrho(x') \, \delta(x - f(x')) \tag{2.68}$$

and from (2.67) we directly find the identity

$$\lambda_L = \int \mathrm{d}x \, \varrho(x) \, \ln |f'(x)| \,. \tag{2.69}$$

In view of (2.66), this expression strongly resembles the usual definition of an (information theoretic) entropy, $\ln |f'(x)|$, averaged over the probability distribution.

The above discussion of the Lyapunov exponent as a quantitative measure of the average exponential separation of neighboring orbits can be extended to

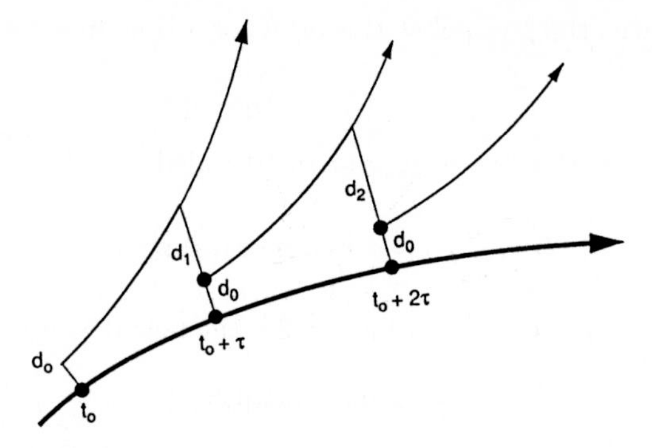

Fig. 2.12. Computation of the Lyapunov exponent by repeated rescaling of the distance of a displaced trajectory from a reference orbit.

higher dimensional discrete mappings and to continuous flows. Let us discuss, for example, a Hamiltonian system in $2N$-dimensional phase space $(\mathbf{p}(t), \mathbf{q}(t))$ evolving from an initial point $(\mathbf{p}(t_0), \mathbf{q}(t_0))$. Let us denote this orbit as the 'reference orbit'. We now follow a slightly displaced orbit, whose starting point is shifted by a small amount $(\Delta\mathbf{p}(t_0), \Delta\mathbf{q}(t_0))$. The time evolution of the separation from the reference orbit can be expressed in terms of the stability matrix $\mathbf{M}(t, t_0)$ connecting the distance in $(\Delta\mathbf{p}(t), \Delta\mathbf{q}(t))$ at time t with the separation at time t_0:

$$(\Delta\mathbf{p}(t), \Delta\mathbf{q}(t)) = \mathbf{M}(t, t_0)(\Delta\mathbf{p}(t_0), \Delta\mathbf{q}(t_0)) . \tag{2.70}$$

The Lyapunov exponent is defined as the long time limit

$$\lambda_L = \lim_{t\to\infty} \frac{\ln \|\mathbf{M}(t, t_0)\|}{t - t_0} , \tag{2.71}$$

where $\|\mathbf{M}\|$ denotes a matrix norm. For a regular trajectory the separation of neighboring trajectories grows as a (low) power of t, i.e. less than exponentially, and the Lyapunov exponent is zero. A positive Lyapunov exponent characterizes a chaotic trajectory. Typically, such a trajectory covers a certain phase space region and the Lyapunov exponent characterizes this region, regardless of the initial condition.

In many practical problems, however, a computation of λ_L using this equation is not practicable because of two problems: first, the exponential growth (2.71) may lead to numerical errors and computer overflow. Second, it cannot hold for all times, simply because in many cases the accessible phase space is bounded.

These problems can be overcome by using a repeated rescaling of the offset from the reference trajectory (see Fig. 2.12). Starting at time $t_0 = 0$ with a displaced orbit at a distance

$$d_0 = |(\Delta\mathbf{p}(0), \Delta\mathbf{q}(0))| \tag{2.72}$$

from the reference orbit, we follow this orbit for a time interval τ, compute the new distance

$$d_1 = |(\Delta\mathbf{p}(\tau), \Delta\mathbf{q}(\tau))|\,, \tag{2.73}$$

and start a new displaced trajectory at the rescaled initial point

$$(\mathbf{p}(\tau), \mathbf{q}(\tau)) + \frac{d_0}{d_1}(\Delta\mathbf{p}(\tau), \Delta\mathbf{q}(\tau))\,. \tag{2.74}$$

The trajectory is followed up to time $t = 2\tau$, the new deviation

$$d_2 = |(\Delta\mathbf{p}(2\tau), \Delta\mathbf{q}(2\tau))|\,, \tag{2.75}$$

is computed, and a second rescaled trajectory is started. This process is continued, yielding a sequence of scaling factors d_0, d_1, d_2, ... from which the Lyapunov exponent can be computed as

$$\lambda_L = \lim_{n\to\infty} \frac{1}{n\tau} \sum_{n=1}^{n} \ln \frac{d_n}{d_0}\,. \tag{2.76}$$

2.4.4 Fixed Points of One-Dimensional Maps

The dynamics of one-dimensional mappings

$$x_{n+1} = f(x_n, r)\,, \tag{2.77}$$

where r is a parameter, may converge in the long-time limit to the fixed points

$$x^* = f(x^*, r)\,. \tag{2.78}$$

The stability of a fixed point x^* can be obtained from linearization of the map (2.77) in the vicinity of the fixed point:

$$x_{n+1} - x^* = f\,'(x^*, r)\,(x_n - x^*) \tag{2.79}$$

with $f' = \mathrm{d}f/\mathrm{d}x$. This implies that deviations from the fixed point shrink for

$$\left|f\,'(x^*, r)\right| < 1 \qquad \text{(stable fixed point)} \tag{2.80}$$

and magnify for

$$\left|f\,'(x^*, r)\right| > 1 \qquad \text{(unstable fixed point)}\,. \tag{2.81}$$

In the case $\left|f\,'(x^*, r)\right| = 1$ we have neutral stability. It is important to realize that this simple criterion also applies to the stability of periodic k-cycles x_0, x_1, ..., $x_k = x_0$ with $x_n = f(x_{n-1}, r)$. Each member of this k-cycle is a fixed point x^* of the k-times iterated map

$$x^* = f^k(x^*, r) = \underbrace{f(f(\ldots f}_{k\,-\text{times}}(x^*, r)\ldots)) \tag{2.82}$$

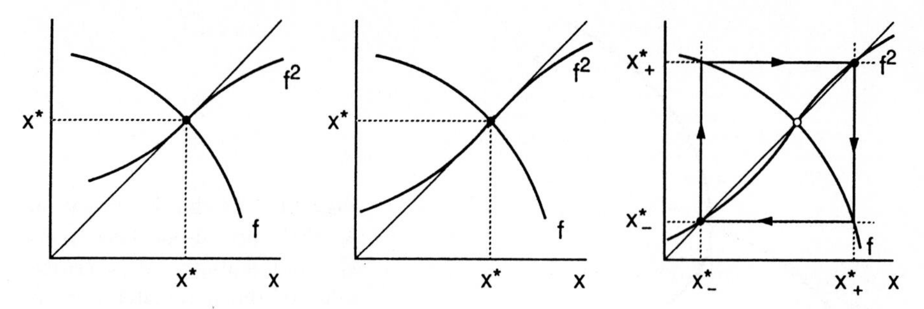

Fig. 2.13. Pitchfork bifurcation: A stable period-one fixed point x^* loses stability at a critical value of the parameter $r = r_1$, where the slope $|f'(x^*)|$ is unity, and a pair of period-two fixed points x^*_- and x^*_+ is born.

and the chain rule for differentiation yields

$$\left|\frac{\mathrm{d}f^k}{\mathrm{d}x}\right|_{x^*} = \prod_{n=0}^{k-1} |f'(x_n, r)| \quad \begin{cases} < 1 & \text{stable fixed point} \\ > 1 & \text{unstable fixed point} \end{cases}, \tag{2.83}$$

for $x^* = x_n$, $n = 0, \ldots, k-1$.

It is evident from (2.83) that the slope of f^k is identical for all k members x_n , $n = 0, \ldots, k-1$ of the k-periodic orbit. Graphically, this means that the function $f^k(x)$ becomes simultaneously tangential to the bisector $y(x) = x$ for the whole chain of fixed points when this orbit loses stability.

Let us look at the loss of stability of a fixed point $x^* = f(x^*, r)$ in more detail. Increasing the parameter r from a stable region ($|f'(x^*, r)| < 1$) to an unstable region ($|f'(x^*, r)| > 1$) crossing a critical value r_1 with $|f'(x^*, r_1)| = 1$, the fixed point $x^* = x^*(r)$ becomes unstable at $r = r_1$. We note that any fixed point of f is also a fixed point of the iterated map f^2 having the same stability properties. However, at $r = r_1$ a stable period-two orbit (x^*_-, x^*_+) with

$$x^*_- \xrightarrow{f} x^*_+ \xrightarrow{f} x^*_- , \tag{2.84}$$

is born. Both points, x^*_- and x^*_+, are stable fixed points of the iterated map f^2, i.e. the slope of f^2 is smaller than unity:

$$\left|\frac{\mathrm{d}f^2}{\mathrm{d}x}\right|_{\{x^*_-\}} = \left|\frac{\mathrm{d}f^2}{\mathrm{d}x}\right|_{\{x^*_+\}} = \left|f'(x^*_-, r)\right| \left|f'(x^*_+, r)\right| < 1 , \tag{2.85}$$

provided that r is close enough to r_1. This bifurcation of fixed points — a so-called *'pitchfork bifurcation'* — is illustrated in Figs. 2.13 and 2.14. It is investigated numerically in Chap. 9.

For increasing values of the parameter r, the fixed points of f^2 can also lose their stability at r_2 and bifurcate again into period-four orbits (fixed points of f^4), and so on.

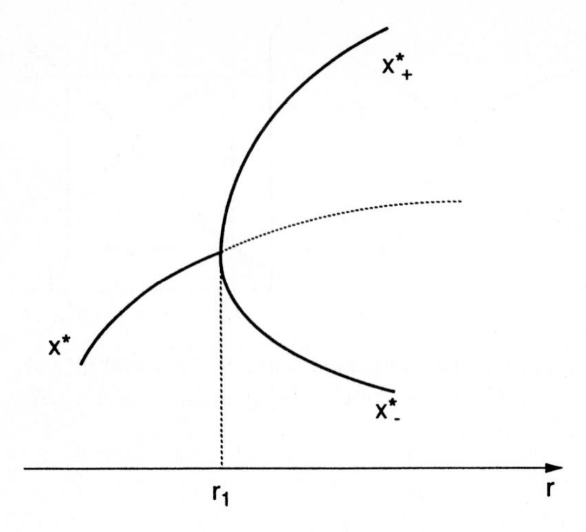

Fig. 2.14. Pitchfork bifurcation: A stable period-one fixed point x^* loses stability at a critical value of the parameter $r = r_1$ and a pair of period-two fixed points x^*_- and x^*_+ is born. The dashed curve marks the position of the unstable fixed point x^*.

Another bifurcation of fixed points of one-dimensional maps can be observed when the mapping function $f(x, r)$ becomes tangential to the bisector at a critical parameter value r_c, as illustrated in Fig. 2.15. For, say, $r > r_c$ we have two intersections with the bisector, $x^* = f(x^*, r)$, i.e. two fixed points of the map. At the critical parameter, the slope of f is unity and therefore one of the fixed points is stable (slope smaller than unity) and one is unstable (slope larger than unity). In the limit $r \to r_c$ the fixed points approach each other. They coalesce at r_c, and disappear for $r > r_c$. This bifurcation is called a *'tangent bifurcation'*.

For $r \gtrsim r_c$ there is a narrow channel between the mapping function and the bisector. When the iterates enter this channel, it takes a large number of iterations until the iterates are ejected from it again. During this process, the behavior of the iterates is very regular, but outside the channel the iteration may be irregular until the channel is re-entered. This behavior is called *'intermittency'* [2.16, 2.17].

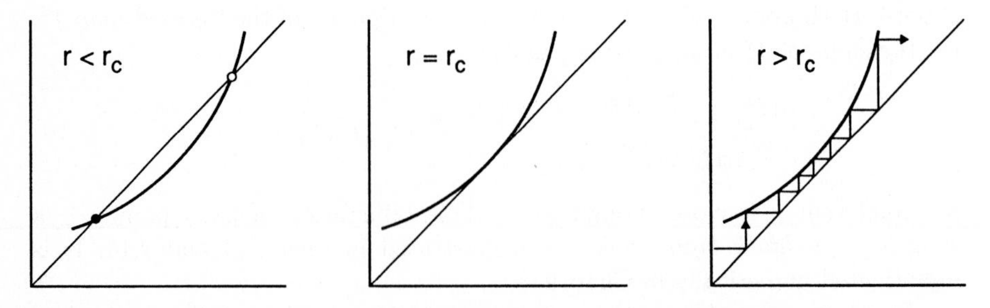

Fig. 2.15. Tangent bifurcation and intermittency: A stable (•) and an unstable (∘) fixed point coalesce and disappear at a critical parameter value r_c.

2.4.5 Fixed Points of Two-Dimensional Maps

Two-dimensional discrete maps

$$
\begin{aligned}
x_{n+1} &= f(x_n, y_n) \\
y_{n+1} &= g(x_n, y_n)
\end{aligned}
\tag{2.86}
$$

are important models for studying chaotic dynamics. A popular example is the Mandelbrot map explored in detail in Chap. 11. Such maps constitute interesting dynamical systems, which may be analyzed without any obvious physical interpretation. They may appear in a direct way, modeling realistic systems as in population dynamics, or they can appear as Poincaré maps of higher dimensional flows. An example are Hamiltonian systems with two degrees of freedom, whose dynamics in four-dimensional phase space is often reduced to the study of area-preserving two-dimensional Poincaré maps, which form an important subclass of two-dimensional maps.

Dissipative Maps. The stability and bifurcation properties of two-dimensional maps are much richer than in the one-dimensional case. Let us consider a fixed point (x^*, y^*) of the map (2.86), so that

$$
\begin{aligned}
x^* &= f(x^*, y^*) \\
y^* &= g(x^*, y^*) \,.
\end{aligned}
\tag{2.87}
$$

We can investigate the stability properties at the fixed point by means of a two-dimensional Taylor expansion

$$
\begin{aligned}
x_{n+1} &\approx f(x^*, y^*) + \left.\frac{\partial f}{\partial x}\right|_* (x_n - x^*) + \left.\frac{\partial f}{\partial y}\right|_* (y_n - y^*) \\
y_{n+1} &\approx g(x^*, y^*) + \left.\frac{\partial g}{\partial x}\right|_* (x_n - x^*) + \left.\frac{\partial g}{\partial y}\right|_* (y_n - y^*)
\end{aligned}
\tag{2.88}
$$

(here, a subscript $*$ denotes that the derivative is evaluated at the fixed point). We rewrite the linearized map in matrix form as

$$
\begin{pmatrix} x_{n+1} - x^* \\ y_{n+1} - y^* \end{pmatrix} \begin{pmatrix} \partial f/\partial x|_* & \partial f/\partial y|_* \\ \partial g/\partial x|_* & \partial g/\partial y|_* \end{pmatrix} \begin{pmatrix} x_n - x^* \\ y_n - y^* \end{pmatrix}
\tag{2.89}
$$

or, by introducing the deviation vector $\boldsymbol{\xi}$ with components $\xi_x = x - x^*$ and $\xi_y = y - y^*$, as

$$
\boldsymbol{\xi}_{n+1} = \mathbf{L}\, \boldsymbol{\xi}_n
\tag{2.90}
$$

with

$$
\mathbf{L} = \begin{pmatrix} \partial f/\partial x/|_* & \partial f/\partial y/|_* \\ \partial g/\partial x/|_* & \partial g/\partial y/|_* \end{pmatrix} = \left.\frac{\partial(f,g)}{\partial(x,y)}\right|_* .
\tag{2.91}
$$

The eigenvalues of the matrix $\mathbf{L}$ are given by

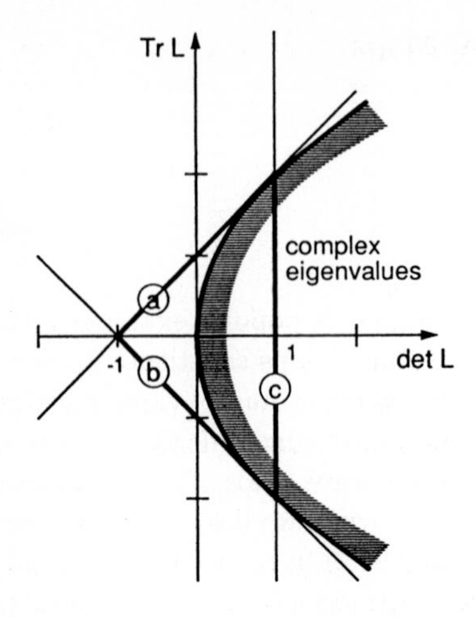

Fig. 2.16. Stability of a two-dimensional map **L**. Crossing the boundary of the stability triangle leads to characteristic bifurcations. In the convex region bounded by the dashed line the eigenvalues are complex conjugate.

$$\lambda_\pm = \tfrac{1}{2}\left\{\mathrm{Tr}\,\mathbf{L} \pm \sqrt{(\mathrm{Tr}\,\mathbf{L})^2 - 4\det\mathbf{L}}\right\} \tag{2.92}$$

(with $\mathrm{Tr}\,\mathbf{L} = L_{11} - L_{22}$ and $\det\mathbf{L} = L_{11}L_{22} - L_{12}L_{21}$), which are interrelated by

$$\lambda_+ + \lambda_- = \mathrm{Tr}\,\mathbf{L} \quad , \quad \lambda_+\lambda_- = \det\mathbf{L}\,. \tag{2.93}$$

The eigenvalues therefore depend on two real valued parameters $\det\mathbf{L}$ and $\mathrm{Tr}\,\mathbf{L}$, and the character of the fixed point can be conveniently related to different regions in the $(\det\mathbf{L}\,,\ \mathrm{Tr}\,\mathbf{L})$-plane. This is discussed below.

There are two possibilities, as is obvious from (2.93): both eigenvalues $\lambda_\pm$ can be real or complex conjugate. These two cases are separated by the parabola

$$(\mathrm{Tr}\,\mathbf{L})^2 = 4\det\mathbf{L} \tag{2.94}$$

in the $(\det\mathbf{L}\,,\ \mathrm{Tr}\,\mathbf{L})$-plane.

If both eigenvalues are inside the unit circle $|\lambda_\pm| < 1$ the magnitude of the difference vector contracts, the iterates of the linearized map converge to $\boldsymbol{\xi} = 0$, and the fixed point is stable. The stability region in the $(\det\mathbf{L}\,,\ \mathrm{Tr}\,\mathbf{L})$-plane is a triangle bounded by the three straight lines

$$\mathrm{Tr}\,\mathbf{L} - \det\mathbf{L} = 1\,,\ \ \mathrm{Tr}\,\mathbf{L} + \det\mathbf{L} = -1\,,\ \text{ and }\ \det\mathbf{L} = 1\,. \tag{2.95}$$

When parameters of the system are varied so that one or two eigenvalues cross the stability boundary $\lambda = 1$, characteristic bifurcations occur. One can distinguish three possibilities, depending on the crossed boundary line of the stability triangle:

(a) a so-called *'divergence'* occurs on the line $\mathrm{Tr}\,\mathbf{L} - \det\mathbf{L} = 1$,

(b) a *'flip'* on the line $\mathrm{Tr}\,\mathbf{L} + \det\mathbf{L} = -1$, and

(c) a *'flutter'* or *'Neimark bifurcation'* on the line $\det\mathbf{L} = 1$.

In the first two cases, (a) and (b), the eigenvalues are real and — because of the inequality $|\lambda_+\lambda_-| < 1$ — only a single eigenvalue, e.g. λ_+, can cross the unit circle and the other eigenvalue, λ_-, must be smaller than one in magnitude. Therefore, the dynamics of the iterates is basically one-dimensional: the iterates of the map first approach a line (contraction because of $|\lambda_-| < 1$) and then converge along this line to the center $\boldsymbol{\xi} = 0$ in the case of stability, or diverge to infinity in the case of instability. The details depend on the character of the crossing, $\lambda = +1$ for a linear divergence or $\lambda = -1$ for a linear flip, as well as on the sign of the second eigenvalue.

For case (c) — a linear 'flutter' or 'Neimark' bifurcation — the behavior is more involved. Here, a pair of complex conjugate eigenvalues crosses the unit circle $|\lambda_\pm| = 1$ simultaneously. The real part of the eigenvalues are identical and two cases can be distinguished:

— For a crossing with a positive real part, the iterates spiral inward in the stable region and outward in the case of instability. Exactly at the crossing we have neutral stability and the iterates trace out an elliptic orbit.

— For a negative real part, the iterates spiral and oscillate simultaneously. In addition, the behavior can be sensitively dependent on the phase of the eigenvalues at the crossing point of the unit circle. A rational phase $2\pi r/s$ leads to an s-periodic rotation (because of $\lambda^S = 1$) tracing out an elliptic orbit, which slowly contracts (inside the stability region) or expands (outside the stability region).

The interested reader can investigate these bifurcations numerically in some of the computer experiments described below (most directly for the two-dimensional maps studied in Chap. 11). More details about the theoretical description can be found in Ref. [2.3, Sect. 8.4].

Area Preserving Maps. In the remainder of this section we will discuss the special case of Hamiltonian dynamics. Here, the phase space volume is preserved, we find area-preserving maps and the linearized mapping (2.90) is restricted by

$$\det \mathbf{L} = 1 . \tag{2.96}$$

Therefore, the eigenvalues $\lambda_\pm$ of $\mathbf{L}$ must satisfy $\lambda_+\lambda_- = 1$ and the stability properties depend on the value of $\mathrm{Tr}\,\mathbf{L}$. Two cases can be distinguished:

(a) $|\mathrm{Tr}\,\mathbf{L}| > 2$: In this case the eigenvalues are real. Let us first assume $\mathrm{Tr}\,\mathbf{L} > 2$. We then have $\lambda_+ > \lambda_- > 0$ and the eigenvalues can be written as

$$\lambda_\pm = \mathrm{e}^{\pm\gamma} , \tag{2.97}$$

where γ is the 'stability exponent'. The eigenvectors

$$\mathbf{L}\,\boldsymbol{\xi}_\pm = \lambda_\pm \boldsymbol{\xi}_\pm \tag{2.98}$$

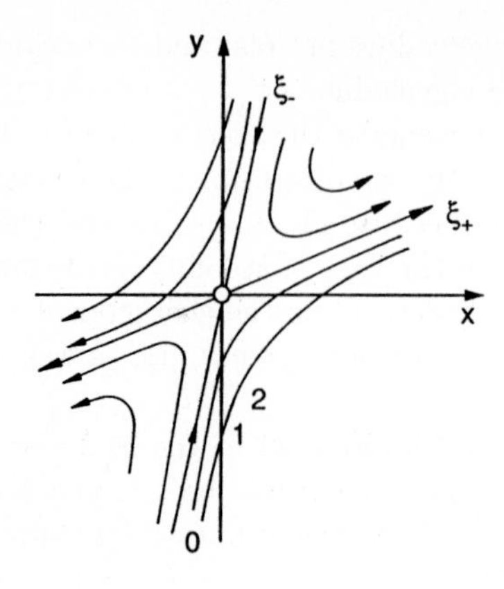

Fig. 2.17. Mapping properties of a hyperbolic fixed point of a two-dimensional linear area-preserving map.

describe the unstable ($\boldsymbol{\xi}_+$) and stable ($\boldsymbol{\xi}_-$) directions of the fixed point. Iterating a point which is initially on the ray in the unstable direction, the iterates stay on this line, moving outwards according to

$$\boldsymbol{\xi}_n = \mathbf{L}^n \boldsymbol{\xi}_0 = \mathrm{e}^{n\gamma} \boldsymbol{\xi}_0 \,. \tag{2.99}$$

Iterates started in the stable direction finally converge to the fixed point. All other points move on a branch of a hyperbola with asymptotes given by the unstable and stable directions, as shown below, iterating finally to infinity in the unstable direction, as illustrated in Fig. 2.17. Such an unstable fixed point is therefore called a *'hyperbolic fixed point'*.

For the case $\operatorname{Tr} \mathbf{L} < -2$, the eigenvalues can be written as $\lambda_\pm = -\mathrm{e}^{\mp\gamma}$, and the behavior is very similar except that the iterates alternate between both branches of the hyperbola. This is a *'hyperbolic fixed point with reflection'*.

(b) $|\operatorname{Tr} \mathbf{L}| < 2$: The eigenvalues are complex conjugate and of unit absolute value. They are conveniently expressed as

$$\lambda_\pm = \mathrm{e}^{\pm i\beta} \,, \tag{2.100}$$

where β is called the *'stability angle'*. In terms of the complex eigenvectors $\boldsymbol{\xi}_\pm$, any initial vector

$$\boldsymbol{\xi}_0 = a_+ \boldsymbol{\xi}_+ + a_- \boldsymbol{\xi}_- \tag{2.101}$$

is mapped to

$$\boldsymbol{\xi}_n = \mathbf{L}^n \boldsymbol{\xi}_0 = a_+ \mathrm{e}^{+in\beta} \boldsymbol{\xi}_+ + a_- \mathrm{e}^{-in\beta} \boldsymbol{\xi}_- \,. \tag{2.102}$$

These iterates are all restricted to an ellipse, which is rotated in the (x, y)-plane (see below) and the fixed point is called an *'elliptic fixed point'*, as illustrated in Fig. 2.18. The fixed point is stable, and any point close to it stays in its neighborhood.

It remains to show that the iterates of the map (2.90) with $\det \mathbf{L} = 1$ trace out a conic section. This can easily be seen by first observing that $\mathbf{L}$ is symplectic, i.e. it satisfies the relation

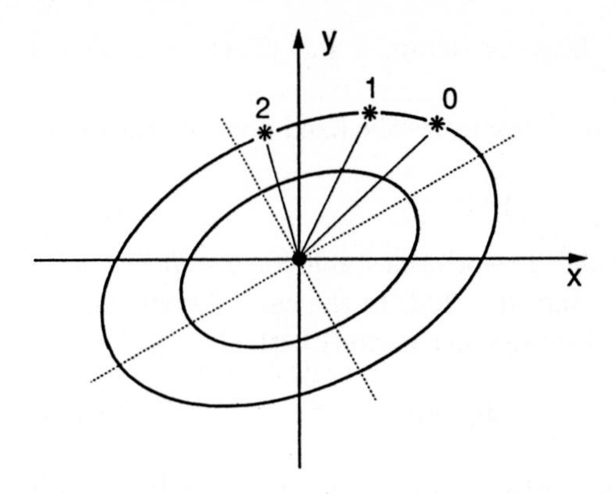

Fig. 2.18. Mapping properties of an elliptic fixed point of a two-dimensional linear area-preserving map.

$$\mathbf{L}^t\,\mathbf{J}\,\mathbf{L} = \mathbf{J} \tag{2.103}$$

with

$$\mathbf{J} = \begin{pmatrix} 0 & -1 \\ 1 & 0 \end{pmatrix}. \tag{2.104}$$

Secondly, the set

$$\Sigma_Q = \left\{ \boldsymbol{\xi} \,|\, \boldsymbol{\xi}^{\,t}\,\mathbf{J}\,\mathbf{L}\,\boldsymbol{\xi} = Q \in \mathbb{R} \right\} \tag{2.105}$$

is invariant under the map $\mathbf{L}$. To show this, we take $\boldsymbol{\xi} \in \Sigma_Q$ with image $\boldsymbol{\xi}' = \mathbf{L}\boldsymbol{\xi}$. We then have

$$\boldsymbol{\xi}'^{t}\,\mathbf{J}\,\mathbf{L}\,\boldsymbol{\xi}' = \boldsymbol{\xi}^{\,t}\,\mathbf{L}^{\mathbf{t}}\,\mathbf{J}\,\mathbf{L}\,\mathbf{L}\,\boldsymbol{\xi} = \boldsymbol{\xi}^{\,t}\,\mathbf{J}\,\mathbf{L}\,\boldsymbol{\xi} = Q\,, \tag{2.106}$$

which shows that $\boldsymbol{\xi}' \in \Sigma_Q$. Finally, the set Σ_Q is a quadratic form in the variables x and y:

$$\begin{aligned} Q = \boldsymbol{\xi}^{\,t}\,\mathbf{J}\,\mathbf{L}\,\boldsymbol{\xi} &= (x, y) \begin{pmatrix} 0 & -1 \\ 1 & 0 \end{pmatrix} \begin{pmatrix} L_{11} & L_{12} \\ L_{21} & L_{22} \end{pmatrix} \begin{pmatrix} x \\ y \end{pmatrix} \\ &= L_{12}\,y^2 + (L_{11} - L_{22})\,xy - L_{21}\,x^2\,. \end{aligned} \tag{2.107}$$

This determines a conic section. The reader can easily show that a rotation

$$\mathbf{D}(\chi) = \begin{pmatrix} \cos\chi & -\sin\chi \\ \sin\chi & \cos\chi \end{pmatrix} \tag{2.108}$$

with

$$\tan 2\chi = \frac{L_{22} - L_{11}}{L_{12} + L_{21}} \tag{2.109}$$

and $\mathbf{L}' = \mathbf{D}^{\mathbf{t}}\mathbf{L}\mathbf{D}$ brings the conic section to the standard form

$$Q = L'_{12}y'^2 - L'_{21}x'^2\,. \tag{2.110}$$

Furthermore, we have

$$L'_{12} - L'_{21} = L_{12} - L_{21} \quad \text{and} \quad L'_{12}L'_{21} = (\mathrm{Tr}\,\mathbf{L}/2)^2 - 1 \tag{2.111}$$

and one can again verify that an ellipse is obtained for $|\mathrm{Tr}\,\mathbf{L}| < 2$ and a hyperbola for $|\mathrm{Tr}\,\mathbf{L}| > 2$.

It is sometimes more convenient to characterize the fixed point by the quantity

$$R = (2 - \mathrm{Tr}\,\mathbf{L})/4\,, \tag{2.112}$$

the so-called *'residue'* of the fixed point. The orbit is stable for $0 < R < 1$ with the exception of the values $R = 3/4$ and $R = 1/2$. In the case of stability, the points move on ellipses at a rate ν (rotations per period) with

$$R = \sin^2(\beta/2) = \sin^2(\pi\nu)\,. \tag{2.113}$$

For $R > 1$ or $R < 0$, the orbit is unstable (hyperbolic or hyperbolic with reflection, respectively).

In the cases $R = 0$, 1, 3/4, 1/2, where the eigenvalues are low order roots of unity ($\lambda^k = 1$ with $k = 1, 2, 3, 4$), the linearized map is not sufficient to determine the stability of the fixed point (see [2.20]).

When a parameter of the (area-preserving) map is varied, the periodic orbits, i.e. the fixed points of the map (2.86) or its iterates, may bifurcate whenever the residue passes a value

$$R_{l,m} = \sin^2(\pi l/m)\,. \tag{2.114}$$

A more detailed discussion of the bifurcation properties of area-preserving maps can be found in the article by Green et al. [2.21].

2.4.6 Bifurcations

In many situations the structural properties of the dynamics are preserved when parameters of the system are (slowly) varied. There are, however, important exceptions: a bifurcation describes a rapid change in the type of dynamics when parameters of the system cross a critical value. Here, we discuss some important examples.

Pitchfork Bifurcation. Pitchfork bifurcation has been discussed above in the context of one-dimensional maps $x' = f(x, r)$. Let us recall its basic properties: a period-one fixed point of the map becomes unstable when the slope of the derivative $|f'(x, r)|$ passes through unity. At the same time, a pair of stable fixed points of period two appears, as illustrated in Figs. 2.13 and 2.14 above.

Tangent Bifurcation. In a tangent bifurcation a stable and an unstable fixed point approach each other and disappear into the complex plane. In the vicinity of the bifurcation point the system still feels the influence of the previous fixed points; they appear as 'ghosts', attracting the orbits for a while until they are ejected again. This leads to so-called *'intermittency'*, where an irregular wandering of the orbit is interrupted at irregular intervals by regular dynamics

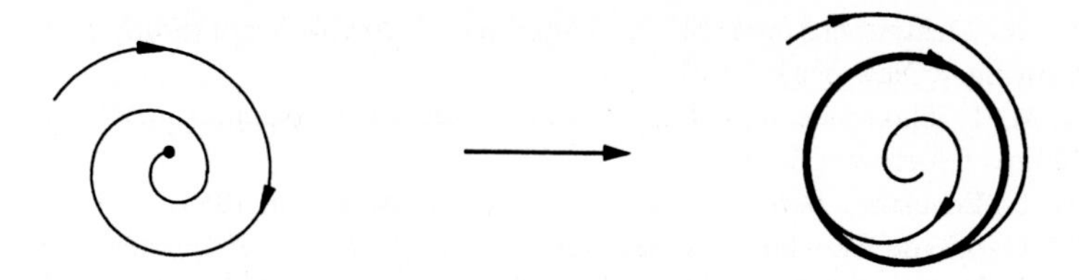

Fig. 2.19. Hopf bifurcation: A point attractor bifurcates into a limit cycle.

in the vicinity of the (disappeared) fixed points. As an example, such behavior is discussed for one-dimensional maps in Sect. 2.4.4 (compare also the numerical experiments in Chap. 9).

Hopf bifurcation. The Hopf bifurcation describes the transition from a point attractor to a limit cycle [2.22]. It can be modeled by the differential equations

$$\begin{aligned} \frac{\mathrm{d}r}{\mathrm{d}t} &= -(g+r^2)r \\ \frac{\mathrm{d}\phi}{\mathrm{d}t} &= \omega \end{aligned} \tag{2.115}$$

in polar coordinates, which can be solved in closed form:

$$r^2(t) = \frac{g\, r_0^2\, \mathrm{e}^{-2gt}}{r_0^2\,(1-\mathrm{e}^{-2gt})+g} \quad , \quad \phi(t) = \omega t \tag{2.116}$$

with $r(0) = r_0$ and $\phi(0) = 0$. The solution approaches a fixed point at $r = 0$ for $g \geq 0$. When the parameter g is decreased to negative values, a limit cycle of radius $\sqrt{-g}$ appears, which attracts all solutions from inside or outside (see Fig. 2.19).

It is furthermore instructive to linearize the differential equations (2.115) in Cartesian coordinates, yielding

$$\frac{\mathrm{d}}{\mathrm{d}t}\begin{pmatrix} x \\ y \end{pmatrix} = \begin{pmatrix} -g & -\omega \\ \omega & -g \end{pmatrix}\begin{pmatrix} x \\ y \end{pmatrix} . \tag{2.117}$$

The eigenvalues of the matrix are $\lambda_\pm = -g \pm i\omega$. At the bifurcation point $g = 0$, a pair of complex conjugate eigenvalues crosses the imaginary axis, which is characteristic for a Hopf bifurcation. A numerical study of a Hopf bifurcation can be found in the computer experiment in Sect. 12.3.2.

References

[2.1] J. Guckenheimer and P. Holmes, *Nonlinear Oscillations, Dynamical Systems, and Bifurcations of Vector Fields. Springer, New York 1983*

[2.2] A. J. Lichtenberg and M. A. Lieberman, *Regular and Stochastic Motion* (Springer, New York 1983)

[2.3] J. M. T. Thompson and H. B. Stewart, *Nonlinear Dynamics and Chaos* (John Wiley, Chichester 1986)

[2.4] H. G. Schuster, *Deterministic Chaos* (VCH, Weinheim 1988)

[2.5] M. Ozorio de Almeida, *Hamiltonian Systems – Chaos and Quantization* (Cambridge University Press, Cambridge 1988)

[2.6] M. Tabor, *Chaos and Integrability in Nonlinear Dynamics* (John Wiley, New York 1989)

[2.7] G. L. Baker and J. P. Gollub, *Chaotic Dynamics – An Introduction* (Cambridge Univ. Press, Cambridge 1990)

[2.8] J. Frøyland, *Introduction to Chaos and Coherence* (IOP Publishing, Bristol 1992)

[2.9] B.-L. Hao, *Chaos* (World Scientific, Singapore 1984)

[2.10] P. Cvitanović, *Universality in Chaos* (Adam Hilger, Bristol 1984)

[2.11] R. S. MacKay and J. D. Meiss, *Hamiltonian Dynamical Systems* (Adam Hilger, Bristol 1987)

[2.12] D. Ruelle, *Strange attractors*, Math. Intelligencer **2** (1980) 126 (reprinted in: P. Cvitanović, *Universality in Chaos*, Adam Hilger, Bristol 1984).

[2.13] S. Smale, *Diffeomorphisms with many periodic points*, in: S. S. Cairns, editor, *Differential and Combinatorial Topology*, Princeton University Press, Princeton 1963

[2.14] D. Ruelle and F. Takens, *On the nature of turbulence*, Commun. Math. Phys. **20** (1971) 167 (reprinted in: B.-L. Hao, *Chaos*, World Scientific, Singapore) 1984.

[2.15] S. E. Newhouse, D. Ruelle, and F. Takens, *Occurrence of strange axiom A attractors near quasiperiodic flows on* T^m, $m \geq 3$, Commun. Math. Phys. **64** (1978) 35

[2.16] P. Manneville and Y. Pomeau, *Intermittency and the Lorenz model*, Phys. Lett. A **75** (1979) 1

[2.17] Y. Pomeau and P. Manneville, *Intermittent transition to turbulence in dissipative dynamical systems*, Commun. Math. Phys. **74** (1980) 189 (reprinted in: B.-L. Hao, *Chaos*, World Scientific, Singapore) 1984 and P. Cvitanović, *Universality in Chaos*, Adam Hilger, Bristol 1984.

[2.18] G. D. Birkhoff, *Nouvelles recherches sur les systèms dynamiques*, Mem. Pont. Acad. Sci. Novi Lyncaei **1** (1935) 85

[2.19] M. V. Berry, *Regular and irregular motion*, in: S. Jorna, editor, *Topics in Nonlinear Dynamics*, page 16. Am. Inst. Phys. Conf. Proc. Vol, 46 1978 (reprinted in R. S. MacKay and J. D. Meiss, *Hamiltonian Dynamical Systems*, Adam Hilger, Bristol 1987.

[2.20] K. R. Meyer, *Generic bifurcations of periodic points*, Trans. AMS **149** (1970) 95

[2.21] J. M. Greene, R. S. MacKay, F. Vivaldi, and M. J. Feigenbaum, *Universal behaviour in families of area–preserving maps*, Physica D **3** (1981) 468 (reprinted in: R. S. MacKay and J. D. Meiss, *Hamiltonian Dynamical Systems*, Adam Hilger, Bristol 1987.

[2.22] J. E. Marsden and M. McCracken, *The Hopf Bifurcation and Its Applications* (Springer, New York 1976)

3. Billiard Systems

An extremely simple example for demonstrating chaotic dynamics in conservative systems numerically is that of Birkhoff's billiard [3.1], i.e. the frictionless motion of a particle on a plane billiard table bounded by a closed curve [3.2]–[3.7]. The limiting cases of strictly regular (*'integrable'*) and strictly irregular (*'ergodic'* or *'mixed'*) systems can be illustrated, as well as the typical case, which shows a complicated mixture of regular and irregular behavior. The onset of chaos follows the so-called Poincaré scenario, i.e. the consecutive destruction of invariant tori for increasing deviation from integrability as described by the KAM-theory and the Poincaré-Birkhoff theorem discussed in Chap. 2.

Billiard systems are particularly suitable as models for educational purposes, since billiard motion is easy to grasp. The numerical treatment, as opposed to other systems, does not require numerical integration of differential equations. This is an important advantage, because such computation is comparatively time-consuming, especially since chaotic phenomena are exhibited in the long-term time behavior of the orbits.

At points P_n, $n = 0, 1, \ldots$, the particle is reflected elastically at impacts with the boundary according to the reflection law (angle of incidence equals angle of reflection). We shall deal with convex billiards, i.e. a straight line has, at most, two intersection points with the boundary curve. In the billiard system, the concept of a Poincaré section evolves quite naturally from the boundary curve. This means that the full dynamics is represented by the orbit's data at impact with the boundary. Let the boundary curve be given as $r = r(\varphi)$ in polar coordinates (this is always possible for convex curves). The point of

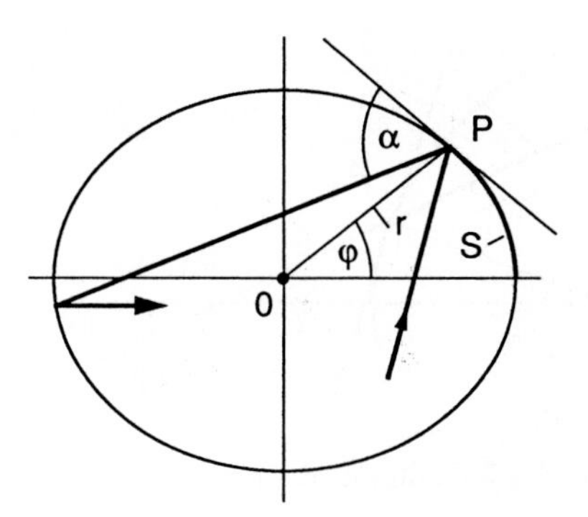

Fig. 3.1. Boundary curve $r(\varphi)$ and coordinates of the billiard system.

impact and the direction of the trajectory can be recorded by the two values of φ and $p = \cos\alpha$, where α is the angle between the velocity vector after reflection and the tangent to the boundary curve (see Fig. 3.1). For convenience, instead of the angle φ, we prefer to use the arc length

$$S(\varphi) = \frac{1}{L}\int_0^{\varphi} \sqrt{r^2(\varphi') + (\mathrm{d}r/\mathrm{d}\varphi')^2}\,\mathrm{d}\varphi' \tag{3.1}$$

of the boundary curve, measured in units of the total perimeter L of the curve. The Poincaré map relates the data of a point of impact (S_n, p_n) to that of the data of the succeeding impact (S_{n+1}, p_{n+1}) :

$$(S_n, p_n) \xrightarrow{T} (S_{n+1}, p_{n+1}) \,. \tag{3.2}$$

The mapping T is determined in a simple geometric manner by the reflection laws, as described in detail in Sect. 3.2 (see Fig. 3.2). Using the variables S and p, the mapping T is area-preserving, i.e.

$$\left| \frac{\partial(S_{n+1}, p_{n+1})}{\partial(S_n, p_n)} \right| = 1\,, \tag{3.3}$$

which can be seen by linearizing the billiard map $P_0 \xrightarrow{T} P_1$

$$\begin{pmatrix} \mathrm{d}S_1 \\ \mathrm{d}p_1 \end{pmatrix} = \mathbf{M}_{10} \begin{pmatrix} \mathrm{d}S_0 \\ \mathrm{d}p_0 \end{pmatrix} \tag{3.4}$$

with

$$\mathbf{M}_{10} = \frac{\partial(S_1, p_1)}{\partial(S_0, p_0)} \,. \tag{3.5}$$

'After a little algebra' (Berry [3.3]), one obtains an explicit expression for the deviation matrix $\mathbf{M}_{10}$:

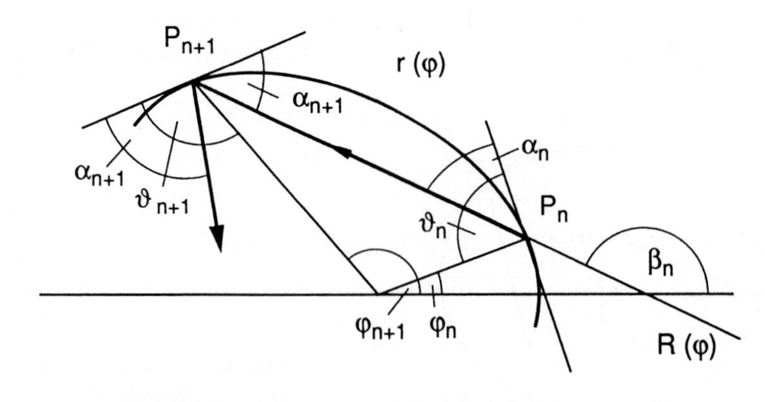

Fig. 3.2. Geometry and angles for billiard mapping.

$$\mathbf{M}_{10} = \begin{pmatrix} -\frac{q_0}{q_1} + \frac{l_{10}}{q_1 \rho_0} & -\frac{l_{10}}{q_0 q_1} \\ -\frac{l_{10}}{\rho_0 \rho_1} + \frac{q_1}{\rho_0} + \frac{q_0}{\rho_1} & -\frac{q_1}{q_0} + \frac{l_{10}}{q_0 \rho_1} \end{pmatrix} \tag{3.6}$$

with $q_i = \sin \varphi_i$. Here, the length of the straight line segment from P_0 to P_1 is denoted by l_{10} , and ρ_i is the radius of curvature at φ_i given by

$$\rho(\varphi) = \frac{(r^2 + r'^2)^{3/2}}{r^2 + 2r'^2 - rr''} , \tag{3.7}$$

with $r = r(\varphi)$, $r' = dr/d\varphi$ and $r'' = d^2r/d\varphi^2$. The determinant of $\mathbf{M}_{10}$ is equal to unity, i.e. the mapping T is area-preserving.

The linearization of the iterated map $T^n(S_0, p_0)$ is

$$\mathbf{M}_{n0} = \prod_{i=1}^{n} \mathbf{M}_{i,i-1} , \tag{3.8}$$

where the $\mathbf{M}_{i,i-1}$ have the form (3.6). For the special case of a periodic n-bounce orbit, we have $P_n = P_0$, and the deviation map

$$\mathbf{M}_n = \mathbf{M}_{0,n-1} \mathbf{M}_{n-1,n-2} \cdots \mathbf{M}_{1,0} \tag{3.9}$$

determines its stability (see Sect. 2.4.5 and also the numerical experiments in Sect. 3.4.4 below).

In the numerical simulation, the presentation on the screen is a two-dimensional Poincaré section of phase space (S, p) in addition to the trajectory in coordinate space. Three typical phase space pictures result:

1. A finite number of fixed points corresponding to periodic orbits in coordinate space.
2. Invariant curves: each initial condition evolves on a curve, which is invariant with respect to the dynamics.
3. A higher dimensional fraction of phase space is filled by the iterated points corresponding to chaotic dynamics.

For a typical billiard, all these different types of motion coexist, and are entangled in a very complicated manner.

3.1 Deformations of a Circle Billiard

It is instructive to discuss the simplest billiard form, the circle, and some of its deformations in some detail:

(a) Circle: This is the simplest form of a boundary curve. The system is rotationally symmetric and, therefore, integrable as a consequence of the existence

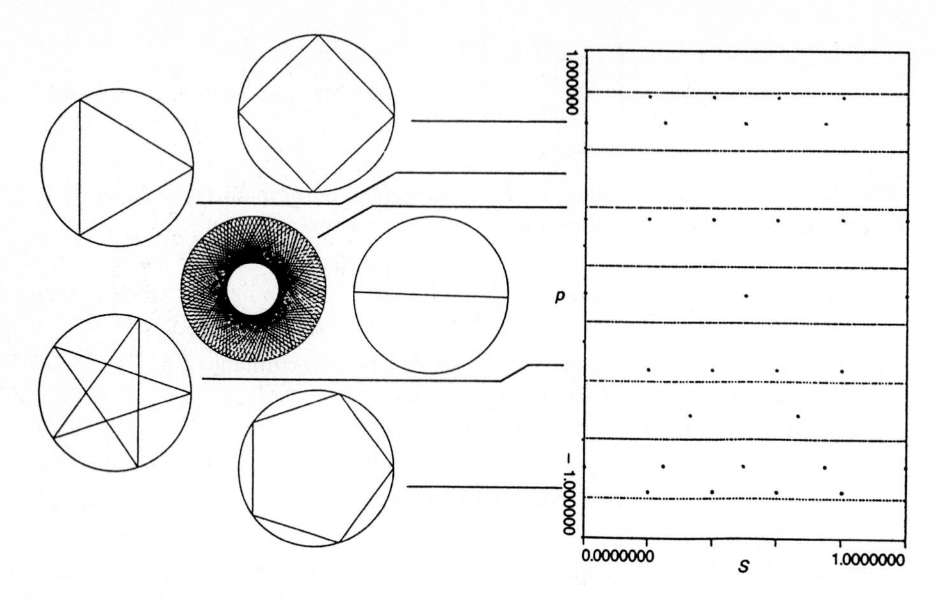

Fig. 3.3. Circular billiard: Trajectories and Poincaré section for periodic and non-periodic orbits. (Reproduced, with permission, from [3.7].)

of a second constant of motion in addition to the energy, namely, the angular momentum. The particle's motion is regular for all initial conditions and the impact points are simply given by

$$\varphi_n = \varphi_0 + 2\alpha n\,, \tag{3.10}$$

and $\alpha_n = \alpha_0 = \alpha$ is constant along the orbit, which constitutes conservation of angular momentum.

In phase space, only points or lines with the same angular momentum (i.e. the same value of p) appear. When choosing an angle α which is a rational multiple of π, $\alpha = \pi\, m/k$, where the integers m and k have no common divisor, one obtains a closed (periodic) curve in coordinate space, and a finite set of k fixed points of the Poincaré map T. The motion along the orbit can be divided into an oscillation between the inner and outer boundary curve, and a rotation about the center where the two frequencies have the ratio m/k. This is illustrated in Fig. 3.3, where trajectories are shown for $m/k = 1/2,\ 1/3,\ 1/4,\ 1/5,\ 2/5$, i.e. $p = \cos\alpha = 0,\ 1/2,\ 1/\sqrt{2}, \ldots$, as well as for an irrational ratio.

(b) Ellipse: An elliptical billiard, which can be described in polar coordinates by the curve

$$r(\varphi) = \frac{R}{\sqrt{1+\varepsilon^2\cos^2\varphi}}\,, \tag{3.11}$$

is also integrable [3.3, 3.6, 3.7], where the second constant of motion (compare Sect. 2.2.1) is the product of the angular momenta of the particle relative to

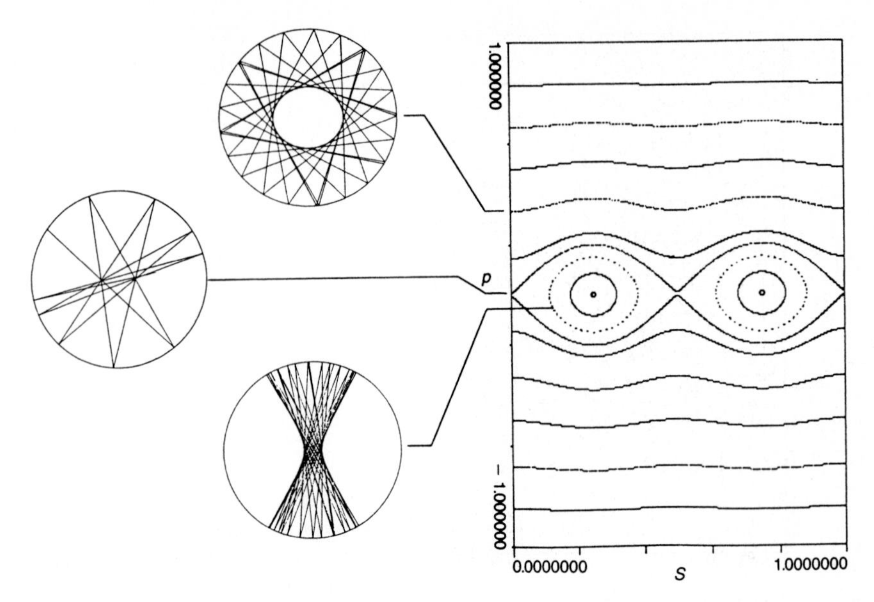

Fig. 3.4. Elliptical billiard with eccentricity $\varepsilon = 0.2$: Trajectories and Poincaré phase space section. (Reproduced, with permission, from [3.7].) This figure is stored (files ELLIPSE.PIC, –.DAT).

the two foci. Obviously, in the limit $\varepsilon \to 0$, this invariant is in agreement with angular momentum conservation for the circular billiard. After some algebra, the invariant can be rewritten as

$$F(\varphi, \alpha) = \frac{1}{2}\left\{r^2 - \varepsilon^2 a^2 + \left(a^2 - \varepsilon^2 r^2 \cos^2\varphi\right)\cos(2\alpha)\right\} , \qquad (3.12)$$

where $r = r(\varphi)$ is given in (3.11) and a is the long halfaxis of the ellipse.

As illustrated in Fig. 3.4, all phase space orbits lie on invariant curves. Two different regions can be distinguished, which are separated by the curve $F = 0$, the so-called separatrix. Invariants inside the separatrix consist of two disjoint closed curves in phase space, which are alternately visited by the trajectory. Invariant curves outside are simply connected (note that opposite boundaries of the (S, p)-plane are identical). In coordinate space the former fill a hyperbolically bounded region, whereas the latter an annular region bounded by an elliptical caustic. A more detailed consideration [3.3] shows that these caustic curves are indeed conic sections, confocal with the boundary curve. For increasing values of $|p|$, the area covered by the trajectory contracts to a tiny annular region close to the boundary, and we observe behavior similar to that of a whispering gallery. Trajectories on the separatrix pass exactly through the focal points.

Another interesting aspect of the ellipse is revealed by examining closed diametric two-bounce orbits. The impacts occur at opposite sides of the boundary curve at normal incidence. Two such orbits can be distinguished. The first, along the short diameter, is stable. This means that an orbit whose initial conditions

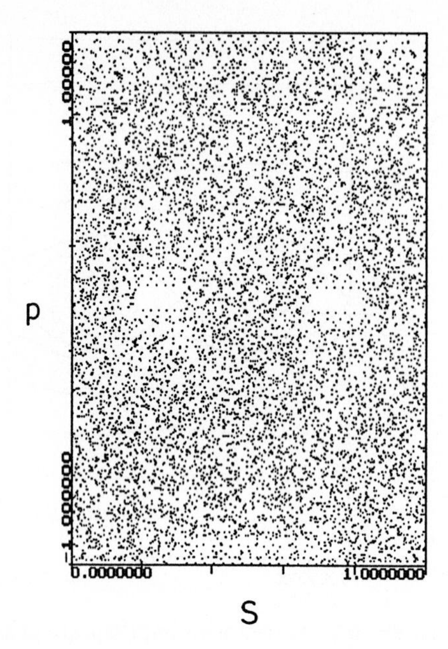

Fig. 3.5. Stadium billiard: Poincaré section of a single trajectory. This figure is stored (files STADIUM.PIC, –.DAT).

vary slightly from the closed orbit always remains close to it. The second orbit, along the long diameter, is unstable. Even the smallest deviation from the initial condition results in an orbit that strays far from the closed orbit. Since the computer is restricted to finite digit precision, small deviations are unavoidable. In the phase space diagram at $p = 0$ these two-periodic orbits appear as elliptic (stable) or hyperbolic (unstable) fixed points of the Poincaré map. They can also be identified with the minima or saddle points of the invariant (3.12). More details regarding stability properties can be found in the article by Berry [3.3].

(c) Stadium: This billiard system consists of two parallel straight lines joined by two semicircles. The ratio between the side length and the diameter of the semicircles can be varied. It has been proven that the stadium billiard is ergodic for all non-zero values of this ratio. Almost every initial condition evolves into chaotic motion, coming arbitrarily close to any other given point. The entire phase space is uniformly filled. Fig. 3.5 shows the Poincaré section of a single trajectory for a large number of iterations for a stadium with a side length/diameter ratio of 0.5. The two empty regions are due to the family of neutrally stable non-isolated two-bounce orbits between the parallel straight line sections of the stadium boundary. Ultimately, these regions will also be filled.

(d) General ovals: Elliptical billiards are the only known integrable billiards with differentiable boundary curve (see, e.g., the article by Poritsky [3.8]). Typical 'oval' billiards are chaotic, showing an intimate mixture of regular and irregular behavior. The type of motion is dependent upon the initial conditi-

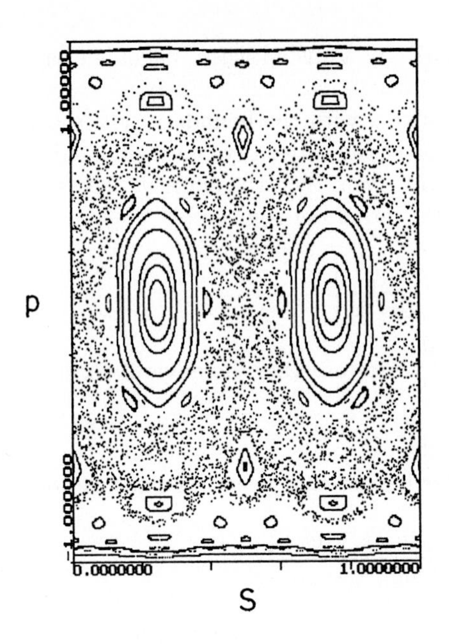

Fig. 3.6. Oval billiard: Poincaré section for several trajectories. This figure is stored (files OVAL.PIC, –.DAT).

ons. In phase space, some domains are filled, while others show invariant curves or fixed points. By way of example, Fig. 3.6 shows the Poincaré section for a boundary curve

$$r(\varphi) = c + \cos(2\varphi) \tag{3.13}$$

with a constant c chosen to be $c = 10$. One observes a complicated island structure within a chaotic sea. The centers of these islands of stability are stable periodic trajectories, and inside the islands we find regular quasiperiodic trajectories as well as more subtle irregular behavior (see the computer experiment 3.4.1). Some of the islands are arranged along chains. The large chaotic sea is produced by iterations of a single trajectory.

3.2 Numerical Techniques

The program computes the boundary curve of a billiard system given in polar form (radius as function of the polar angle φ). Boundary curves can be defined interactively by the user (see 'parameter menu'). The program checks that they are both closed and convex. The first derivative $dr(\varphi)/d\varphi$, which is used in equation (3.1) for the arc length and in a Newton iteration scheme described below, is evaluated analytically by the program. The calculation of the total arc length

$$L = \int_0^{2\pi} \sqrt{r^2(\varphi') + (\mathrm{d}r/\mathrm{d}\varphi')^2}\,\mathrm{d}\varphi' \tag{3.14}$$

of the boundary curve, as well as the subsequent computation of the integral $S(\varphi)$ defined in equation (3.1), are carried out by numerical integration. The accuracy of this integration is automatically adjusted when magnified parts of the phase space are considered.

The trajectory is a sequence of bounces with the boundary. The sequence of the points of impact and the direction of the trajectory can easily be derived from the reflection law. If φ_n and α_n are given, then the angle ϑ_n between the positive direction of the tangent and the radial ray is given by

$$\tan \vartheta_n = \left. \frac{r(\varphi)}{\mathrm{d}r/\mathrm{d}\varphi} \right|_n \tag{3.15}$$

and, therefore, the direction β_n of the trajectory (i.e. the angle it forms with the $\varphi = 0$ direction) is

$$\beta_n = \pi + \varphi_n + \alpha_n - \vartheta_n \, . \tag{3.16}$$

Hence, the straight line trajectory after impact n is given in polar form by

$$R(\varphi) = r(\varphi_n) \frac{\sin{(\beta_n - \varphi_n)}}{\sin{(\beta_n - \varphi)}} \, . \tag{3.17}$$

The next impact coordinate φ_{n+1} is determined by the intersection of the line (3.17) with the boundary $r(\varphi)$, i.e. the solution of the equation

$$R(\varphi) - r(\varphi) = 0 \, . \tag{3.18}$$

When $r(\varphi)$ is convex, there exists only one further solution, φ_{n+1}, in addition to φ_n, which is numerically extracted by the Newton iteration scheme. The angle of the trajectory with respect to the tangent in φ_{n+1} is

$$\alpha_{n+1} = \varphi_{n+1} - \varphi_n + \vartheta_n - \vartheta_{n+1} - \alpha_n \, , \tag{3.19}$$

as illustrated in Fig. 3.2. The succeeding boundary reflections are computed by repeating these steps.

For phase space pictures, at each of these iteration steps the φ and α coordinates are converted to $S(\varphi)$ and p just prior to display.

3.3 Interacting with the Program

The program BILLIARD can be controlled by means of two menus: in the *Main Menu*, shown in Fig. 3.7, one can store and load pictures, select the presentation mode, access various printers, and start the iteration. On-line information can be obtained by pressing the ⟨F1⟩ key. In the *Parameter Menu*, the boundary form and the initial conditions of the simulation are set.

Main Menu

- **Clear graph/initialize** — clears the buffered graphics.

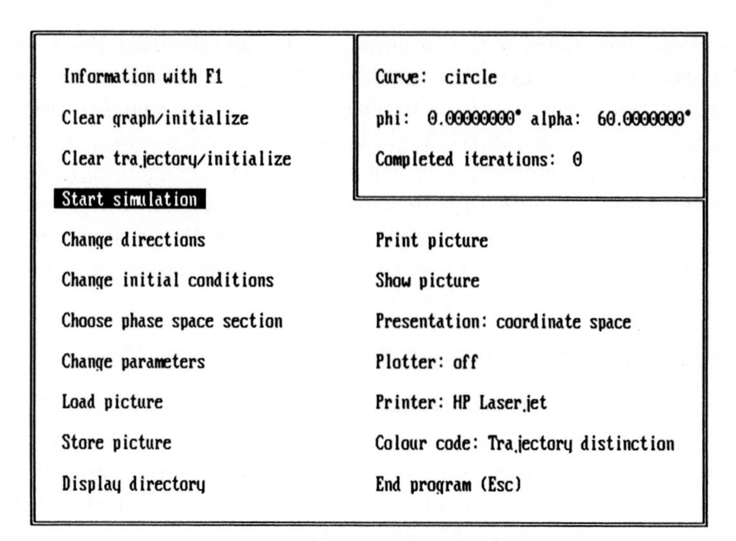

Fig. 3.7. Main menu of program BILLIARD.

- **Clear trajectory/initialize** — clears the coordinate space orbit in the buffered graphics. This menu item does not effect the phase space (in contrast to the above menu item).

- **Start simulation** — switches to graphics and starts simulation of the billiard system with the present parameters. The simulation can be interrupted by pressing any key. If an important parameter has been altered since the last simulation (for example, a new boundary curve has been introduced), the existing graph is cleared. Otherwise drawing continues on the original graph.

- **Change directions** — changes the direction of the particle, which should return to its previous path. This serves as a control of the numerical accuracy.

- **Change initial conditions** — is only possible in presentation mode *phase space* and *coordinate/phase space*. In the phase space plot, a cross, which can be controlled by means of cursor keys, appears. Pressing the ⟨TAB⟩ key switches between slow and fast motion of the cross. If the cross is in the desired position, then pressing the ⟨ENTER⟩ key fixes the corresponding phase space coordinates as new initial conditions, and the computation of the trajectory is started. Pressing ⟨ESC⟩ returns to the presentation mode without any changes.

- **Choose phase space section** — is only possible in presentation mode *phase space* or *coordinate/phase space* and allows a magnification of phase space sections. A rectangular window appears in the center of the screen.

The boundary lines of this window can be moved using the cursor keys. One can switch between the upper and lower, or right and left boundary, by using the ⟨SPACE BAR⟩. Pressing the ⟨TAB⟩ key switches between fast and slow motion. If the rectangle is in the desired position, then pressing the ⟨ENTER⟩ key starts the iteration. After leaving this item and using the ⟨ENTER⟩ key, the chosen phase space section is magnified and the iteration starts. If the rectangle is enlarged until it agrees with the shown phase space boundary, the program returns to the entire phase space $|p| < 1$, $0 \leq S \leq 1$. Pressing ⟨ESC⟩ returns without any changes.

- **Change parameters** — switches to the *Parameter Menu.*
- **Load picture** — reads the data from files *.DAT and *.PIC. The results of the present simulation are then lost. The program is in the same state as when the loaded picture was stored. Pictures for the four cases discussed in Sect. 3.1(a-d) above are already given, and can be loaded (files CIRCLE.DAT, ELLIPSE.DAT, STADIUM.DAT, and OVAL.DAT).
- **Store picture** — when this menu item is called, one is asked under which name the picture is to be stored. Suffixes are ignored. After the input of a valid file name, two files under this name, with the suffixes *.PIC and *.DAT , are created on the logged drive. The file *.PIC contains the last compiled picture, while *.DAT stores the necessary data for this picture (i.e. bounds of phase space section, momentary curve, presentation mode).
- **Display directory** — shows the content of the active directory.
- **Print picture** — A hardcopy of the recent graphics is printed. One must verify that the printer type chosen in the menu item *printer* coincides with the type linked to the computer. *Laser printer* is pre-set.
- **Show picture** — displays the present graph until the user presses any key.
- **Presentation** — offers the option of three presentation modes:
  - *Coordinate space:* The orbit of the particle is drawn in coordinate space. The last segment of the trajectory is shown in different color.
  - *Phase space:* The phase space of the system is displayed.
  - *Coordinate/phase space:* The left side of the screen is used for coordinate space, while the right is used for phase space presentation.

  Coordinate space is pre-set.
- **Plotter** — switches the plotter 'on' or 'off', and is only appropriate in the presentation mode *Coordinate space*. Verification should be made that the plotter is actually linked to the computer. At the start of the plotting routine, the user is requested to define the plotter's coordinate axes. For

this purpose, the ⟨ENTER⟩ key must be pressed three times. *Plotter: off* is pre-set.

- **Printer** — informs the program which printer type is linked to the computer. This is only of interest when a hardcopy of the graph is to be made. Options: *Laser printer* (HP-Laserjet compatible), *Postscript printer*, and *Matrix printer* (Epson 9 pins compatible). *Laser printer* is pre-set.

- **Color Code** — by pressing the space bar, one can choose between two schemes for coloring the phase space points: *Trajectory distinction* uses different colors for different trajectories, *Frequency distinction* monitors the number of recurrences of the phase space pixels, i.e. the phase space density.

- **End program** — terminates program. Data not saved are lost.

Parameter Menu

- **Initial conditions**

  - *alpha, phi* — determine the coordinate and direction of the particle at the start of the simulation. The coordinate is parameterized by *phi* $= \varphi$, the polar angle of the point of impact on the boundary. The distance to the origin is determined by the boundary curve. The direction of the trajectory is labeled by its angle *alpha* $= \alpha$ with respect to the tangent to the boundary curve at the given coordinate ($0^0 < \alpha < 180^0$). Angles are measured in degrees. The values $\alpha = 60^0$, $\varphi = 0^0$ are pre-set.

- **Function selection** — different boundary curves can be chosen:

  - *Circle*
  - *Ellipse* — the variable eccentricity is requested. Pre-set value: 0.8 .
  - *Stadium* — the ratio of side length to diameter of semicircles is requested. All values greater than 0 are valid. Pre-set value: 0.5 .
  - *Oval* — The equation of an oval (in polar form): $r(\varphi) = const + \cos(2\varphi)$ (see equation (3.13)) The constant *const* (parameter of the oval) is requested; for values *const* > 5, the oval is convex.
  - *Self-defined* — The user has the option of defining a boundary curve in polar form. To this effect, the pre-set function must be overwritten. On-line input of mathematical formulae is possible, as described in Appendix B.3. If one line is not sufficient, ⟨ENTER⟩ or ⟨↓⟩ must be pressed and two further lines are at one's disposal. Instead of the variable *'phi'*, one can also use *'x'*, while instead of *'cos(phi)'* one can use *'C'*. The computer converts this to φ or $\cos\varphi$, respectively. To edit the expressions, the ⟨INS⟩ and ⟨DEL⟩ keys are quite convenient (for more details see Appendix B.3). Upon calling the menu item *accept*, the function

is proof-read. If a mistake occurs, a message is displayed and the old function remains valid. For non-convex boundary curves, the next intersection of a straight line (the trajectory) with the boundary is not unique and errors are possible. Pressing ⟨Esc⟩ returns without changing the boundary curve.

3.4 Computer Experiments

3.4.1 From Regularity to Chaos

In this computer experiment we will study a 'typical' billiard with the boundary curve

$$r(\varphi) = 1 + \varepsilon \cos\varphi \,. \tag{3.20}$$

It can be shown that this cos-billiard is identical to the analytical billiard studied by Robnik [3.5]). Figure 3.8 shows the deformation of the boundary curve for increasing values of ε. Between $\varepsilon = 0$ (circle) and $\varepsilon = 0.5$, the boundary curve is convex. In the following discussion, we shall limit ourselves to this ε-interval.

Poincaré sections for $\varepsilon = 0.1, 0.2, \dots, 0.5$ are displayed in Figs. 3.9 (a) through 3.9 (e). One encounters behavior completely different from that exhibited by the elliptical billiard (see Fig. 3.4). The differences become more distinct with increasing ε. For small perturbation, the picture strongly resembles that of an elliptical deformation of the circle-billiard: the two-periodic orbits for $p = 0$ separate into a stable and unstable orbit, which appear as elliptic and hyperbolic fixed points of the Poincaré map. It should, however, be noted that the stability, or instability, of these orbits is exchanged as compared to the elliptical billiard. A more detailed discussion can be found in Sect. 3.4.4 below.

The resemblance with an elliptical billiard for small values of ε is merely superficial. Even for $\varepsilon = 0.1$, some clear island chains can be discerned, a first indication for the non-integrability of the cos-billiard. If one studies the island structures more closely, one finds in the centers of these islands an n-periodic

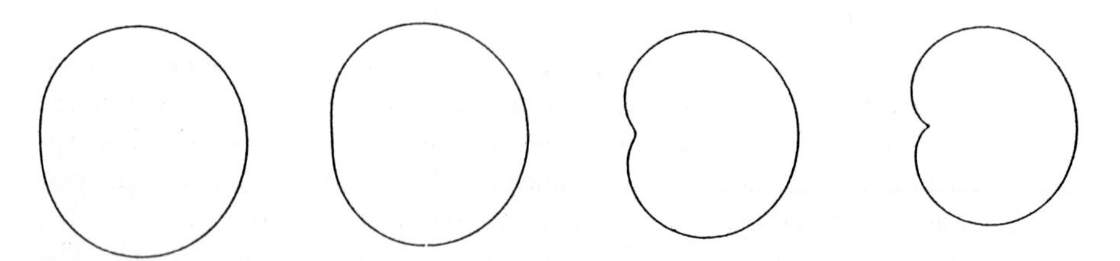

Fig. 3.8. The 'typical' cos-billiard for increasing deformation parameters $\varepsilon = 0.4$, 0.5, 0.8, and 1.0.

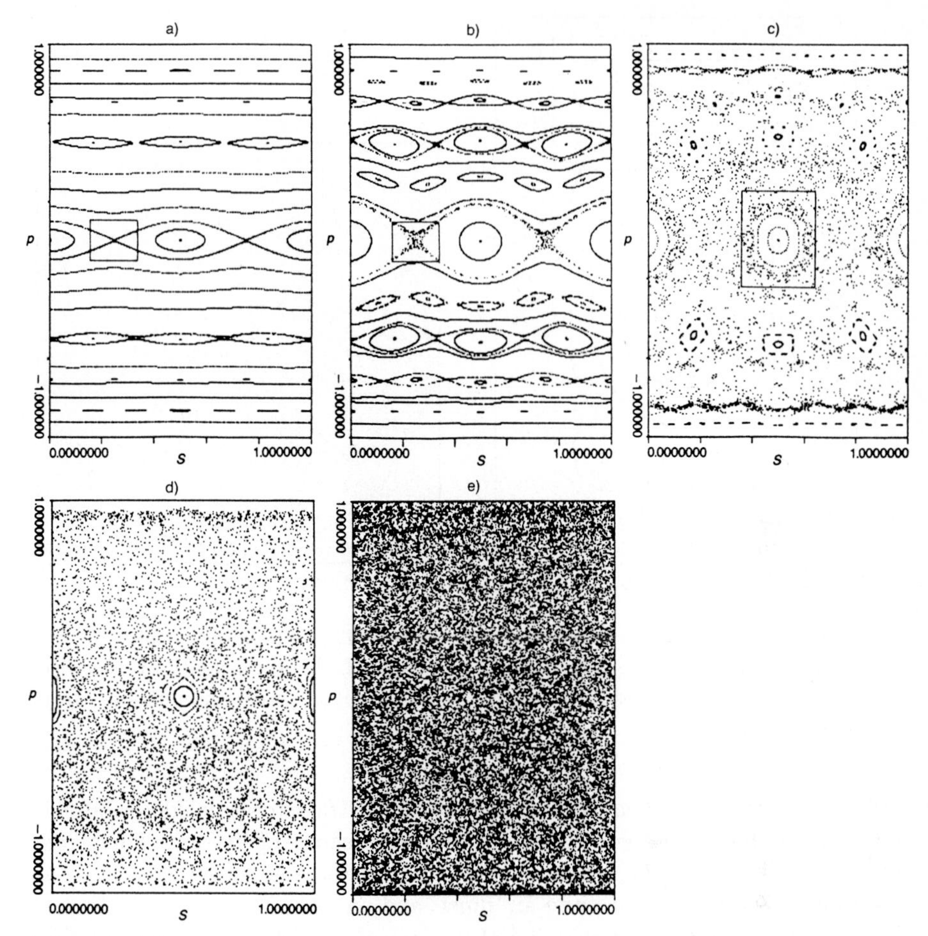

Fig. 3.9. Phase space diagram for the 'typical' cos-billiard for deformation parameters $\varepsilon = 0.1$ (a), 0.2 (b), 0.3 (c), 0.4 (d), 0.5 (e). (Reproduced, with permission, from [3.7].) Figure (a) is stored on the disk.

orbit, for example the 3-periodic orbit in Fig. 3.9 (a) for $\alpha = \pi/3$ $(p = 0.5)$, the 4-periodic orbit for $\alpha = \pi/4$ $(p = 0.707)$ or both 5-periodic orbits at $\alpha = \pi/5$ $(p = 0.809)$ and $\alpha = 2\pi/5$ $(p = 0.309)$. The centers of the island chains in Fig. 3.9 (a) consist of stable fixed points of T^n for $n = 2, 3, \ldots$. Between these stable fixed points, one finds unstable fixed points of T^n.

The second major difference to the Poincaré plots of the ellipse is the appearance of irregular orbits. In the vicinity of the unstable fixed points one can recognize orbits whose Poincaré sections no longer lie on a curve. The iterated points fill an area in the two-dimensional Poincaré map. For small values of ε, most of the phase space is still filled with invariant curves, as predicted by the KAM theorem (see Sect. 2.2.3), and the chaotic bands are trapped between the

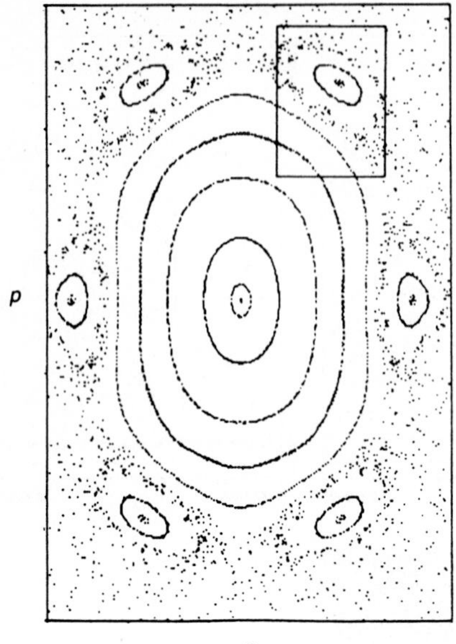

Fig. 3.10. Magnification of the neighborhood of the central fixed point in Fig. 3.9 (c). (Reproduced, with permission, from [3.7].)

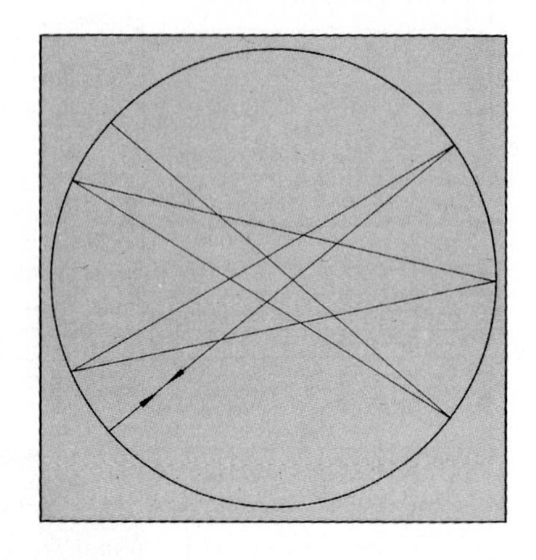

Fig. 3.11. Periodic orbit in the center of the six islands of Fig. 3.10. (Reproduced, with permission, from [3.7].)

invariant curves. This confinement of the chaotic trajectories inside or outside an invariant torus is only valid for one and two degrees of freedom. For higher dimensional systems, the n-dimensional invariant torus no longer separates the $2n$-dimensional phase space into an 'inner' and an 'outer' part.

With increasing ε, a growing number of invariant curves are destroyed. The case $\varepsilon = 0.2$ in Fig. 3.9 (b) gives a good illustration of the KAM-theorem, showing destroyed periodic orbits, where the widths of these zones decrease as the period increases. For $\varepsilon = 0.3$, we find a wide, continuous chaotic region. All points in this wide chaotic sea $|p| < 0.7$ are created from a single orbit. Only small stability islands remain. For $\varepsilon = 0.4$, these regions are further diminished, and for $\varepsilon = 0.5$ we find no more remnants of invariant curves. All points in Fig. 3.9 (e) stem from a single orbit. This system is presumably ergodic.

3.4.2 Zooming In

In this experiment, we will divert our attention to the fine structure of phase space dynamics. We saw in Sect. 3.4.1 that invariant curves with rational frequency ratios break up into chains of stable and unstable fixed points (compare the discussion of the Poincaré-Birkhoff theorem in Sect. 2.4.1). Each of these stable fixed points is now itself a center of a system of invariant curves, which

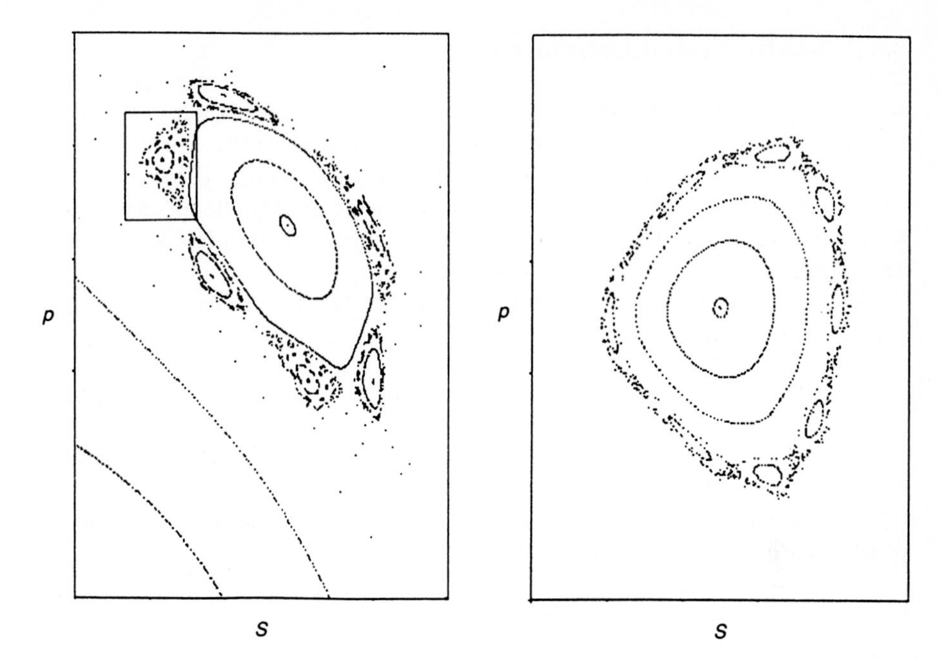

Fig. 3.12. Magnification of the island in the upper right corner of Fig. 3.10. (Reproduced, with permission, from [3.7].)

Fig. 3.13. Magnification of the island in the upper left corner of Fig. 3.12. (Reproduced, with permission, from [3.7].)

can be further broken apart. Figure 3.10 shows a magnification of the vicinity of the central two-periodic fixed point of Fig. 3.9 (c). The neighborhood of this fixed point shows a system of invariant curves and narrow regions, which are due to destroyed rational orbits. This becomes especially evident for the six outer islands: they belong to the periodic orbit with period 12 in the center of these islands, i.e. to a fixed point of T^{12}. Figure 3.11 shows this orbit in coordinate space.

Magnifying, for example, the island at the right upper corner once more, one finds the same structure as that displayed in Fig. 3.12. A fixed point of T^{12} is located in the center, which stems from the 12-periodic in Fig. 3.11. This fixed point is again surrounded by a system of invariant curves, and by regions broken up into stable and unstable fixed points. Figure 3.13 shows a further magnification of an island in Fig. 3.12. Theoretically, one can continue this magnification to the deepest level. Each fixed point "is a microcosmos of the whole, down to arbitrary small scales" [3.9].

3.4.3 Sensitivity and Determinism

Regular and irregular (chaotic) orbits can be distinguished by their phase space behavior. There exists yet another characteristic of chaotic orbits: in the regular case, initially neighboring orbits remain 'close' to one another. However, this is not true for irregular, chaotic orbits, which separate quickly. Here, very important questions concerning the predictability of strictly deterministic processes are raised.

For the case of a circular billiard, the effects of an initial uncertainty $(\Delta\varphi_0, \Delta\alpha_0)$ is immediately found to be

$$\Delta\varphi_n = \Delta\varphi_0 + 2\Delta\alpha_0\, n\,, \tag{3.21}$$

i.e. the errors grow linearly. Similar relations are valid for other integrable cases, as well as for regular orbits in general. The irregular case is completely different: here, the orbits are extremely sensitive to small deviations in the initial conditions, and the uncertainties grow exponentially:

$$|\Delta\varphi_n| = |\Delta\varphi_0|\, \mathrm{e}^{\lambda n}\,. \tag{3.22}$$

As an example, the angular separation of two orbits is computed for the chaotic billiard (3.20) [3.7] for a parameter value of $\varepsilon = 0.5$, starting at $\varphi = 0$ and $\alpha = 70.0^0$ or $\alpha = 70.1^0$. The numerical values of φ after each iteration can be read from the box shown in the main menu of the program, where φ, α and the number n of iterations are displayed (see Fig. 3.7).

In Fig. 3.14, the angular separations $\Delta\varphi_n$ for the first six collisions with the boundary are shown as a function of n on a logarithmic scale. The exponential law (3.22) is approximately satisfied for $\lambda = 0.7$. For comparison, the separation

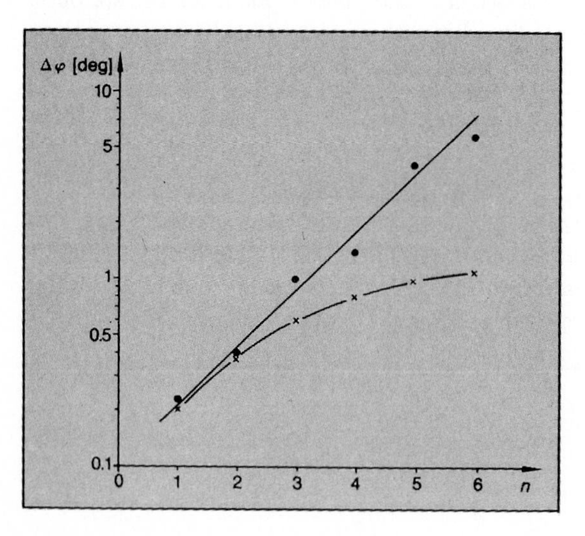

Fig. 3.14. Angle difference $\Delta\varphi_n$ for two neighboring irregular trajectories of the chaotic cos-billiard (•) as compared with regular orbits for the circular billiard (x).

of two regular orbits for the circle-billiard is also shown, for which the error grows linearly.

An immediate consequence of this exponential error growth is the complete unpredictability of the trajectory (error exceeds 90^0) after only 10 collisions with the boundary, when the initial error is 0.1^0. This should be compared to the value of 450 predictable impacts with the boundary for the regular circle trajectory. Reducing the initial error by a factor of two doubles the predictability in the regular case to 900, whereas in the irregular case it only increases additively by $(\ln 2)/\lambda \approx 1$ to 11 impacts! A predictability of 100 impacts requires a precision of some 29 decimal digits in the initial angle. A similar precision is needed for all parameters of the system, and for all unavoidable external perturbations. This characteristic uncertainty of chaotic orbits makes long term predictions impossible, despite strictly deterministic dynamics.

3.4.4 Suggestions for Additional Experiments

Stability of Two-Bounce Orbits. The periodic orbits, and in particular their stability properties, play a very important role in the organization of the billiard map. The analysis of diametric two-bounce orbits $P_0 \to P_1 \to P_0$ is straightforward: from equation (3.6), using $\alpha_1 = \alpha_0 = \pi/2$, the deviation matrix is obtained as

$$\mathbf{M}_2 = \begin{pmatrix} -1+\dfrac{l}{\rho_1} & -l \\ -\dfrac{l}{\rho_0\rho_1}+\dfrac{1}{\rho_0}+\dfrac{1}{\rho_1} & -1+\dfrac{l}{\rho_0} \end{pmatrix} \begin{pmatrix} -1+\dfrac{l}{\rho_0} & -l \\ -\dfrac{l}{\rho_0\rho_1}+\dfrac{1}{\rho_0}+\dfrac{1}{\rho_1} & -1+\dfrac{l}{\rho_1} \end{pmatrix}, \tag{3.23}$$

($l = l_{10}$ = diameter of the billiard) and we find stability for $|\mathrm{Tr}\,\mathbf{M}_2| < 2$, i.e.

$$l < \rho_0 + \rho_1 \,. \tag{3.24}$$

Figure 3.15 illustrates stable and unstable situations. (Note, that the stability condition allows a simple interpretation in terms of the focusing and defocusing of optical rays in a spherical mirror system.) For the special case $\rho_0 = \rho_1 = \rho$, (3.23) simplifies to

$$\mathbf{M}_2 = \begin{pmatrix} 2\left(\dfrac{l}{\rho}-1\right)^2 -1 & 2\,l\left(1-\dfrac{l}{\rho}\right) \\ \dfrac{2}{\rho}\left(\dfrac{l}{\rho}-1\right)\left(2-\dfrac{l}{\rho}\right) & 2\left(\dfrac{l}{\rho}-1\right)^2 -1 \end{pmatrix} \tag{3.25}$$

with stability condition $l < 2\rho$ (note the neutral stability for the circular billiard).

An immediate consequence of these stability considerations is the stability of the two-bounce orbits along the short diameter of an elliptical billiard, and

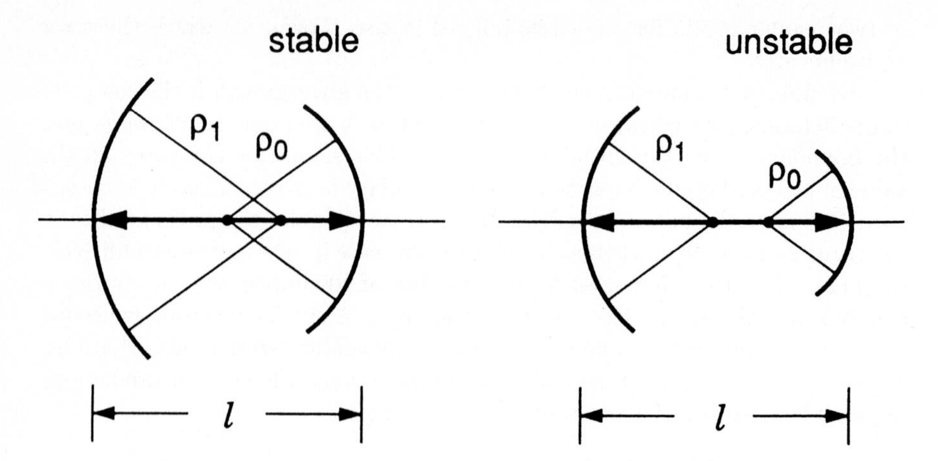

Fig. 3.15. Stable and unstable diametric periodic two-bounce orbits.

the instability of the orbit on the long diameter. These periodic orbits appear as elliptic and hyperbolic fixed points of the Poincaré map (see Fig. 3.4). For the stadium billiard (Fig. 3.5), the horizontal two-bounce orbit is always unstable, the vertical one has neutral stability and is non-isolated (note that this is a non-generic property of the stadium billiard). For the cos-billiard (3.20), the horizontal orbit is stable and the vertical one unstable (note the differences between the Poincaré maps in Figs. 3.4 and 3.5).

These stability considerations can be carried over to n-periodic orbits, but the analytic treatment is more complicated. A numerical study is, of course, possible, e.g. a numerical computation and classification of all stable and unstable four-bounce orbits (a short discussion can be found in an article by Berry [3.3] — see also Fig. 3.16 and the following discussion of bifurcation properties). For more general results on the stability of periodic orbits, see articles [3.10, 3.11].

Bifurcations of Periodic Orbits. On varying a parameter of the billiard curve, the periodic orbits show interesting bifurcation properties (birth and

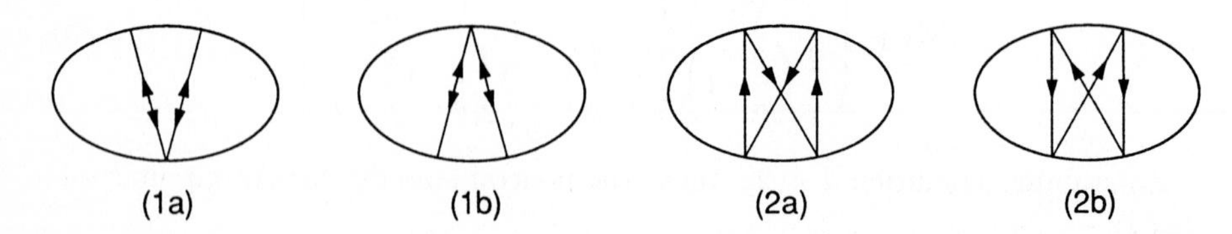

Fig. 3.16. Stable and unstable four-periodic orbits: (1a, b) are unstable, (2a, b) stable.

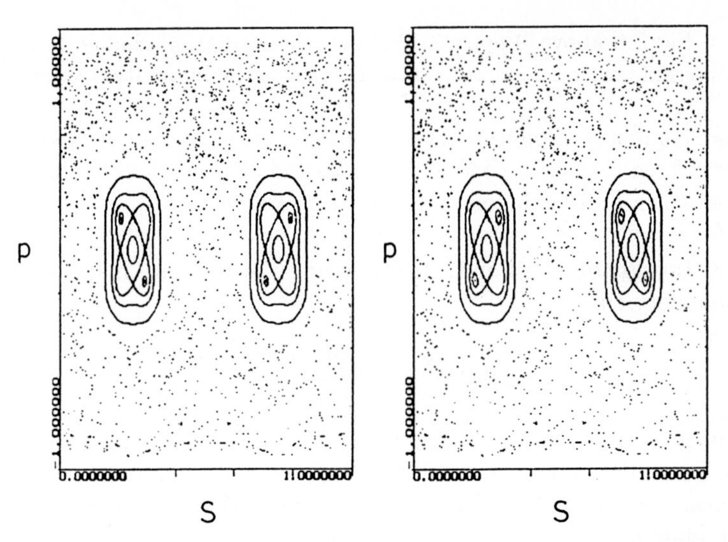

Fig. 3.17. Phase space diagrams for $\varepsilon = 0.12$ illustrating the birth of new four-periodic orbits. The new stable $(2a)$- and $(2b)$-type orbits are shown in the left and right diagrams, respectively. This figure is stored on the disk.

disappearance). In analogy to the study by Berry [3.3], an example of such bifurcation can be found for the billiard with boundary curve

$$r(\varphi) = 1 + \varepsilon \cos(2\varphi)\,, \tag{3.26}$$

which agrees with the 'oval' billiard of equation (3.13). Here the diametric two-bounce orbit $\varphi_0 = \pi/2 \to \varphi_1 = -\pi/2 \to \varphi_0$ is stable for $0 < \varepsilon < 1$ (compare the stability considerations above) because

$$\frac{l}{2\rho} = 1 - \frac{4\varepsilon}{1-\varepsilon} < 1\,, \tag{3.27}$$

where l is the diameter of the billiard and the curvature ρ is given in (3.7)). Trivially, this two-bounce orbit is also a stable four-bounce orbit, i.e. a fixed point of T^4. The stability properties of the four-periodic orbits are determined by the deviation matrix $\mathbf{M}_4 = \mathbf{M}_2\mathbf{M}_2$. Using equation (3.25) for $\mathbf{M}_2$ in the case of a symmetric two-bounce orbit, one obtains

$$\mathrm{Tr}\,\mathbf{M}_4 = 2 - 16\frac{l}{\rho}\left(2 - \frac{l}{\rho}\right)\left(\frac{l}{\rho} - 1\right)^2 \tag{3.28}$$

(see Berry [3.3], (42)). For the billiard (3.26), this four-bounce orbit along the small diameter is always stable. It is neutrally stable ($|\mathrm{Tr}\,\mathbf{M}_4| = 2$) for $\varepsilon = 1/9$. On increasing ε beyond this value, four new four-periodic orbits are born, two of which are stable and two unstable (see Fig. 3.16). For $\varepsilon = 0.1$, we come close to the bifurcation point. The corresponding phase space diagram can be found in Fig. 3.6 above. Figure 3.17 shows the structural change in the phase space diagrams when ε is increased to 0.12 across the bifurcation value of $1/9 = 0.111\ldots$.

More complicated bifurcation phenomena appear almost everywhere in the transition from regular to chaotic dynamics and offer a variety of different phenomena, which are worthy of study (see, e.g. [3.12]).

A New Integrable Billiard? Investigating the Poincaré map for a billiard with boundary curve

$$r(\varphi) = \frac{\cos\varphi + \sqrt{a(5\cos^2\varphi + 3)}}{1 + \cos^2\varphi} \tag{3.29}$$

one finds phase space plots showing typical mixed regular and chaotic behavior. For the particular case of $a = 0.5$, however, the phase space seems to be filled by regular orbits only. Is the billiard integrable in this case? Is this a new case of integrability?

Non-Convex Billiards. Billiards with a non-convex boundary curve can also be studied using the program and may show new and unexpected effects. An example is the cos-billiard (3.20) for $\varepsilon > 0.5$, or the oval billiard (3.13) for values of the constant $c > 10$.

3.5 Suggestions for Further Studies

Analytic billiards, i.e. billiards which are quadratic conformal images of the unit disc (in the complex plane), have been investigated [3.5]. An example is the cos-billiard (3.20). Various interesting results concerning the existence of caustics and invariant tori (e.g. Lazutkin's theorem [3.13]–[3.15]) and related topics are discussed in [3.5].

Billiards in magnetic fields have been studied as examples for simple Hamiltonian systems without time reversal symmetry. This symmetry breaking gives rise to new and interesting phenomena [3.16, 3.17].

Gravitational billiards are billiards (particles elastically reflected from hard boundaries) in constant force fields. A characteristic case is the gravitational wedge-billiard [3.18], which is discussed in detail in Chap. 4 (see also [3.19, 3.20]).

Polygon Billiards are billiards whose boundaries are polygons. The dynamics of these systems is different from that of those having 'smooth' boundaries and they are referred to as 'pseudointegrable'. Typically, polygon billiards allow for far reaching analytical analysis. Details can be found in the literature [3.21]–[3.24].

3.6 Real Experiments and Empirical Evidence

When playing billiard, precision is mandatory; it is recommended that the student rereads Sect. 3.4.3 on sensitivity and determinism within this context. The probability of hitting a further billiard ball after few collisions ($n > 6$) is pretty small (see Fig. 3.14). A predictability of 100 impacts requires a precision of about 28 decimal digits in the initial angle. In reality, a player can, at best, achieve an angle accuracy of about 10^{-2} degree with his cue. It should, however, be noted that the almost inevitable friction may, to some extent, serve to reduce the sensitivity to the initial conditions.

An interesting view of the billiard problem is given by Stöckmann and Stein [3.25, 3.26, 3.27], who determined the eigenvalue spectrum and wave functions of a stadium billiard in a microwave cavity experiment. They demonstrated experimentally and theoretically the correspondence between classical periodic orbits and the eigenvalue spectrum. Some well known periodic orbits – bouncing ball, double diamond, whispering galley – can be clearly recognized as 'scars' in the eigenfunction patterns. The famous Chladni figures, the nodal pattern of vibrating plates, for regular, and especially for irregular shapes, are also commented upon.

References

[3.1] G. D. Birkhoff, *On the periodic motions of dynamical systems*, Acta Math. **50** (1927) 359 (reprinted in R. S. MacKay and J. D. Meiss, *Hamiltonian Dynamical Systems*, Adam Hilger, Bristol 1987.

[3.2] G. Benettin and J.-M. Strelcyn, *Numerical experiments on the free motion of a mass point moving in a plane convex region: Stochastic transition and entropy*, Phys. Rev. A **17** (1978) 773

[3.3] M. V. Berry, *Regularity and chaos in classical mechanics, illustrated by three deformations of a circular 'billiard'*, Eur. J. Phys. **2** (1981) 91

[3.4] N. Saito, H. Hirooka, J. Ford, F. Vivaldi, and G. H. Walker, *Numerical study of billiard motion in an annulus bounded by non–concentric circles*, Physica D **5** (1982) 273

[3.5] M. Robnik, *Classical dynamics of a family of billiards with analytic boundaries*, J. Phys. A **16** (1983) 3971

[3.6] A. Ramani, A. Kalliterakis, B. Grammaticos, and B. Dorizzi, *Integrable curvilinear billiards*, Phys. Lett. A **115** (1986) 25

[3.7] H. J. Korsch, B. Mirbach, and H.-J. Jodl, *Chaos und Determinismus in der klassischen Dynamik: Billard–Systeme als Modell*, Praxis d. Naturwiss. (Phys.) 36(7) (1987) 2

[3.8] H. Poritsky, *The billiard ball problem on a table with a convex boundary — an illustrative dynamical problem*, Ann. Math. **51** (1950) 446

[3.9] M. V. Berry, *Regular and irregular motion*, in: S. Jorna, editor, *Topics in Nonlinear Dynamics*, page 16. Am. Inst. Phys. Conf. Proc. Vol. 46 (1978) (reprinted in R. S. MacKay and J. D. Meiss, *Hamiltonian Dynamical Systems*, Adam Hilger, Bristol 1987.

[3.10] R. S. MacKay and J. D. Meiss, *Linear stability of periodic orbits in Lagrangian systems*, Phys. Lett. A **98** (1983) 92 (reprinted in: R. S. MacKay and J. D. Meiss, *Hamiltonian Dynamical Systems*, Adam Hilger, Bristol 1987.

[3.11] R. S. MacKay, J. D. Meiss, and I. C. Percival, *Transport in Hamiltonian systems*, Physica D **13** (1984) 55 (reprinted in: R. S. MacKay and J. D. Meiss, *Hamiltonian Dynamical Systems*, Adam Hilger, Bristol 1987.

[3.12] J. M. Greene, R. S. MacKay, F. Vivaldi, and M. J. Feigenbaum, *Universal behaviour in families of area–preserving maps*, Physica D **3** (1981) 468 (reprinted in: R. S. MacKay and J. D. Meiss, *Hamiltonian Dynamical Systems*, Adam Hilger, Bristol 1987.

[3.13] V. F. Lazutkin, *The existence of caustics for a billiard problem in a convex domain*, Math. Izv. USSR **7** (1973) 185

[3.14] V. F. Lazutkin, *Asymptotics of the eigenvalues of the Laplacian and quasimodes. A series of quasimodes corresponding to a system of caustics close to the boundary of the domain*, Math. Izv. USSR **7** (1973) 439

[3.15] V. F. Lazutkin, *The existence of an infinite number of elliptic and hyperbolic periodic trajectories for a convex billiard*, Funct. Anal. Appl. **7** (1973) 103

[3.16] M. Robnik and M. V. Berry, *Classical billiards in magnetic fields*, J. Phys. A **18** (1985) 1361

[3.17] M. Robnik, *Regular and chaotic billiard dynamics in magnetic fields*, in: S. Sakar, editor, *Nonlinear Phenomena and Chaos*, page 303, Hilger, Bristol 1986.

[3.18] H. E. Lehtihet and B. N. Miller, *Numerical study of a billiard in a gravitational field*, Physica D **21** (1986) 93

[3.19] H. J. Korsch and J. Lang, *A new integrable gravitational billiard*, J. Phys. A **24** (1990) 45

[3.20] B. Grammaticos and V. Papageorgiou, *Integrable bouncing–ball models*, Phys. Rev. A **37** (1988) 5000

[3.21] A. Hobson, *Ergodic properties of a particle moving inside a polygon*, J. Math. Phys. **16** (1975) 2210

[3.22] P. J. Richens and M. V. Berry, *Pseudointegrable systems in classical and quantum systems*, Physica D **2** (1981) 495

[3.23] P. J. Richens, *Unphysical singularities in semiclassical level density expansions for polygon billiards*, J. Phys. A **15** (1983) 3961

[3.24] B. Mirbach and H. J. Korsch, *Long–lived states and irregular dynamics in inelastic collisions: Analysis of a polygon billiard model*, Nonlinearity **2** (1989) 327

[3.25] H.-J. Stöckmann and J. Stein, *"Quantum" chaos in billiards studied by microwave absorption*, Phys. Rev. Lett. **64** (1990) 2215

[3.26] J. Stein and H.-J. Stöckmann, *Experimental determination of billiard wave functions*, Phys. Rev. Lett. **68** (1992) 2867

[3.27] H.-D. Gräf, H. L. Harney, H. Lengeler, C. H. Lewenkopf, C. Rangacharyulu, A. Richter, P. Schardt, and H. A. Weidenmüller, *Distribution of eigenmodes in a superconducting billiard with chaotic dynamics*, Phys. Rev. Lett. **69** (1992) 1296

4. Gravitational Billiards: The Wedge

The program WEDGE studies the dynamics of a billiard in a gravitational field, or more precisely, a falling body in a symmetric wedge. The boundary of this billiard (compare the discussion of billiard systems in Chap. 3) consists of two planes symmetrically inclined with respect to a constant (e.g. gravitational) force field. The particle is reflected elastically from these planes. For simplicity, we consider the motion to be two-dimensional. We use Cartesian coordinates (u, h) in coordinate space (see Fig. 4.1). In contrast to the billiard systems discussed in Chap. 3, the velocity of the particle changes here, and the trajectory between the reflections is curved (a parabola). An example of such a trajectory is shown in Fig. 4.1.

We are interested in the long-term time behavior of the dynamics, which will be found to be critically dependent on the opening angle 2θ of the wedge (in the following, we always assume $0 < \theta < 90°$). This system is very well suited for numerical studies, since the dynamics reduces to a simple set of iterated equations, as shown in Sect. 4.1, i.e. the Poincaré mapping can be constructed analytically. Moreover, it should also be possible to examine the motion experimentally. For theoretical studies of the gravitational wedge billiard, see the

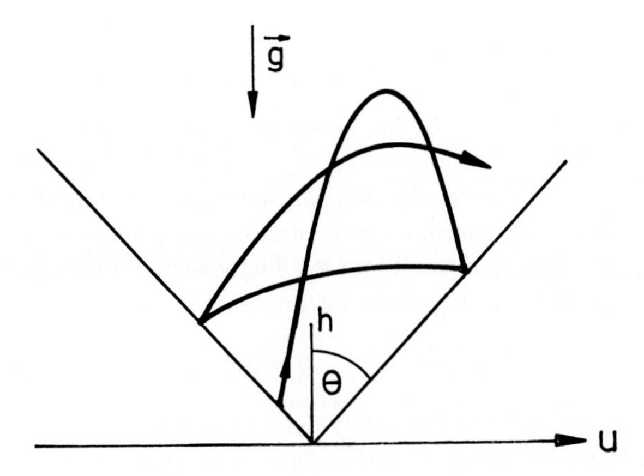

Fig. 4.1. A mass point hopping in a two-dimensional wedge of opening angle 2θ in a gravitational field in a vertical direction.

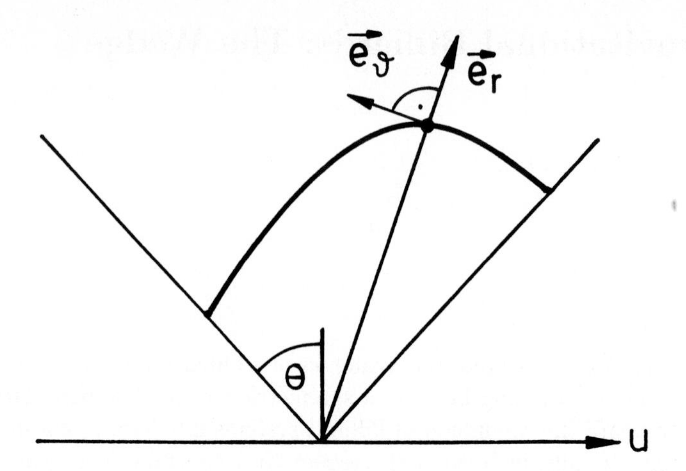

Fig. 4.2. Configuration space coordinates.

excellent articles by Lehtihet and Miller [4.1], and Richter et al. [4.2] (see also [4.3] for a short discussion of a computer program related to that described here).

4.1 The Poincaré Mapping

In order to study the dynamics of the system, it is sufficient to monitor the sequence of impact points with the wall. Let us denote the horizontal coordinates of these points as $u_0,\ u_1,\ u_2,\ldots$ (see Fig. 4.1) and the corresponding velocity components as $\dot{u}_0,\ \dot{u}_1,\ \dot{u}_2,\ldots$. From the geometry of the wedge we can obtain the corresponding heights h_n, and from conservation of the energy

$$E_{\text{kin}} + E_{\text{pot}} = \frac{m}{2}\left(\dot{u}^2 + \dot{h}^2\right) + mgh = E \tag{4.1}$$

the velocity component $\dot{h}_n \geq 0$. The phase space motion therefore reduces to a two-dimensional Poincaré mapping $(u_n, \dot{u}_n) \to (u_{n+1}, \dot{u}_{n+1})$.

Following Ref. [4.1] we now choose coordinates, which simplify this mapping. First we introduce polar coordinates: $\mathbf{e}_r$ and $\mathbf{e}_\vartheta$ are orthogonal unit vectors in radial and angular direction (see Fig. 4.2). The velocity of the particle is $\mathbf{v}$, and the velocity components x and y are introduced by

$$x = \mathbf{v} \cdot \mathbf{e}_r \qquad y = |\mathbf{v} \cdot \mathbf{e}_\vartheta|\,. \tag{4.2}$$

Furthermore, we use units where the gravitational acceleration g, as well as the ratio between the total energy E and mass m, have both the value 1/2. In these units, the energy conservation is written as

$$x^2 + y^2 + h = 1\,. \tag{4.3}$$

A straightforward calculation now provides the Poincaré mapping T in terms of x and y. (Hint: using the parabolic trajectory, the point of next impact as well as the velocity can easily be determined in the coordinates u and h. One then transforms into polar coordinates and applies the elastic reflection law $x \to -x$.) It should be noted that, in doing this, we have to make a distinction between two cases: T_A describes a collision with the same wall and $T_{\rm B}$ a collision with the opposite wall. Further, using the abbreviations

$$X = x/\cos\theta\,, \quad Y = y/\sin\theta\,, \tag{4.4}$$

$$\alpha = \tan\theta\,, \quad \xi = \frac{1-\alpha^2}{(1+\alpha^2)^2}\,, \tag{4.5}$$

the final result for the Poincaré mapping in terms of X and Y is

$$\begin{aligned} T_{\rm A}\ :\quad X_{n+1} &= X_n - 2Y_n \\ Y_{n+1} &= Y_n \\ \\ T_{\rm B}\ :\quad X_{n+1} &= Y_n - X_n - Y_{n+1} \\ Y_{n+1}^2 &= 2 + 2\xi(Y_n - X_n)^2 - Y_n^2\,. \end{aligned} \tag{4.6}$$

The condition

$$(X-2Y)^2\cos^2\theta + Y^2\sin^2\theta \le 1 \tag{4.7}$$

implies A-motion as long as the energy constraint (4.3) is still satisfied after being mapped by $T_{\rm A}$. In the case of an equality in (4.7) the next collision will be exactly at the wedge vertex. In this case, the mapping $T_{\rm A}$ is applied for convenience.

In many applications, one can make use of the linearization of the mapping $T_{\rm B}$:

$$\begin{pmatrix} {\rm d}X_{n+1} \\ {\rm d}Y_{n+1} \end{pmatrix} = \mathbf{B}_{n+1,n} \begin{pmatrix} {\rm d}X_n \\ {\rm d}Y_n \end{pmatrix}, \tag{4.8}$$

with

$$\begin{aligned} \mathbf{B}_{n+1,n} &= \left(\frac{\partial(X_{n+1}, Y_{n+1})}{\partial(X_n, Y_n)}\right) = \\ &= \frac{1}{Y_{n+1}} \begin{pmatrix} 2\xi(Y_n - X_n) - Y_{n+1} & (2\xi-1)Y_n - 2\xi X_n + Y_{n+1} \\ -2\xi(Y_n - X_n) & (2\xi-1)Y_n - 2\xi X_n \end{pmatrix}. \end{aligned} \tag{4.9}$$

The mapping $T_{\rm A}$ is already linear, i.e. $T_{\rm A} = \mathbf{A}$ with

$$\mathbf{A} = \mathbf{A}_{n+1,n} = \left(\frac{\partial(X_{n+1}, Y_{n+1})}{\partial(X_n, Y_n)}\right) = \begin{pmatrix} 1 & 0 \\ -2 & 1 \end{pmatrix}. \tag{4.10}$$

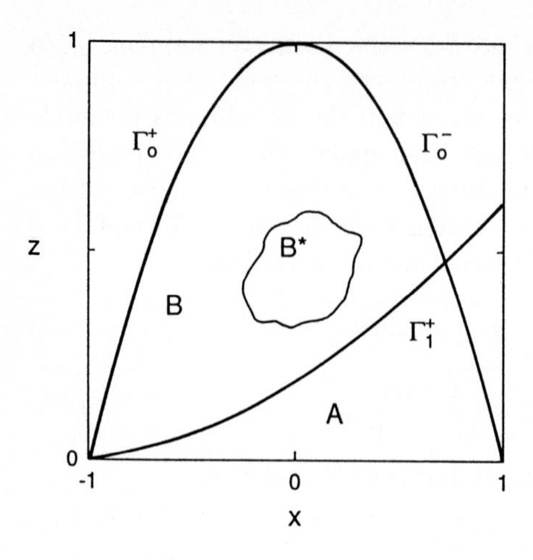

Fig. 4.3. Phase space region in (x, z) variables. The parabolic region is classically allowed. Points in the A-region are mapped by T_{A}, those in the B-region by T_{B}.

It should be noted that the nth iteration of $\mathbf{A}$ is simply

$$\mathbf{A}^n = \begin{pmatrix} 1 & 0 \\ -2n & 1 \end{pmatrix}. \tag{4.11}$$

The equations (4.6) are used for numerical iteration. It is more illustrative, however, to display the mapping in terms of the transformed coordinates (x, z), where z is defined by

$$z = y^2 , \tag{4.12}$$

and x and y are obtained from (4.4). Furthermore, we use the notation

$$Z = Y^2 = (1 + \alpha^{-2})\, z . \tag{4.13}$$

The Jacobi matrices (4.9) and (4.10) can be directly rewritten in terms of the (X, Z) coordinates by noting that

$$\left(\frac{\partial(X_{n+1}, Z_{n+1})}{\partial(X_n, Z_n)}\right) = \begin{pmatrix} \dfrac{\partial X_{n+1}}{\partial X_n} & \dfrac{1}{2Y_n}\dfrac{\partial X_{n+1}}{\partial Y_n} \\ 2Y_{n+1}\dfrac{\partial Y_{n+1}}{\partial X_n} & \dfrac{Y_{n+1}}{Y_n}\dfrac{\partial Y_{n+1}}{\partial Y_n} \end{pmatrix}. \tag{4.14}$$

It can be verified that, in terms of the (x, z)-coordinates, the Poincaré mapping $(x_n, z_n) \to (x_{n+1}, z_{n+1})$ is area-preserving with the measure $\mathrm{d}x\,\mathrm{d}z$, i.e.

$$\left|\frac{\partial(x_{n+1}, z_{n+1})}{\partial(x_n, z_n)}\right| = 1 . \tag{4.15}$$

The same is valid for the coordinates (X, Z).

It should be noted that the Poincaré mapping in terms of the variables (x, z) is actually a phase space section in momentum space. This mapping is ideally suited for displaying the essential features of the dynamics. It is, however, somewhat abstract, and it may be helpful to recall some physical properties:

- The variable x is the radial component of the velocity immediately after reflection at the wall.
- The variable z is the kinetic energy for angular motion (in energy units $E = m/2$).
- Energy conservation is given by $z = 1 - h - x^2$.
- The motion is restricted to the parabolic section (see Figure 4.3)

$$0 \leq z \leq 1 - x^2 \,. \tag{4.16}$$

- For points on the z-axis, the velocity is orthogonal to the wall.
- Points on the x-axis represent a sliding motion along the walls of the wedge.
- Points on the parabolic boundary $x = \pm\sqrt{1-z}$ describe trajectories originating from the wedge vertex. In the following, this boundary will be denoted by $\Gamma_0^\pm$.
- Points on the curve $x = 2\sqrt{z}/\alpha - \sqrt{1-z}$ map in the next iteration onto the vertex, those in the A-region below this curve are mapped by T_A, and those in the B-region by T_B.

It is, furthermore, helpful to discuss the pre-images (+) and images (−) of the wedge vertex, i.e. those points that will map after k iterations onto the vertex, or those that move away from the vertex. It can easily be shown from the mapping (4.6) that these points are given by the curves $\Gamma_k^\pm$ defined by

$$x = \pm\left(\frac{2k}{\alpha}\sqrt{z} - \sqrt{1-z}\right) \,. \tag{4.17}$$

These curves separate the (x, z)-plane into different regions. Points in two such regions separated by a curve Γ_k^+ will be split after k iterations. This beam splitting is very important for a detailed understanding of the intricate properties of the mapping. For further details, see the original article by Lehtihet and Miller [4.1].

Another important set of points is given by the curve $y = \alpha x$, i.e.

$$x = \pm\sqrt{z}/\alpha \,. \tag{4.18}$$

Points on this curve correspond to trajectories in coordinate space moving vertically and falling back on themselves. Under certain conditions, such trajectories are periodic. It is also important to observe that pure A-type motion is impossible, i.e. the particle cannot hop indefinitely on the same side of the wedge. This can be directly shown from the mapping T_A given in (4.6), which is linear. We immediately obtain

$$X_n = X_0 - 2nY_0\,, \qquad Y_n = Y_0\,. \tag{4.19}$$

This iteration is n-periodic only in the case $Y_0 = 0$, i.e. for a sliding motion along the wall, which will reach the vertex within a finite time. Indefinite B-motion is, however, possible within a subregion B^* of the (x, z)-plane containing the two-bounce orbit $\tilde{P}$:

$$\begin{aligned}(\tilde{x}, \tilde{z}) &= (0\,, \alpha^2\,(1+\alpha^2)^{-1}(1-\xi)^{-1}) \\ &= (0\,, (1+\alpha^2)\,(3+\alpha^2)^{-1}) \\ &= (0\,, (2+\cos 2\theta)^{-1})\,, \end{aligned} \tag{4.20}$$

which can be directly obtained as a period-one fixed point of the mapping T_{B}

$$(\tilde{X}, \tilde{Z}) = (0\,, (1-\xi)^{-1})\,, \tag{4.21}$$

which transforms to (4.20). It is a simple exercise to derive the same result from a symmetric parabolic motion orthogonal to the two sides of the wedge. The shape of the region B^* depends sensitively on the half-wedge angle θ. This is studied in the numerical experiments in sections 4.3.2 and 4.3.4 below.

The properties of the map in the neighborhood of the fixed point (4.21) can be found from a linearization of the mapping T_{B}, which leads to

$$\begin{pmatrix} X_1 \\ Z_1 - \tilde{Z} \end{pmatrix} = \mathbf{M} \begin{pmatrix} X_0 \\ Z_0 - \tilde{Z} \end{pmatrix}, \tag{4.22}$$

where the deviation matrix $\mathbf{M}$ is given by [4.1]

$$\mathbf{M} = \begin{pmatrix} 2\xi - 1 & (1-\xi)^{3/2} \\ -4\xi(1-\xi)^{-1/2} & 2\xi - 1 \end{pmatrix} \tag{4.23}$$

with determinant one.

The stability condition for the fixed point (4.21) is

$$|\mathrm{Tr}\,\mathbf{M}| = 2\,|2\xi - 1| < 2\,, \tag{4.24}$$

i.e. the fixed point is elliptic for $\xi > 0$ ($\theta < 45°$) and hyperbolic for $\xi < 0$ ($\theta > 45°$). The residue R and the rotation number ϵ (compare Sect. 2.4.5) are, in this case, given by

$$R = \sin^2 \pi\epsilon = \tfrac{1}{4}(2 - \mathrm{Tr}\,\mathbf{M}) = 1 - \xi\,, \tag{4.25}$$

or simply,

$$\cos^2 \pi\epsilon = \xi\,. \tag{4.26}$$

Some other low order fixed points of ${T_{\mathrm{A}}}^m {T_{\mathrm{B}}}^n$ (m integer and $n = 1, 2, 3$) are:

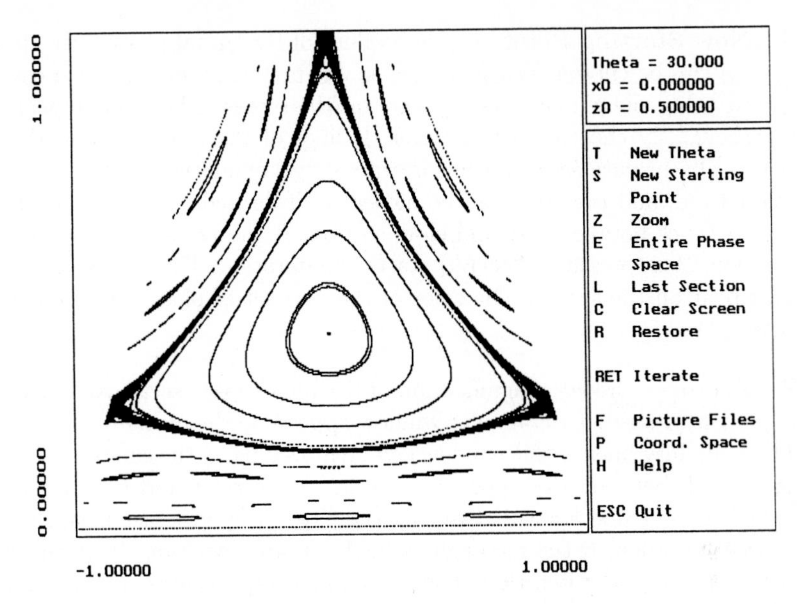

Fig. 4.4. Main menu of program WEDGE.

$$
\begin{aligned}
T_\mathrm{A}{}^m T_\mathrm{B} \quad &: \quad Z = Y^2 = \frac{1}{1-(m+1)^2\xi} \quad , \quad X = mY \\
T_\mathrm{A}{}^m T_\mathrm{B}{}^2 \quad &: \quad Z = Y^2 = \frac{1}{\frac{1}{2}\left[1+(m+1)^2\right]-(m+1)^2\xi} \quad , \quad X = mY \\
T_\mathrm{A}{}^m T_\mathrm{B}{}^3 \quad &: \quad Z = Y^2 = \frac{1}{(m+1)^2\left[\frac{1}{2}(\xi/(\frac{1}{8}-\xi))^2-\xi\right]} \quad , \quad X = mY\,.
\end{aligned}
\tag{4.27}
$$

More details on periodic motion are given in Sect. 4.3.1.

4.2 Interacting with the Program

On accessing the program WEDGE, the *Main Menu* appears. To the left, we see the empty phase space section in the (x, z)-variables described in Sect. 4.1. Here, x varies between -1 and $+1$, z between 0 and 1. In the small rectangle on the upper right, the values of the half-wedge angle θ and the starting point (x_0, z_0) are displayed; 30° and $(0, 0.5)$ are pre-set.

The action of the program can be controlled by means of the menu listed in the box on the right side of the screen. In order to activate the various points, the key letter must be pressed.

- **T New Theta** — a new value of the half-wedge angle θ (in degrees) can be chosen after pressing ⟨T⟩. To start the iteration, press ⟨ENTER⟩.

- **S New Starting Point** — new values of the initial phase space point (x_0, z_0) can be chosen. When the ⟨S⟩ key is pressed, a cross, which can be moved using the cursor keys, appears on the screen. The cross appears at the previously chosen starting point. If lines of the cross are outside the shown phase space section, the corresponding boundaries of the screen are colored red, and the cross can be moved in, if desired. Pressing the ⟨TAB⟩ key switches between fast and slow motion. If the cross is in the desired position, then pressing ⟨ENTER⟩ starts the iteration. The values of x_0 and z_0 appear on the screen. The color of the new set of phase space points is changed.

- **Z Zoom** — allows magnification of the phase space sections. When the ⟨Z⟩ key is pressed, a rectangular window appears in the center of the screen. The boundary lines of this window can be moved using the cursor keys. One can switch between the upper and lower, or the right and left boundary, by using the ⟨SPACE BAR⟩. Pressing the ⟨TAB⟩ key switches between fast and slow motion. If the rectangle is in the desired position, then, pressing ⟨ENTER⟩ starts the iteration. The chosen phase space section is magnified. After leaving this item, the iteration starts as soon as the ⟨ENTER⟩ key is pressed.

- **E Entire Phase Space** — pressing ⟨E⟩ switches back to the entire phase space. The iteration continues with the chosen starting conditions.

- **L Last Section** — pressing the ⟨L⟩ key switches back to the previously chosen phase space section. The iteration continues with the chosen starting conditions.

- **C Clear Screen** — pressing the ⟨C⟩ key clears the screen.

- **R Restore** — pressing the ⟨R⟩ key restores the phase space section and the results of the previous picture.

- **RET Iterate** — pressing ⟨ENTER⟩ starts the iteration. The iteration can be stopped by pressing the ⟨ESC⟩ key, and restarted by ⟨ENTER⟩.

- **F Picture Files** — upon pressing the ⟨F⟩ key, a submenu appears. The entries can be activated by pressing the corresponding key:

  - P *Print Picture* — a hardcopy of the entire screen is printed.
  - S *Save Picture* — one is asked under which name the phase space picture is to be stored. Suffixes are ignored. After the input of a valid file name, two files under this name with the suffixes *.PIC and *.PAR are created on the logged drive. The file *.PIC contains the last compiled picture, while *.PAR stores the necessary data for this picture. If these files are loaded, simulation using these exact data can be continued.

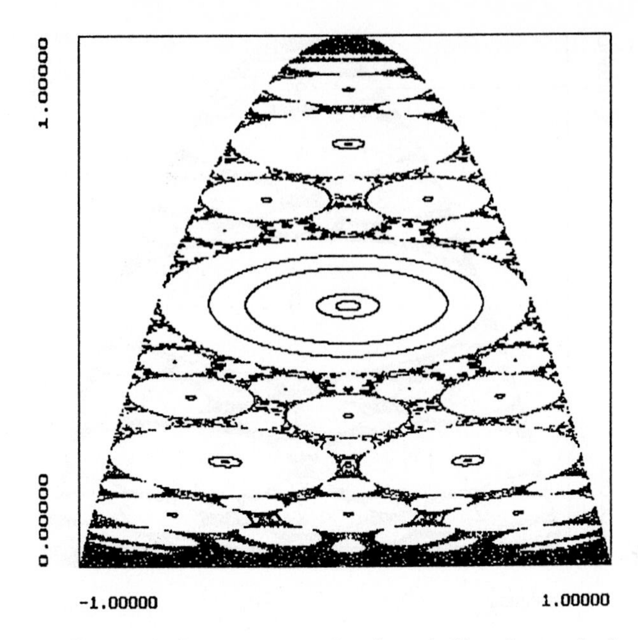

Fig. 4.5. Poincaré phase space section for a half-wedge angle $\theta = 44°$.

- L *Load Picture* — requests the name under which a picture has been stored and reads the data from files *.PIC and *.PAR. The results of the present computation are lost. The program is in the same state as when the loaded picture was stored. Four examples of half-wedge angles $\theta = 20°$, $30°$, $40°$, and $46°$ have already been computed and can be loaded from files 20.PIC , 30.PIC etc.

- **P Coord. Space** — upon calling this item, the trajectory in coordinate space is shown. When any key – except ⟨ESC⟩ – is pressed, the particle jumps to the next impact with a wall. Pressing the ⟨ESC⟩ key returns to the phase space mode.

- **H Help** — pressing the ⟨H⟩ key provides on-line information.

- **ESC Quit** — pressing the ⟨ESC⟩ key terminates the program.

4.3 Computer Experiments

4.3.1 Periodic Motion and Phase Space Organization

In this computer experiment, we study some of the properties of periodic orbits for the gravitational wedge billiard. Let us first start the program with the preset values of $\theta = 30°$ and $(x_0, z_0) = (0, 0.5)$. We observe that the phase space

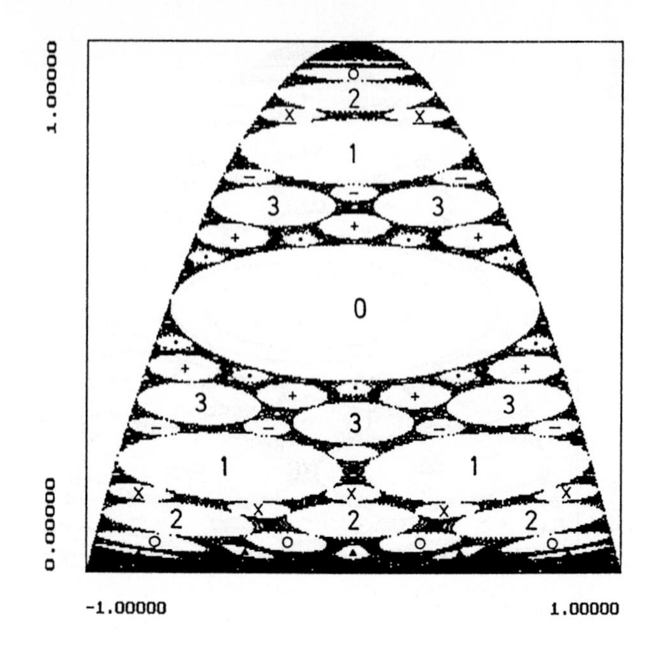

Fig. 4.6. As in Fig. 4.5, the dominant stability islands are marked by the symbol s given in Table 4.1 .

points arrange themselves on a seemingly smooth closed curve. When new initial conditions closer to the center of phase space are chosen, the curve traced out by the iterated points shrinks and, in the limiting case, approaches a point, namely $(x_0, z_0) = (0, 0.4)$. This point can be identified with the central fixed point of the Poincaré map given in (4.20). Switching to coordinate space by activating the menu item T the trajectory is indeed found to be a two-bounce orbit orthogonal to the wall at the moment of impact.

In the following, we will study higher periodic orbits and phase space organization in more detail. We choose a half-wedge angle of $\theta = 44°$. The phase space picture is shown in Fig. 4.5 (the same phase space section is stored as a picture file (44.PIC), which can be loaded by the program). One observes a complex organization of bigger and smaller islands, where the motion is nearly integrable. Let us first examine the centers of these islands. In the middle of the big central island, we again find the stable two-bounce orbit, which is shown in Fig. 4.7 in coordinate space. Also shown are the trajectories for two of the regular orbits inside the central island. These trajectories only cover a small part of the accessible coordinate space, bounded by a smooth envelope of the trajectory field (a caustic).

Some of the periodic orbits appearing at the centers of the stability islands are shown in Fig. 4.8. (Note that the periodic orbits within the very small islands can be more easily located using the zoom-feature of the program.) Such periodic orbits follow increasingly complicated sequences of T_A and T_B mappings, and connect a number of the stability islands. In these periodic trajectories, there are two special phenomena which deserve attention:

1. For perpendicular incidence at the boundary, the trajectory is exactly reversed and follows the opposite direction after collision. It should be noted, that for such points $x = \mathbf{v} \cdot \mathbf{e}_r = 0$, i.e. they appear on the z-axis in phase space.

2. The same reversion of the orbit occurs when the particle bounces vertically after reflection. It reaches a summit – visible as a spike in the coordinate space plot (this point does not appear in the phase space plot) – and falls back. Such vertical loops appear after (and before) iterates on line $x = \sqrt{z}/\alpha$ in phase space (see (4.18)).

Table 4.1 lists some properties of the (stable) periodic orbits shown in Figs. 4.5–4.8. We use a short-hand notation to record the topology of an orbit: '0' stands for B-motion (collision with the opposite wall) and '1' for A-motion (re-collision with the same wall). In this way, we have mapped the topology of the orbits onto the binary numbers. Furthermore a '·' denotes an orthogonal impact. Listed in the table are period and symmetry $\sigma = \pm$, as well as the symbol s used in Figs. 4.6–4.8 (the last two orbits are somewhat harder to detect in Fig. 4.5). All these orbits start with an orthogonal impact. For even periods, we find a second orthogonal impact per period, for odd values a vertical spike. As illustrated in Fig. 4.7, nearby trajectories are quasiperiodic and have the same collision sequence. Different regimes are separated by trajectories hitting the vertex, i.e. separated by the lines $\Gamma_k^{\pm}$ defined in (4.17).

Table 4.1. Short periodic orbits

Symbolic Code	Period	σ	$n^{\uparrow}$	$n^{\perp}$	$\bar{z}$	s
·0·0·	2	+	0	2	0.4914	0
·010·	3	−	1	1	0.7950	1
·01·10·	4	+	0	2	0.0998	2
·00100·	5	−	1	1	0.2837	3
·01110·	5	−	1	1	0.9403	○
·110·011·	6	+	0	2	0.0385	△
·0001000·	7	−	1	1	0.6536	+
·1001001·	7	−	1	1	0.1513	×
·0010·0100·	8	+	0	2	0.2267	−
·000010000·	9	−	1	1	0.2493	·
·100111001·	9	−	1	1	0.0571	
·01001·10010·	10	+	0	2	0.0185	

The analysis of a system by such a reduced code is known as *'symbolic dynamics'* and provides useful mathematical techniques, as well as valuable insight into the behavior of a dynamical system.

On considering Table 4.1 above, one observes:

- The number $n^{\perp}$ counts the number of orthogonal collisions with the boundary. In all these cases there is at least one of these points. This implies that there is at least one point on the z-axis in phase space for each orbit. Such z-values are listed in the table as $\bar{z}$.
- The number $n^{\uparrow}$ counts the number of 'spikes', which is either zero or one.
- The sum $n^{\perp} + n^{\uparrow}$ must be zero or two. In the present cases, we always have a sum of two. There exist, however, periodic trajectories with sum zero (see, e.g., the unstable trajectory in Figure 4.18 below).
- All these symbolic codes are symmetric with respect to reflection. This is due to the reflection symmetry of these orbits at a point of vertical incidence.
- The number of 0's must be even, i.e. there must be an even number of hops from one wall to the other. This implies that the central figure in the odd symbolic codes must be '1'.

The symbolic code shows directly that some conceivable sequences are missing. Some of these missing sequences (such as the six-periodic orbit ·010·010·) simply possess a smaller period (three), while others are identical to a listed one (· 110 · 011 · = · 011 · 110 ·) . Some, such as · 01010 ·, have already been excluded above. Missing in the list are, e.g., ·0011000· or ·101· . Such orbits can be hard to detect (small surrounding stability island, or unstable) or dynamically forbidden. It is suggested that the reader carry out further numerical experiments, which will help to clarify these questions.

For a global understanding of the dynamics, it is important to describe the set of all periodic trajectories (this set is countable), i.e. their topology in

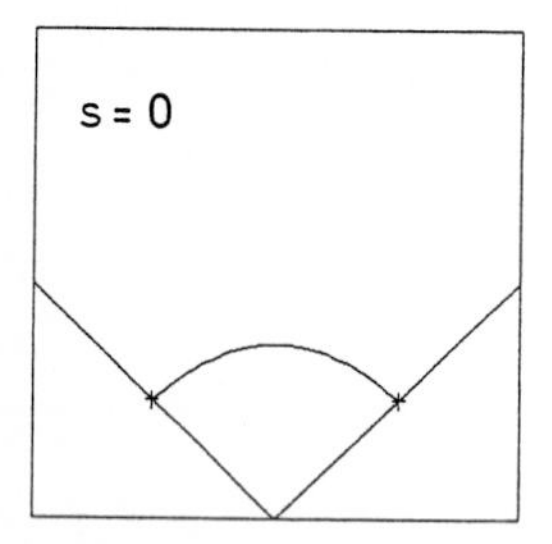

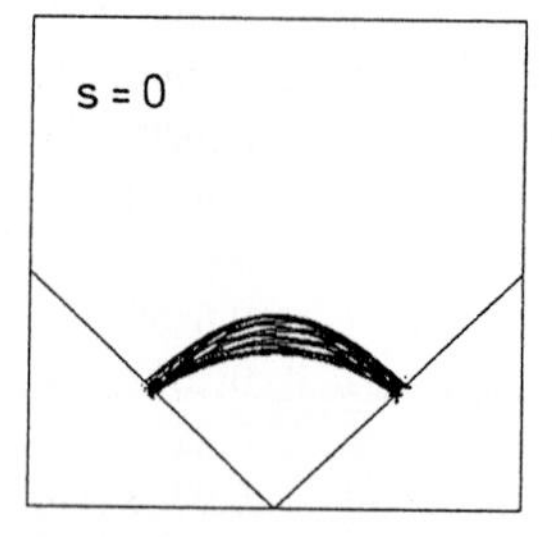

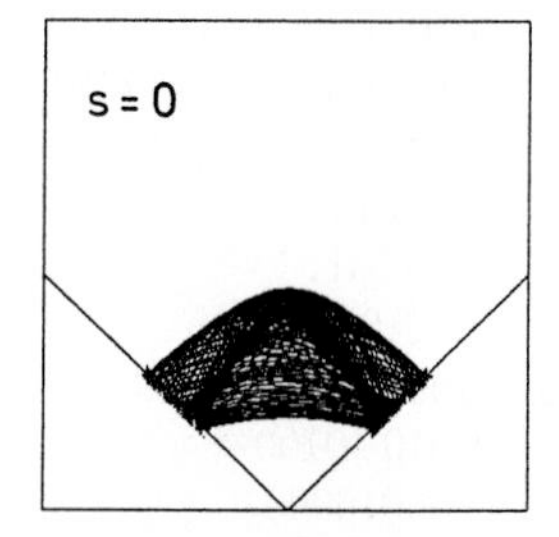

Fig. 4.7. Coordinate space plot of three trajectories within the central stability island $s = 0$ of Fig. 4.6. The left figure shows the periodic two-bounce orbit at the center.

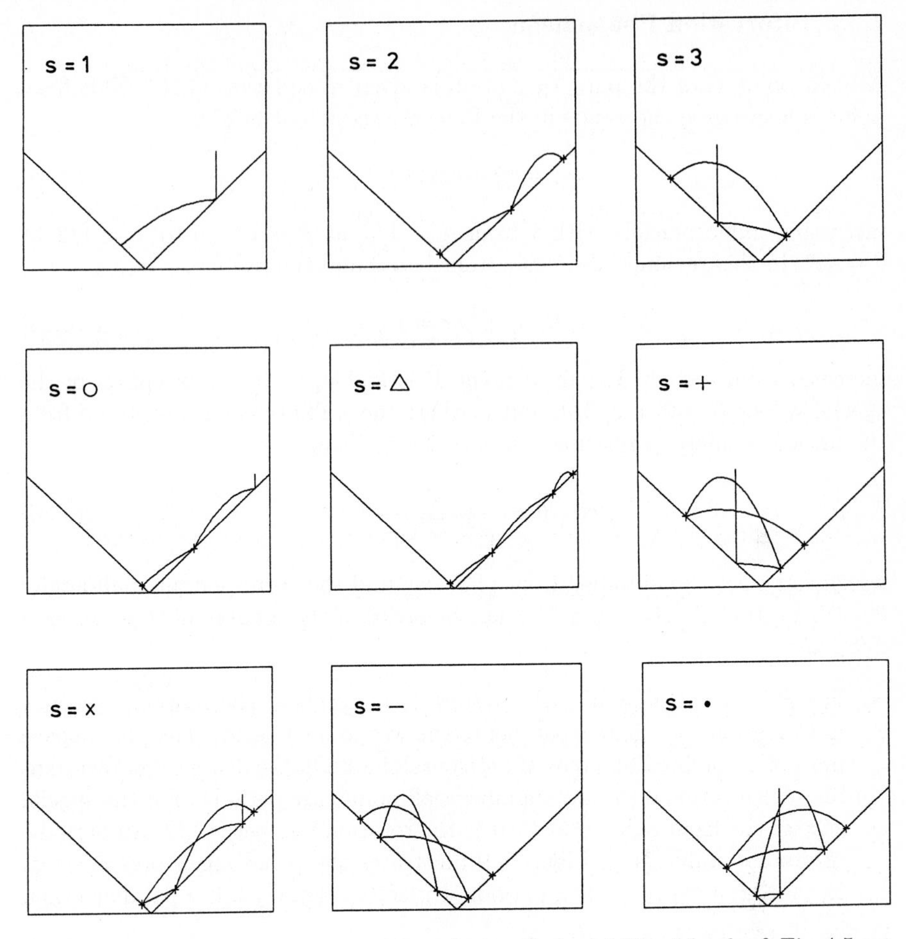

Fig. 4.8. Periodic trajectories within the marked stability island of Fig. 4.5 as described in the text. The mark s agrees with that in Fig. 4.6 and Table 4.1.

terms of a symbolic code, their stability parameters, their number as a function of the period (typically, there is an exponential proliferation of this number with increasing period), etc. Such data form the basic input for a semiclassical description of the quantum dynamics involved (see, for instance, the work by Gutzwiller [4.4] and references given there).

4.3.2 Bifurcation Phenomena

A fixed point $\tilde{P}$ of the pure T_{B} motion is given in equation (4.20). This fixed point is located on the z-axis in the Poincaré section at height

$$\tilde{z} = (2 + \cos 2\theta)^{-1}, \tag{4.28}$$

increasing monotonically with θ from $\tilde{z} = 1/3$ at $\theta = 0°$ up to $\tilde{z} = 1/2$ at $\theta = 45°$. In this region, ξ decreases from 1 to 0 and the residue,

$$R = \sin^2 \pi\epsilon = 1 - \xi, \tag{4.29}$$

increases from 0 to 1. The fixed point $\tilde{P}$ is stable, with the exception of the special values $R = 0$, $1/2$, $3/4$, and 1, where the stability cannot be found from the linearized map, as discussed in Sect. 2.4.5. Using

$$\cos 2\theta = \frac{4\xi}{\sqrt{8\xi + 1} - 1} - 1, \tag{4.30}$$

which can easily be obtained from (4.5), we find the corresponding half-angles $\theta = 0°$, $25.914°$, $34.265°$, and $45°$. The behavior of the map in all these cases is different:

- For $R = 1$, i.e. $\theta = 45°$, the system is integrable. The motion separates in Cartesian coordinates parallel to the wedge boundaries. The phase space motion is confined to a pair of horizontal lines. On the dense set of invariant lines with rational rotation number ϵ all points are periodic. For the special case of the fixed point $\tilde{P}$ at $(0, 0.5)$ the rotation number is $1/2$. All periodic orbits are stable [4.1]. Figure 4.9 illustrates the phase space organization. In coordinate space, a non-periodic trajectory densely fills a parallelogram.

Fig. 4.9. Poincaré section: $\theta = 45°$

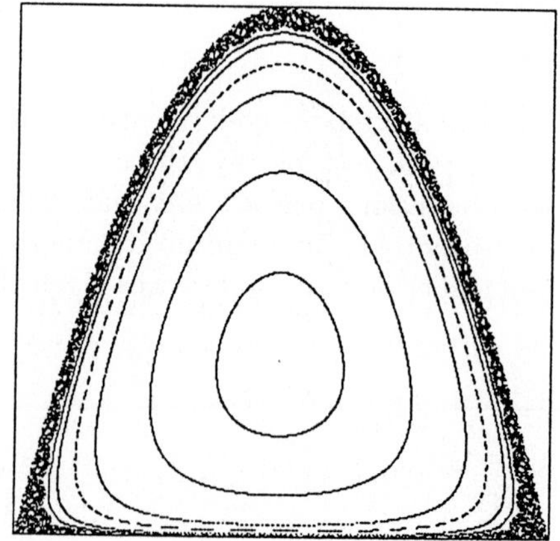

Fig. 4.10. Poincaré section: $\theta = 3°$

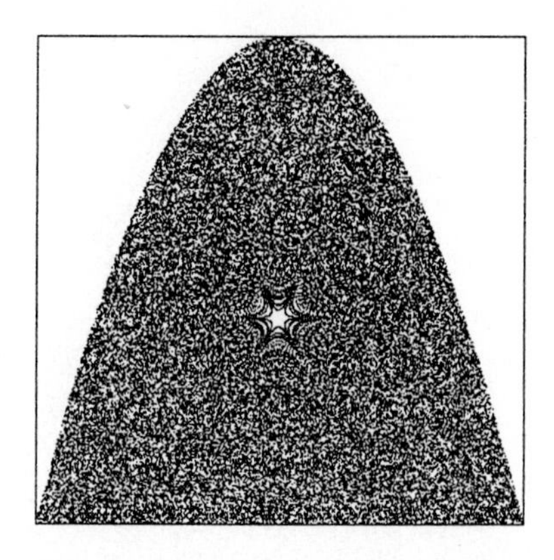

Fig. 4.11. Poincaré section for the critical angle $\theta = 34.265°$.

- For $R \to 0$, i.e. $\theta \to 0°$, the wedge approaches two parallel vertical lines with zero distance. The motion is almost entirely a B-motion, because collisions with the same wall become increasingly improbable. The phase space map consists almost entirely of invariant curves surrounding the fixed point $\tilde{P}$, which approaches $(0, 1/3)$. The rotation number ϵ goes to zero. Figure 4.10 shows the Poincaré map for $\theta = 3°$. Here, the complicated chaotic dynamics is already confined to a thin layer at the phase space boundary.

- For $R = 3/4$, i.e. $\theta = 34.265°$, the rotation number is $\epsilon = 1/3$. A numerical inspection of the Poincaré map shows that the phase space seems to be densely filled with iterations of a single trajectory, as demonstrated in Fig. 4.11. The only point with pure T_{B} motion is the fixed point $\tilde{P}$ in the center of the six-point star, which is still empty. If the iteration is continued, this star-like region will also be filled with points.

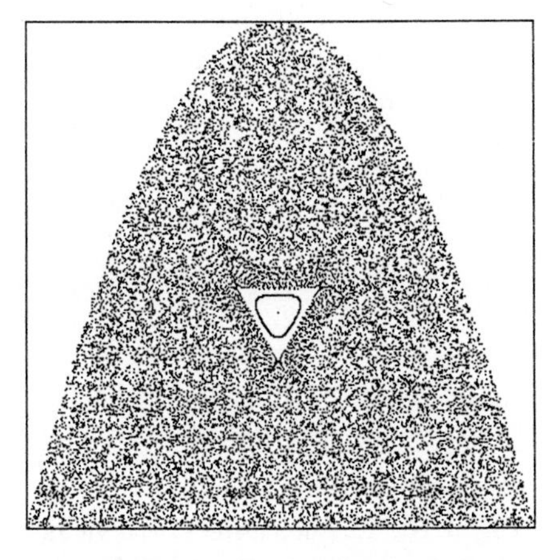

Fig. 4.12. Poincaré section: $\theta = 35°$

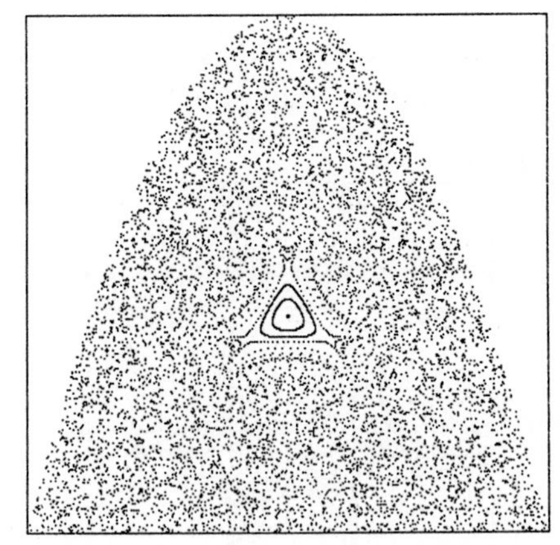

Fig. 4.13. Poincaré section: $\theta = 33.5°$

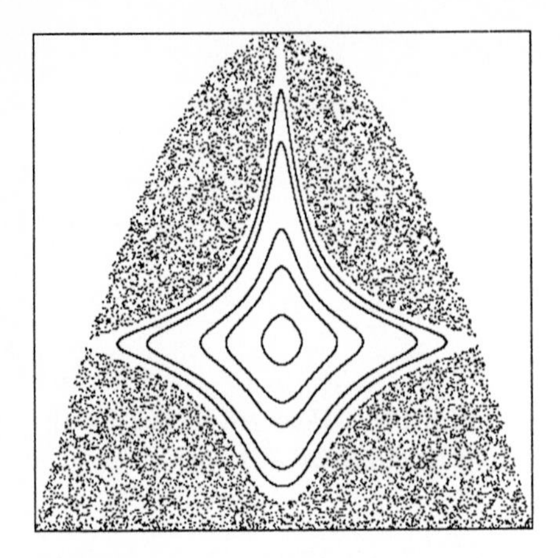

Fig. 4.14. Poincaré section: $\theta = 26°$

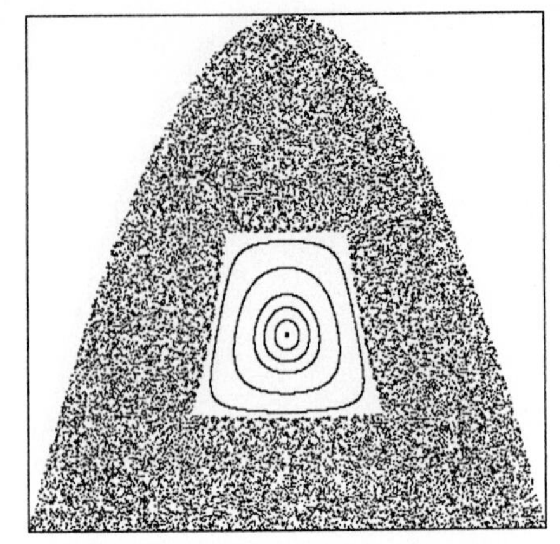

Fig. 4.15. Poincaré section: $\theta = 25°$

Figures 4.12 and 4.13 illustrate the θ-dependence in the vicinity of the critical value $\theta = 34.265°$. In Figures 4.12 and 4.13, the angles are 35° and 33.5°, i.e. slightly above and below the critical value, respectively. In both cases, we observe three hyperbolic period-three ($\epsilon^{-1} = 3$) fixed points, which move towards $\tilde{P}$, when the critical θ is approached. The hyperbolic invariant curves extend outside and form the six-star pattern observed at some distance from $\tilde{P}$ (compare Fig. 4.11). It should be noted that the triangle formed by the three hyperbolic fixed points is reversed when θ passes through the critical value.

- For $R = 1/2$, i.e. $\theta = 25.914°$, the rotation number is $\epsilon = 1/4$ and the behavior of the Poincaré map is similar to that of the preceding case, with the exception that we find four hyperbolic period-four ($\epsilon^{-1} = 4$) fixed points. This is illustrated in Figs. 4.14 and 4.15 for $\theta = 26°$ and 25°, i.e. above and below the critical value. Figure 4.16 finally shows a magnification of the vicinity of the fixed point for an angle slightly below the critical one.

Fig. 4.16. Magnification of the vicinity of the fixed point for $\theta = 25.910°$.

Table 4.2. Critical angles

n	θ [deg]	$\tilde{z}$
2	45.000	0.50000
3	34.265	0.42265
4	25.914	0.38197
5	20.765	0.36382
6	17.313	0.35425
7	14.843	0.34858

A very interesting set of θ values is given by the condition $\epsilon^{-1} = n$, where n is an integer, at the fixed point $\tilde{P}$ (compare the general discussion in Sect. 2.4.6). These values can be obtained from (4.30) using $\xi = \cos^2 \pi\epsilon = \cos^2 \pi/n$. Table 4.2 lists the first of these critical angles, as well as the z-coordinate of $\tilde{P}$.

As the parameter θ passes through one of these critical values, periodic orbits of n times the original period are created or destroyed. There are several ways in which such bifurcations can occur, as discussed in Sect. 2.4.6 . Here, some of these bifurcations can be studied numerically. The cases $n = 2$, 3, and 4 have already been discussed above as exceptional cases.

Let us look in some detail at the case $n = 6$. For values of θ above the critical value of 17.313°, we observe a stable elliptic fixed point at approximately $(\tilde{x}, \tilde{z}) = (0, 0.36)$ (Fig. 4.21 below). As θ crosses the critical values, a new pair of period-six orbits is born — a stable and an unstable one —, which appear as a chain of six stability islands, separated by hyperbolic saddles. Fig. 4.17 shows the evolution of this new stable fixed point for decreasing values of θ. The stable periodic orbit approaches the centers of the six islands in a clockwise manner. For θ values close to the critical angle, we also find a chain of unstable hyperbolic fixed points between the elliptic ones. Fig. 4.18 shows the stable central period-

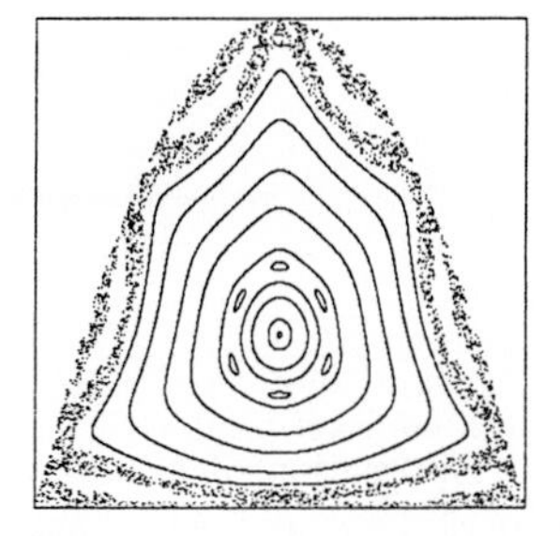
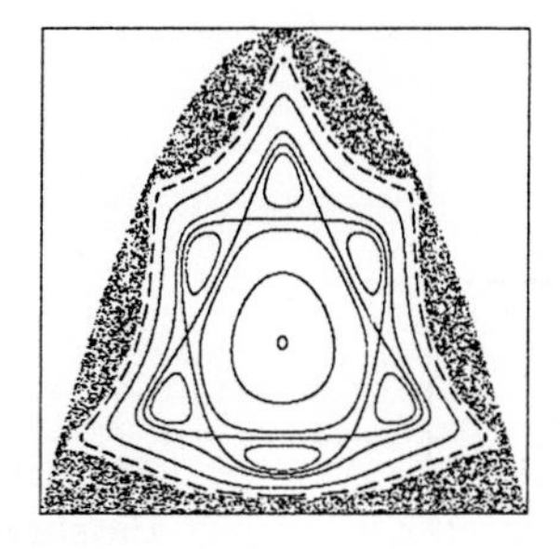
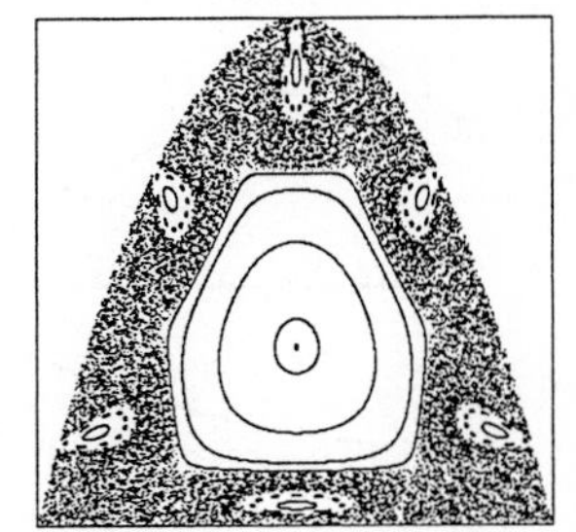

Fig. 4.17. Evolution of the period-six fixed point for $\theta = 17.25°$, 17.0°, and 16.6°. These figure is stored on the disk.

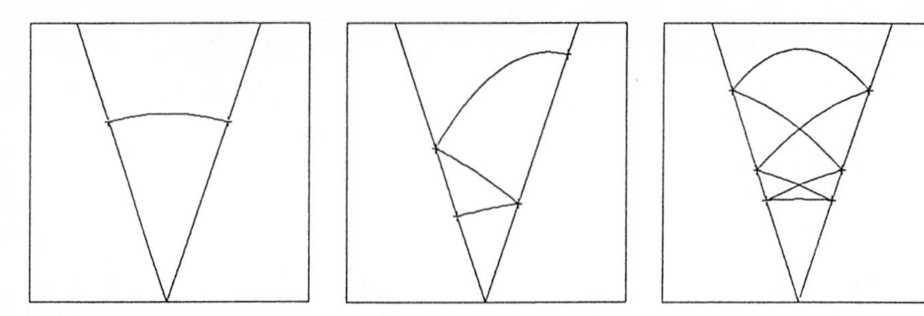

Fig. 4.18. Stable central period-two orbit and stable and unstable period-six orbits in coordinate space ($\theta = 17°$).

two orbit, as well as the stable and unstable period-six orbits in coordinate space. In Fig. 4.19, the location of the coordinates $\tilde{z}$ of the stable fixed points on the z-axis is plotted as a function of θ, which shows the typical pitchfork shape (note, however, that here the central fixed point remains stable). When the angle θ is decreased even further, the six elliptic fixed points move towards the phase space boundary, and the area of their stability islands goes to zero. Finally, they are destroyed at the boundary for a θ-value of about 16.4°.

The above examples nicely illustrate some typical bifurcation properties of stable fixed points of area-preserving maps. Similar observations can be made for other critical angles. A related study of bifurcation properties for the oval billiard has been suggested in Sect. 3.4.4. More details can be found in the literature (see, e.g., [4.5] and references therein).

4.3.3 'Plane Filling' Wedge Billiards

For half-angles

$$\theta = \frac{\pi}{2k}, \quad k \in \mathbb{N} \tag{4.31}$$

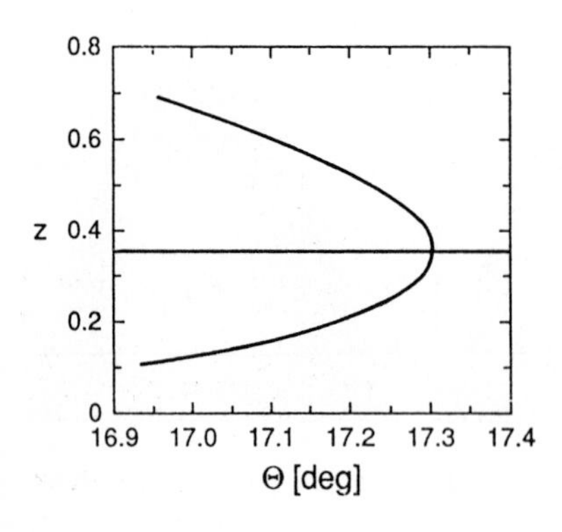

Fig. 4.19. Bifurcation diagram of the stable period-two and period-six fixed points on the z-axis.

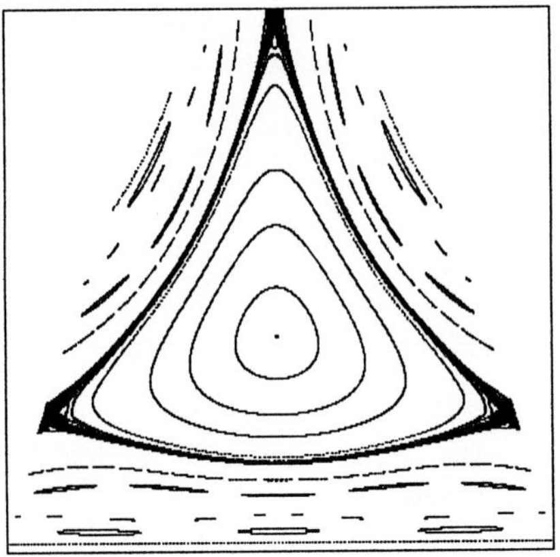

Fig. 4.20. Poincaré section: $\theta = 30°$. This figure is stored (files 10sc.pic, 10sc.par).

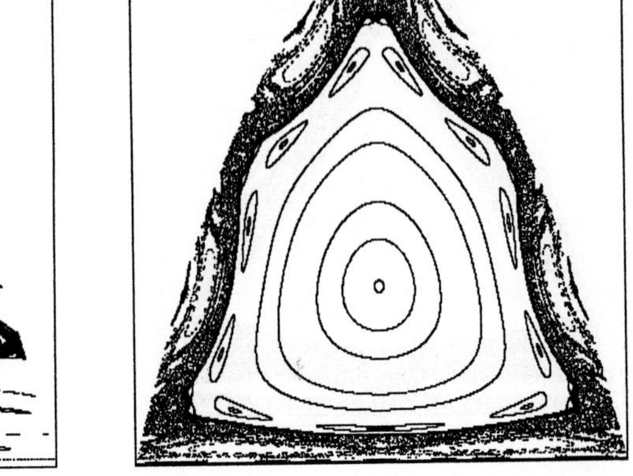

Fig. 4.21. Poincaré section: $\theta = 18°$. This figure is stored on the disk.

the wedge billiard exhibits a special behavior [4.1]: $2k$ copies of the wedge can be joined at the boundaries and fill the plane. The constant acceleration in each of these segments acts in the direction of the segment's bisector. The walls are transparent and the coordinate and velocity of the particle are continuous at the boundaries.

The first of these cases are $\theta = 45°$, $30°$, $22.5°$, and $18°$. Figures 4.20 and 4.21 show the Poincaré sections for $30°$ and $18°$. No simply connected global chaotic region is observed. Invariant curves exist even in the vicinity of the wedge vertex, as can be seen from a magnification of the chaotic region close to the

Fig. 4.22. Magnification of the lower part of Fig. 4.20. This figure is stored on the disk.

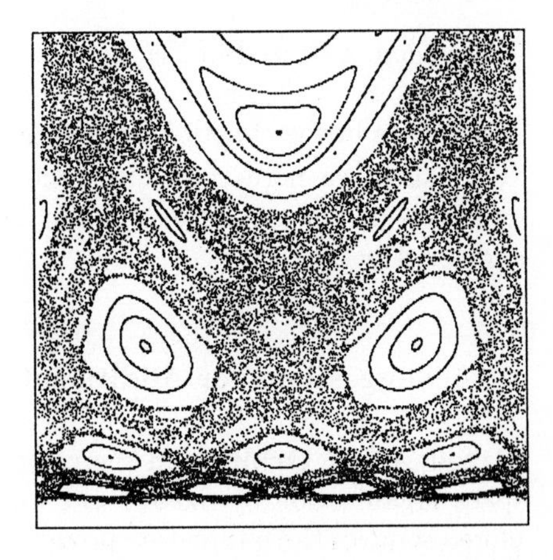

Fig. 4.23. Magnification of the lower part of Fig. 4.21. This figure is stored on the disk.

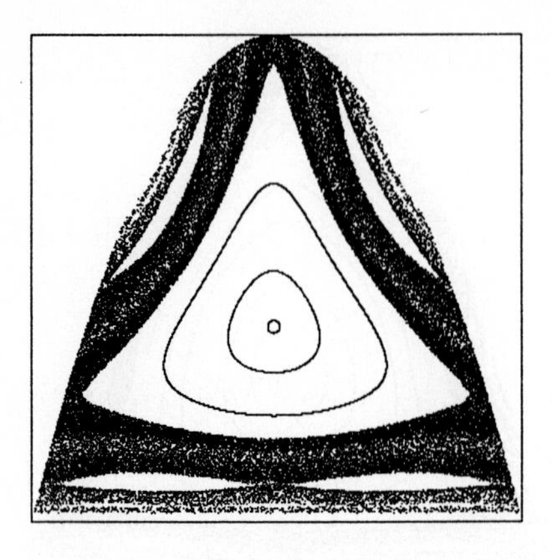

Fig. 4.24. Poincaré section for $\theta = 30.1°$.

top of the phase space boundary. A magnification of the lower part of Figs. 4.20 and 4.21 is shown in Figs. 4.22 and 4.23. Again we see disconnected chaotic regions bounded by invariant curves. A detailed numerical inspection [4.1] shows that some of the invariant curves cross the Γ-lines (see (4.17)), i.e. they cross the discontinuous $A-B$ boundary. This is not observed for θ-values different from $\pi/2k$, where the invariant curves are stopped by the Γ-curves (compare the numerical experiment 4.3.1 above). As observed by Lehtihet and Miller [4.1, p. 98]

> "...every intersection of curves $\Gamma_n^{\pm}$ and $\Gamma_m^{\pm}$ (n and m integer) behaves under iteration like an unstable periodic point. The small chaotic region surrounding it is again bounded by near-integrable curves some of which are crossing the Γ lines. A slight change of the parameter annihilates this phenomenon. Consequently no near-integrable curve will prevent chaos from expanding along the Γ lines ...".

This is demonstrated in Fig. 4.24, where θ is changed to 30.1° as compared to Figs. 4.20 and 4.22. The confinement of the different chaotic regions is destroyed – all points in the chaotic region stem from a single chaotic trajectory.

Finally, we note that a characteristic reduction in the chaoticity of the dynamics at the angles $\theta = \pi/2k$ has been observed (the 'breathing chaos' addressed below).

4.3.4 Suggestions for Additional Experiments

Mixed $A-B$ Orbits. In the computer experiment 4.3.1, we studied phase space organization by fixed points and invariant curves. Experiment 4.3.2 explored some of the bifurcation properties of pure T_B motion. It is now of interest to analyze in addition the low order mixed fixed points of ${T_\mathrm{A}}^m {T_\mathrm{B}}^n$. The coordinates of some of these fixed points are given in (4.27). Guided as much as possible by analytic formulae these fixed points can be localized numerically, and their stability properties, as well as their bifurcation behavior when the wedge half-angle is varied explored.

Pure B Dynamics. B^* was defined above as the phase space subregion with pure B-motion (i.e. there are no re-collisions). Outside B^* the dynamics is much more complicated: we find successive beam splittings at the wedge vertex, and the KAM theory is not applicable. The shape of region B^* is sensitively dependent on the parameter θ.

First, we recall that B^* contains the two-bounce orbit $\tilde{P}$, whose stability properties were studied above. For the two critical values of θ, where the rotation number ϵ equals 1/3 or 1/4, the region B^* is reduced to a point. For small θ values, B^* fills the whole map, thereby confining the chaotic region to a thin layer at the phase space boundary (compare Fig. 4.10).

As illustrated in Fig. 4.17, B^* is not always simply connected: as the period-six fixed points created at the critical angle move away from $\tilde{P}$, global chaos invades the unstable hyperbolic fixed points. It is interesting to investigate the metamorphosis of B^*, as well as the interior dynamics, which is governed by the KAM theory (compare also the discussion in [4.1]).

The Stochastic Region. No stable fixed points can be found in the region $\theta > 45°$. The motion appears to be stochastic. The degree of stochasticity depends, however, on the value of θ. For angles close to 45° and 90°, the stochasticity is weak and the divergence of an initial condition is small. For θ slightly below 90°, we find almost exclusively A-motion, and an adiabatic invariant exists. The strongest stochasticity is obtained for $\theta = 60°$. Further information — including numerical results for the Lyapunov exponents — can be found in Ref. [4.1].

Breathing Chaos. Richter et al. [4.2] used the gravitational wedge billiard to introduce the scenario of 'breathing chaos', i.e. an oscillation in the ratio of chaotic to regular parts of the Poincaré surface of section as a system parameter (here, the wedge half-angle θ) varies. The chaoticity was found to be drastically reduced at the plane-filling θ-angles given in (4.31).

4.4 Suggestions for Further Studies

Smooth gravitational billiards, i.e. gravitational billiards which are bounded by a smooth boundary curve instead of the wedge, have been discussed in Ref. [4.6]. In this article, it has been shown that a parabolic boundary is a new example of an integrable billiard. A second integral of motion is constructed analytically, and the stability and bifurcation properties of fixed points are studied. The results also shed new light on the known integrable non-gravitational elliptical billiards (compare Chap. 3).

Birkhoff's symmetry lines and Birkhoff's decomposition of the Poincaré map into a product of involutions is very helpful for an understanding of the global behavior of the dynamics. For more details, see the article by Richter et al. [4.2].

4.5 Real Experiments and Empirical Evidence

Imagine that a classical particle be replaced by an atom and the form of the wedge altered in a specific way, so that the atoms always bounce between the boundary planes. One can then study the eigenmodes of such a cavity, which is dependent on opening angle and geometry of the boundary. Wallis, Dalihard, and Cohen–Tannoudji [4.7] constructed a single horizontal concave parabolic mirror placed in a gravitational field, i.e. a smooth gravitational billiard such that discussed in Sect. 4.4. This method allows one to study individual or bunches of ions and atoms in other cavity types such as Paul traps. As in a laser resonator, where photons are stored, atoms can be stored and studied experimentally in such a gravitational cavity .

References

[4.1] H. E. Lehtihet and B. N. Miller, *Numerical study of a billiard in a gravitational field*, Physica D **21** (1986) 93

[4.2] P. H. Richter, H.-J. Scholz, and A. Wittek, *A breathing chaos*, Nonlinearity **3** (1990) 45

[4.3] B. N. Miller and H. Lehtihet, *Chaotic dynamics: An instructive model*, in: E.F. Redish and J.S. Risley, editors, *Computers in Physics Instruction*, Addison–Wesley, New York 1990

[4.4] M. C. Gutzwiller, *Chaos in Classical and Quantum Mechanics* (Springer, New York 1990)

[4.5] J. M. Greene, R. S. MacKay, F. Vivaldi, and M. J. Feigenbaum, *Universal behaviour in families of area–preserving maps*, Physica D **3** (1981) 468 (reprinted in: R. S. MacKay and J. D. Meiss, *Hamiltonian Dynamical Systems*, Adam Hilger, Bristol 1987.

[4.6] H. J. Korsch and J. Lang, *A new integrable gravitational billiard*, J. Phys. A **24** (1990) 45

[4.7] H. Wallis, J. Dalibard, and C. Cohen-Tanoudji, *Trapping atoms in a gravitational cavity*, Appl. Phys. B **54** (1992) 407

5. The Double Pendulum

The planar double pendulum consists of two coupled pendula, i.e. two point masses m_1 and m_2 attached to massless rods of fixed lengths l_1 and l_2 moving in a constant gravitational field (compare Fig. 5.1). For simplicity, only a planar motion of the double pendulum is considered. Such a planar double pendulum is most easily constructed as a mechanical model to demonstrate the complex dynamics of nonlinear (i.e. typical) systems in mechanics, in contrast to the more frequently discussed linear (i.e. atypical) harmonic oscillators. Here, numerical experiments are helpful for investigating the complex dynamics, in particular by means of Poincaré sections.

5.1 Equations of Motion

The configuration of the pendulum is most conveniently described by the two angles φ and ψ, which are related to the Cartesian coordinates (x_i, y_i), $i = 1, 2$ of the masses m_i by

$$\begin{aligned} x_1 &= +l_1 \sin\varphi \ , \quad x_2 = +l_1 \sin\varphi + l_2 \sin\psi \\ y_1 &= -l_1 \cos\varphi \ , \quad y_2 = -l_1 \cos\varphi - l_2 \cos\psi \ , \end{aligned} \tag{5.1}$$

and the equations of motion can be derived from the Lagrangian

$$\begin{aligned} L = {} & \tfrac{1}{2} M \, l_1{}^2 \, \dot{\varphi}^2 + \tfrac{1}{2} m_2 \, l_2{}^2 \, \dot{\psi}^2 + m_2 \, l_1 \, l_2 \, \dot{\varphi} \, \dot{\psi} \, \cos(\psi - \varphi) \\ & - M \, g \, l_1 \, (1 - \cos\varphi) - m_2 \, g \, l_2 \, (1 - \cos\psi) \ , \end{aligned} \tag{5.2}$$

where $M = m_1 + m_2$ is the total mass, and the Euler-Lagrange equations are

$$\frac{\mathrm{d}}{\mathrm{d}t} \frac{\partial L}{\partial \dot{\varphi}} - \frac{\partial L}{\partial \varphi} = 0 \qquad \text{and} \qquad \frac{\mathrm{d}}{\mathrm{d}t} \frac{\partial L}{\partial \dot{\psi}} - \frac{\partial L}{\partial \psi} = 0 \ . \tag{5.3}$$

One obtains

$$\begin{aligned} \ddot{\varphi} = {} & \{1 - \mu \cos^2(\psi - \varphi)\}^{-1} \Big[\mu g_1 \sin\psi \cos(\psi - \varphi) \\ & + \mu \dot{\varphi}^2 \sin(\psi - \varphi) \cos(\psi - \varphi) - g_1 \sin\varphi + \frac{\mu}{\lambda} \, \dot{\psi}^2 \sin(\psi - \varphi) \Big] \end{aligned} \tag{5.4}$$

and

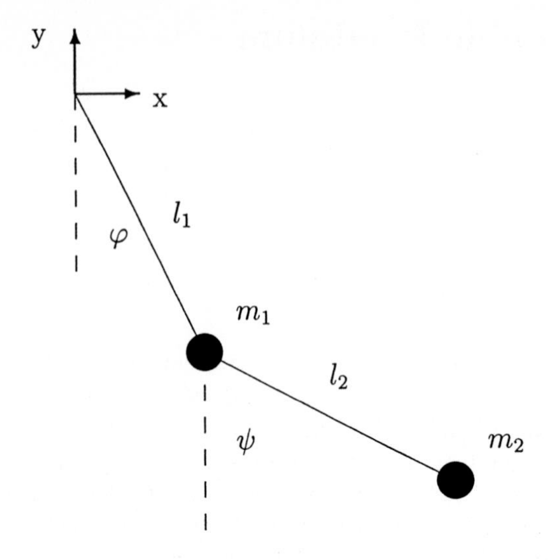

Fig. 5.1. The double pendulum: Parameters and coordinates.

$$\begin{aligned}\ddot{\psi} = \left\{1 - \mu\cos^2(\psi-\varphi)\right\}^{-1} \Big[& g_2 \sin\varphi\cos(\psi-\varphi) \\ & - \mu\dot{\psi}^2 \sin(\psi-\varphi)\cos(\psi-\varphi) - g_2\sin\psi - \lambda\dot{\varphi}^2\sin(\psi-\varphi)\Big]\,, \end{aligned} \tag{5.5}$$

with

$$\lambda = l_1/l_2 \quad , \quad g_1 = g/l_1 \quad , \quad g_2 = g/l_2 \quad , \quad \mu = m_2/M\,. \tag{5.6}$$

The dynamics of the double pendulum can be described by four variables, the two angles and the corresponding angular velocities (momenta), which span the four-dimensional phase space of the system. Conservation of the total energy

$$\begin{aligned} E_{\text{tot}} = & \tfrac{1}{2} M\, {l_1}^2\, \dot{\varphi}^2 + \tfrac{1}{2} m_2\, {l_2}^2\, \dot{\psi}^2 + m_2\, l_1\, l_2\, \dot{\varphi}\, \dot{\psi} \cos(\psi-\varphi) \\ & + M\, g\, l_1\, (1-\cos\varphi) + m_2\, g\, l_2\, (1-\cos\psi) \end{aligned} \tag{5.7}$$

reduces the four-dimensional phase space to a three-dimensional manifold.

We can construct a two-dimensional Poincaré section by looking at the trajectory only at those moments when the outer pendulum passes the vertical position, i.e. for $\psi = 0$. Equation (5.7) then yields a quadratic equation for $\dot{\psi}$, with solutions

$$\begin{aligned} \dot{\psi}_{\pm} = & -\frac{l_1}{l_2}\dot{\varphi}\cos\varphi \\ & \pm\sqrt{\left(\frac{l_1}{l_2}\dot{\varphi}\cos\varphi\right)^2 - \frac{2}{m_2 {l_2}^2}\left(M g\, l_1 (1-\cos\varphi) + \tfrac{1}{2} M {l_1}^2\dot{\varphi}^2 - E_{\text{tot}}\right)}\,. \end{aligned} \tag{5.8}$$

We now plot a point $(\varphi, \dot{\varphi})$ in the phase space of the inner pendulum, when the two conditions

$$\psi = 0 \qquad \text{and} \qquad \dot{\psi} + \lambda\dot{\varphi}\cos\varphi > 0 \tag{5.9}$$

are fulfilled. This leads to a unique definition of a Poincaré phase space section.

One should be aware of the fact that the Poincaré map is plotted as a function of the variables φ and the angular velocity $\dot{\varphi}$, which is *not* proportional to the canonical momentum conjugate to φ. Therefore, the Poincaré map is *not* area-preserving.

The Poincaré section allows fast and informative insight into the dynamics of the double pendulum. The different types of motion appear as a finite number of points for periodic orbits, curve-filling points (*'invariant curves'*) for quasiperiodic motion and area-filling points for chaotic trajectories.

5.2 Numerical Algorithms

The second order differential equations of motion (5.4)–(5.5) can be rewritten as a system of four ordinary differential equations of first order by introducing $\dot{\varphi}$ and $\dot{\psi}$ as additional independent variables. The program solves numerically a system

$$\frac{\mathrm{d}y_k}{\mathrm{d}t} = f_k(y_1, \ldots, y_n, t) \qquad k = 1, \ldots, n\,, \tag{5.10}$$

with $n = 4$ and $(y_1, \ldots, y_4) = (\varphi\,, \psi\,, \dot{\varphi}\,, \dot{\psi}\,)$ for given initial angles $(\varphi_0\,, \psi_0)$ and angular velocities $(\dot{\varphi}_0\,, \dot{\psi}_0\,)$ at $t = 0$. Numerically, the $\mathrm{d}y$'s and $\mathrm{d}t$ are replaced by finite steps Δy and Δt. The numerical algorithm has to predict the values of the variables at the next time step $t_{i+1} = t_i + \Delta t$. Four different numerical methods for solving the differential equations (5.10) can be selected in the program, and these are briefly described below. For more details concerning the numerical methods, see [5.1, 5.2]. The first three methods keep the time step constant for all times, whereas in the last method its value is readjusted during the computation.

In order to simplify the notation, in the following equations we write down the propagation scheme only for the case of one or two variables. A generalization is straightforward.

Euler Method. This is the simplest and most naïve technique. Coordinates and velocities at time step i are used to construct these values at the next step $i+1$ by linear extrapolation

$$y_{i+1} = y_i + \Delta t\, f(y_i, t_i)\,. \tag{5.11}$$

This method is the first-point Euler method – other varieties are known [5.2]. It is conceptually very simple, but it is immediately clear that the algorithm uses exclusively the approximate derivative $\mathrm{d}y/\mathrm{d}t \approx (y_{i+1} - y_i)/\Delta t \approx f(y_i, t_i)$ computed from the first point of the time interval, which is a biased choice. In application, the Euler method has serious drawbacks. First, it is a first order method, i.e. the error term is $O((\Delta t)^2)$, which may require a small time step in order to satisfy the required error bounds. Secondly, the errors may increase systematically and, moreover, the method is not very stable.

Leapfrog Method. The leapfrog or half-step method [5.2] is much more accurate. Its basic idea is a parallel propagation of a variable x and its time derivative v, alternating between x and v propagation. Let us consider a pair of equations of motion

$$\dot{x} = v(x, v, t) \qquad \text{and} \qquad \dot{v} = a(x, v, t) \tag{5.12}$$

appearing typically in Hamiltonian dynamics. The leapfrog method computes the variable at step $i+1$ from the value at step $i-1$ using the velocity at step i by linearization, as in the Euler method:

$$x_{i+1} = x_{i-1} + \Delta t\, v_i \tag{5.13}$$

and the velocity at step $i+2$ is calculated by

$$v_{i+2} = v_i + \Delta t\, a_i \tag{5.14}$$

with initial conditions chosen as x_0 and $v_1 = v_0 + \frac{1}{2}\Delta t\, a_0$. The error term for the Leapfrog method is $O((\Delta t)^3)$, i.e. it is a second order method.

Runge-Kutta Method (simple). The Runge-Kutta method is one of the most popular algorithms for numerically solving ordinary nonlinear differential equations. The success of the fourth-order Runge-Kutta method used here is based on the balance it affords between modest programming effort and high numerical accuracy. The time propagation is given by the equations

$$y_{i+1} = y_i + \frac{\Delta t}{6}\,(k_1 + 2k_2 + 2k_3 + k_4) \tag{5.15}$$

with

$$\begin{aligned} k_1 &= f(y_i, t_i) \\ k_2 &= f(y_i + \tfrac{1}{2}\Delta t\, k_1\,,\ t_i + \tfrac{1}{2}\Delta t) \\ k_3 &= f(y_i + \tfrac{1}{2}\Delta t\, k_2\,,\ t_i + \tfrac{1}{2}\Delta t) \\ k_4 &= f(y_i + \Delta t\, k_3\,,\ t_i + \Delta t)\,. \end{aligned} \tag{5.16}$$

The numerical error of this scheme is $O((\Delta t)^5)$, i.e. the method is the so-called *'fourth order Runge-Kutta method'*. Higher order methods are available (see, however, the remarks in Ref. [5.1]).

Runge-Kutta Method (adapted). This algorithm is simply a fourth order Runge-Kutta method, where the time step Δt is adapted during the computation. The time step is halved or doubled, depending on the results of an accuracy

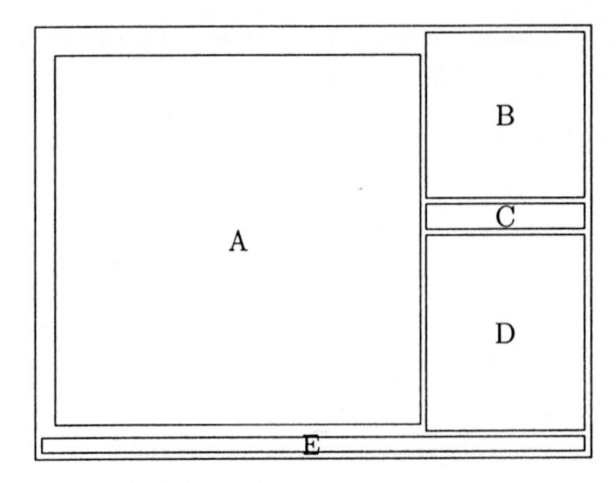

Fig. 5.2. Working screen for the Double Pendulum.

check. The program uses a test of conservation of the total energy E_{tot} as such a check. At each time step i, the difference ΔE_{tot} between the numerical and exact value of the total energy E_{tot} is computed. For $\Delta E_{\text{tot}} > Tol$ the time step is halved, while for $\Delta E_{\text{tot}} < Tol$ it is doubled. In the first case, the last computation step is repeated, i.e. the time step is repeatedly halved, until the energy accuracy criterion is satisfied. However, in this case the computation continues to the next time step. The accepted tolerance Tol can be changed by the user. Clearly this procedure typically increases the computation time. In addition, the user should be warned that conservation of energy does not, of course, guarantee the correctness of the numerical integration.

5.3 Interacting with the Program

The program for the double pendulum displays the numerical results in the different modes of operation on the screen in a common form. The working screen is partitioned into five windows, as shown in Fig. 5.2.

A: Presentation of Motion — The dynamics of the system can be displayed in various modes of presentation.

B: Pendulum Window — The parameters of the pendulum are shown. In addition to the current presentation, the motion of the double pendulum can be displayed.

C: Energy Window — The instantaneous value of the total energy is displayed. This value is computed after each time step and can be used to control the accuracy of the computation.

D : Menu Window — Here, menu items which can be activated during the computation ('hot keys') are listed.

E : Information Box — Messages from the program.

Main Menu

In the *Main Menu*, the user can change the mass and geometry of the pendulum configuration, store and load data, select the presentation, and the numerical integration algorithm. In addition, a brief information about the program can be obtained.

- **Information** — This menu item provides a short on-line description of the geometry of the double pendulum and the program.

- **Pendulum Geometry and Starting Conditions** — Here, the masses m_i, the lengths l_i of the pendula, the initial values of the angles φ, ψ and the angular velocities $\dot{\varphi}$, $\dot{\psi}$, as well as the gravitational acceleration g, can be altered. The parameters can be stored in a file to which the suffix *.PAR is automatically added. Pre-set values are $m_1 = m_2 = 1\mathrm{kg}$, $l_1 = l_2 = 1\mathrm{m}$, $\varphi = \psi = 90°$, $\dot{\varphi} = \dot{\psi} = 0\,\mathrm{s}^{-1}$, and $g = 9.81\mathrm{m\,s}^{-2}$. The pre-set parameters are saved in the file STANDARD.PAR.

  If a Poincaré section has been computed, the initial values of φ and $\dot{\varphi}$ are those of the last chosen trajectory in the Poincaré diagram; the angle ψ of the outer pendulum is set equal to zero and its angular velocity $\dot{\psi}$ is chosen according to energy conservation (5.8), where the positive sign is chosen (compare (5.9)). If desired, the parameters can be reset to the standard values by loading the parameter file STANDARD.PAR.

- **Mode of Presentation** — The program can show the dynamics of the double pendulum in various ways. Here, the desired mode of presentation can be selected:

  - *Double Pendulum:* The trajectory is shown in coordinate space, i.e. in the (φ, ψ)-plane. By pressing ⟨p⟩, the motion of the double pendulum is displayed graphically on the screen in the B-window. It should be noted that this motion is *not* a real-time movie of the pendulum, since the observed 'time' depends on the speed of computation.
  - *Phase Space of Inner/Outer Pendulum:* The phase space trajectory for the inner or outer pendulum is displayed in blue or red, respectively. The angular velocity ($\dot{\varphi}$ or $\dot{\psi}$, respectively) is plotted versus the angle (φ or ψ) as a function of time.
  - *Energies of Inner/Outer Pendulum:* The potential energy of the inner or outer pendulum is plotted (in blue or red, respectively) versus the kinetic energy as a function of time.

– *Individual Energies:* Four different energies of the double pendulum (kinetic and potential energies of the inner and outer pendulum) are shown simultaneously versus time (in blue or red, respectively). This allows a direct investigation of the flow of energy between different degrees of freedom.

– *Comparison of Numerical Algorithms:* The quality of the four available numerical integration schemes differs considerably. The two Runge-Kutta methods are able to compute the trajectory with satisfactory precision. The other two algorithms are faster, but much less accurate. In this mode of presentation, the %-deviation of the instantaneous total energy with respect to the true value (the initial value of the total energy) is simultaneously shown for all four numerical propagation schemes. The curves shown on the screen are scaled to a maximum deviation of 3%. This value (shown in window E) can be increased or decreased using the $\pm$ keys.

– *Poincaré Map:* A two-dimensional Poincaré section can be constructed by looking at the trajectory only at those moments when the outer pendulum passes the vertical position ($\psi = 0$), as described in Sect. 5.1. The program switches to the Poincaré map menu described below.

- **Numerical Algorithm** — The equations of motion for the double pendulum are integrated numerically. Four different numerical algorithms can be selected (Euler, Leapfrog, and simple or adapted Runge-Kutta method) and the time step Δt or the energy tolerance *Tol* can be modified. The chosen method is displayed in the main menu. The simple Runge-Kutta method with time step $\Delta t = 0.003\,\mathrm{s}$ is pre-selected in the program.

Menu Window

Here, some menu entries which can be activated during a program run are listed:

- **Stop** ⟨SPACE BAR⟩ : The motion is interrupted.

- **Delay** ⟨F1⟩ – ⟨F10⟩ : The motion is slowed down. ⟨F1⟩ shows the normal time scale and ⟨F10⟩ the single time step mode (pressing the ⟨SPACE BAR⟩ shows the next step).

- **Pendulum** ⟨p⟩ : The motion of the double pendulum is graphically displayed in the pendulum window. Pressing the key again erases the picture and displays the pendulum parameters.

- **Total Energy** ⟨e⟩ : The instantaneous value of the energy is shown/not shown.

- **Clear** ⟨c⟩ : The graphical presentation in window A is erased. The computation is continued.

- **Start again** ⟨n⟩ : The graphical presentation in window A is erased. The computation starts again with the given initial values.
- **Draw Boundary** ⟨b⟩ : The boundary of the energetically accessible region is plotted.
- **Magnify** ⟨+⟩ The graphical presentation is magnified (the coordinate axes are rescaled by a factor 0.5). In case of a phase space plot, this rescaling affects only the angular velocities.
- **Reduce** ⟨−⟩ The graphical presentation is reduced (the coordinate axes are rescaled by a factor two). In case of a phase space plot, this rescaling affects only the angular velocities.
- **Hardcopy** ⟨h⟩ : Prints the screen. By pressing ⟨M⟩, ⟨L⟩ or ⟨P⟩ a ⟨M⟩atrix-, ⟨L⟩aser- or ⟨P⟩ostscript printer can be selected; ⟨E⟩ or ⟨ESC⟩ returns. Before printing, the screen is inverted. The computation is continued after printing.
- **Main Menu** ⟨ESC⟩ : Returns to the selection of presentation.

Poincaré Map

In this presentation mode, a Poincaré section for the motion of the inner pendulum is constructed, as described in Sect. 5.1 . A colored point appears on the screen when the conditions (5.9) are met. The adapted Runge-Kutta method is used for numerical integration because the other algorithms are not accurate enough for this purpose.

First, a menu appears allowing a *'Configuration of the Poincaré section'*. The total energy of the system and the accuracy of integration (the energy tolerance) can be specified (pre-set values are 10J and 10^{-8}J, respectively). Different trajectories can be distinguished by different colors. A *cyclic* (pre-set) or *manual* color selection can be selected. In case of a manual selection, the color must be chosen by means of the keys ⟨F1⟩ to ⟨F10⟩.

If this menu is left by activating *'Accept configuration'* the Poincaré map can be constructed by manual selection of initial points (see below).

Furthermore, it is possible to choose several initial conditions, compute the entire Poincaré map automatically, and save the result to a file (move the cursor to the entry *'Choose initial conditions and save automatically to file'* and press ⟨ENTER⟩). The number of initial conditions (≤ 50) and the number of computed points for each trajectory (≤ 10000) must be specified. Finally, the name of the file can be changed (POINC is pre-set). Activating now *'Accept configuration'*, the Poincaré section is constructed by automatic selection of the initial points (see below).

Poincaré section: Manual selection — The menu window for the Poincaré section appears. By pressing ⟨i⟩, a cross for selecting the initial conditions by means of the cursor keys appears on the screen. Pressing the ⟨TAB⟩ key toggles the stepsize for the cursor. When the desired initial position is reached (the instantaneous values of the angle and the angular velocity are shown in window D), these values are confirmed by pressing ⟨ENTER⟩. Initial values outside the energetically allowed region are rejected. In case of a manual selection, the color of the plotted points must then be chosen by means of the keys ⟨F1⟩ to ⟨F10⟩ (the color 1, 2, ... corresponding to the keys ⟨F1⟩, ⟨F2⟩, ... is shown in window E). Otherwise the colors are chosen cyclically. During the computation, the following menu items can be activated:

- ⟨i⟩ : New initial values of the phase space point can be chosen.
- ⟨z⟩ : A phase space section can be magnified. The previously computed section is not retained and should be saved, if required. A rectangular window, which can be moved by means of the cursor keys, is shown on the screen. One can switch between the upper left and lower right edge by using the ⟨SPACE BAR⟩. The ⟨TAB⟩ key toggles the stepsize for the cursor. The chosen phase space section is magnified. If the entire zoom box is extended to the window boundaries, the Poincaré section is reset to the maximum possible values.

  ⟨ENTER⟩ leaves this item and chooses the selected phase space section. A new initial condition can then be chosen. ⟨ESC⟩ returns without action.
- ⟨p⟩ : The motion of the double pendulum is graphically displayed in the pendulum window and can be delayed (keys ⟨F1⟩– ⟨F10⟩). Pressing the key again erases the picture.
- ⟨e⟩ : The actual value of the total energy is shown/not shown.
- ⟨b⟩ : Draws the boundary of the energetically allowed phase space section. Initial points can only be chosen inside this boundary.
- ⟨s⟩ : The computed phase space section is saved. Two files are created: *.POP contains the parameters of the pendulum and information concerning the chosen phase space section; *.POI contains the data of the picture.
- ⟨l⟩ : A stored phase space section can be loaded.
- ⟨h⟩ : Hardcopy of the Poincaré section. By pressing ⟨M⟩ or ⟨L⟩, a ⟨M⟩atrix- or ⟨L⟩aserprinter can be selected. Before printing, the screen is inverted. After printing, the screen is re-inverted.
- ⟨ESC⟩ : Returns to the mode of presentation menu. Attention: The Poincaré section is *not* automatically saved!

Poincaré section: Automatic selection — If the Poincaré section is constructed automatically, the user must first specify the number of different initial conditions and the number of computed points in the chosen section in each case. In addition, the name of the file for saving the results must be defined ('POINC' is pre-set). Then, the window for the Poincaré sections appears and the boundary of the energetically allowed phase space region can be drawn, if desired. The specified number of initial conditions must be selected using the cursor, as described above. Initial values outside the energetically allowed region are rejected. The computation then starts. The completed Poincaré section is automatically saved. During the computation, the following menu items can be activated:

- ⟨i⟩ : Allows the selection of new initial conditions.
- ⟨z⟩ : A phase space section can be magnified, as described above.
- ⟨ESC⟩ : Stops the computation and returns to the mode of presentation menu. The instantaneous status of the Poincaré section is saved as specified in the configuration menu.

5.4 Computer Experiments

In the following, we will describe some numerical experiments which explore the dynamics of the double pendulum. In the program, all quantities are measured in standard physical units (mass in kilograms, length in meters, time in seconds, energy in joules). Most of our experiments are performed for the preset parameters $m_1 = m_2 = 1\text{kg}$, $l_1 = l_2 = 1\text{m}$ and $g = 9.81\text{ms}^{-2}$, and we quite often provide formulae which are valid for this standard choice, and data for starting conditions, energies, etc. In order to simplify such expressions, in the following we drop the dimensions of physical quantities whenever parameters of the system are specified.

5.4.1 Different Types of Motion

The dynamics of the double pendulum will turn out to be complicated and it may be helpful to discuss some special cases. First, if the mass of the inner pendulum, m_1, is large compared to m_2, the motion of the inner pendulum decouples from the outer one. The motion of the inner one is integrable (it is a simple pendulum motion), whereas the outer one feels a periodic driving force.

For a small length l_1, the mass m_1 is practically fixed at the origin, and we have only a single pendulum of mass m_2 and length l_2. Similarly, for small l_2 the system reduces to a single pendulum of mass M.

For large energy, the kinetic terms dominate and the dynamics is considerably simplified. The same is true if the gravitational force is zero, which can be achieved experimentally by a pendulum swinging in a horizontal plane. In this case, the Lagrangian (5.2) is

$$L = \tfrac{1}{2} M \, l_1{}^2 \, \dot{\varphi}^2 + \tfrac{1}{2} m_2 \, l_2{}^2 \, \dot{\psi}^2 + m_2 \, l_1 \, l_2 \, \dot{\varphi} \, \dot{\psi} \, \cos(\psi - \varphi) \,, \tag{5.17}$$

or, in terms of the variable $\delta = \psi - \varphi$,

$$L = \tfrac{1}{2} M \, l_1{}^2 \, \dot{\varphi}^2 + \tfrac{1}{2} m_2 \, l_2{}^2 \, (\dot{\varphi} + \dot{\delta})^2 + m_2 \, l_1 \, l_2 \, \dot{\varphi} \, (\dot{\varphi} + \dot{\delta}) \, \cos \delta \, . \tag{5.18}$$

The angle φ is a cyclic variable, and hence the canonical angular momentum

$$\begin{aligned} L_\varphi = \frac{\partial L}{\partial \dot{\varphi}} &= M l_1^2 \dot{\varphi} + m_2 l_2^2 (\dot{\varphi} + \dot{\delta}) + m_2 l_1 \, l_2 (2\dot{\varphi} + \dot{\delta}) \cos \delta \\ &= l_1 \dot{\varphi} \, (M l_1 + m_2 \cos(\psi - \varphi)) + m_2 \, l_2 \dot{\psi} \, (l_2 + l_1 \cos(\psi - \varphi)) \end{aligned} \tag{5.19}$$

is a constant of motion. Hence, this system with two degrees of freedom is integrable because we have two invariants, the total energy E_{tot} and the angular momentum L_φ of the inner pendulum.

Another integrable limit is the small amplitude motion. Linearizing the equations of motion (5.4) and (5.5), or expanding the Lagrangian (5.2) up to second order terms, we obtain

$$\begin{pmatrix} \ddot{\varphi} \\ \ddot{\psi} \end{pmatrix} + \frac{1}{1-\mu} \begin{pmatrix} g_1 & -g_1 \mu \\ -g_2 & g_2 \end{pmatrix} \begin{pmatrix} \varphi \\ \psi \end{pmatrix} = 0 \tag{5.20}$$

and the eigenfrequencies

$$\omega_\pm^2 = \frac{1}{2(1-\mu)} \left\{ g_1 + g_2 \pm \sqrt{(g_1 - g_2)^2 + 4 g_1 g_2 \mu} \right\} \tag{5.21}$$

$$= \frac{1}{2 m_1 l_1 l_2} \left\{ M(l_1 + l_2) \pm \sqrt{M\{\, M(l_1 - l_2)^2 + 4 m_2 l_1 l_2 \}} \right\} . \tag{5.22}$$

The ratio of the components of the corresponding normal coordinates is

$$(\psi/\varphi)_\pm = \frac{1}{2\mu} \left\{ 1 - \lambda \mp \sqrt{(1-\lambda)^2 + 4\lambda\mu} \right\} , \tag{5.23}$$

with $\lambda = l_1/l_2$ and $\mu = m_2/M$ (compare (5.6)). For the pre-set parameter values $m_1 = m_2 = 1$, $l_1 = l_2 = 1$, and $g = 9.81$ this simplifies to

$$\omega_\pm^2 = g \left\{ 2 \pm \sqrt{2} \right\} , \tag{5.24}$$

i.e. $\omega_- = 5.75$, $\omega_+ = 33.49$ and $(\psi/\varphi)_\pm = \mp\sqrt{2}$. Normalized eigenvectors for the two eigenmodes are

$$\mathbf{y}_\pm = \frac{1}{\sqrt{3}} \begin{pmatrix} 1 \\ \mp\sqrt{2} \end{pmatrix} , \tag{5.25}$$

and the general solution vector can be written as

$$\begin{pmatrix} \varphi(t) \\ \psi(t) \end{pmatrix} = \mathrm{Real}\,(\,\mathbf{y}(t)) \tag{5.26}$$

with

$$\mathbf{y}(t) = A_- \mathrm{e}^{i\omega_- t} \mathbf{y}_- + A_+ \mathrm{e}^{i\omega_+ t} \mathbf{y}_+ \, , \tag{5.27}$$

where the complex coefficients $A_\pm$ are determined by the initial conditions.

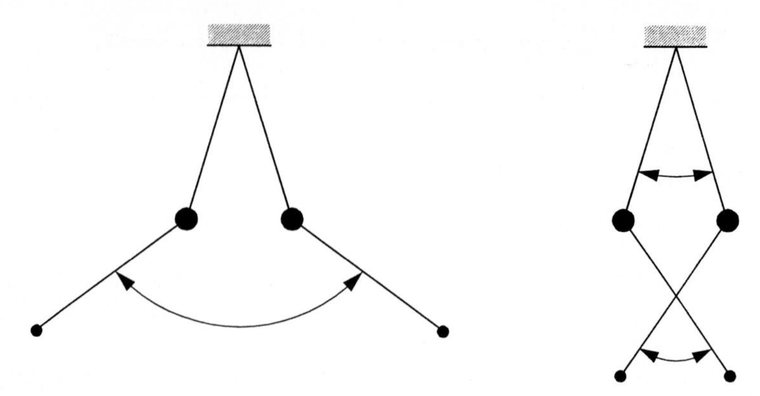

Fig. 5.3. Normal modes of oscillation of the double pendulum ($l_1 = l_2$ and $m_1 = 8m_2$).

In the small amplitude limit, the total energy (5.7) is given by

$$E_{\mathrm{tot}} = \tfrac{1}{2}M{l_1}^2\,\dot{\varphi}^2 + \tfrac{1}{2}m_2{l_2}^2\,\dot{\psi}^2 + m_2 l_1 l_2 \dot{\varphi}\dot{\psi} + \tfrac{1}{2}Mgl_1\varphi^2 + \tfrac{1}{2}m_2 g l_2 \psi^2\,, \quad (5.28)$$

which is equal to

$$E_{\mathrm{tot}} = \dot{\varphi}^2 + \tfrac{1}{2}\dot{\psi}^2 + \dot{\varphi}\,\dot{\psi} + \tfrac{1}{2}g\,(\varphi^2 + \psi^2) \quad (5.29)$$

for the pre-set parameters.

As a first computer experiment, one can start the program, switch to the mode *'Pendulum Geometry and Starting Conditions'*, activate the item *'Change'* and change the initial position of the inner and outer pendulum from the pre-set horizontal position to $\varphi = 10°$ and $\psi = 14.14°$, the conditions for the eigenmode with frequency ω_-. We then return to the main menu, activate *'Mode of Presentation'*, and select *'Double Pendulum'*. We indeed observe a periodic oscillation, where the double pendulum swings in a synchronized manner. When the key $\langle$e$\rangle$ is pressed, the momentary numerical value of the total energy is displayed. This is equal to $E_{\mathrm{tot}} = 0.595$ and stays constant, as it should. The displayed motion of the pendulum is quite slow and can be speeded up by changing the accuracy or the method of the numerical integration.

The higher frequency eigenmode can be excited by starting at an initial configuration $\varphi = 10°$ and $\psi = -14.14°$. The two pendula are now counter-oscillating. The same oscillation can also be excited if the pendulum is in a downward position, $\varphi = \psi = 0°$ and the initial angular velocities are chosen as, for example, $\dot{\varphi} = 1$ and $\dot{\psi} = 1.414$. For other starting conditions with small amplitude, according to (5.27) one observes a quasiperiodic motion of the double pendulum, because the eigenfrequencies $\omega_\pm$ are incommensurable. It is also instructive to display the small amplitude oscillations in phase space of the inner or outer pendulum, where we find closed elliptical curves for the normal modes and quasiperiodic motion otherwise.

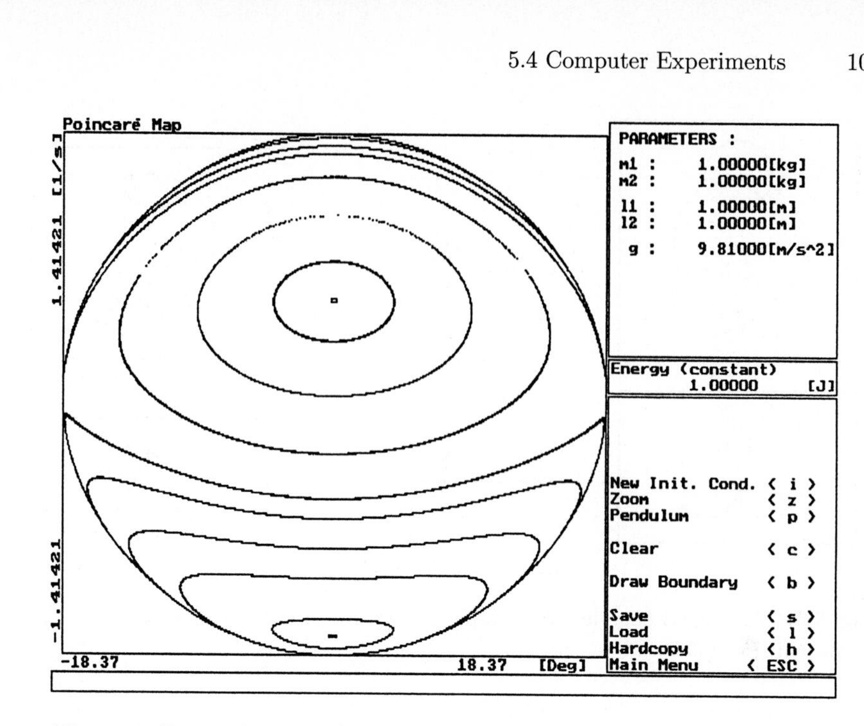

Fig. 5.4. Poincaré section for the double pendulum at low energy $E_{tot} = 1$.

Next we explore the phase space organization of the dynamics in more detail by means of a Poincaré section of the phase space for $\psi = 0$, i.e. when the outer pendulum passes the vertical position with positive values of the quantity $\dot{\psi} + \lambda\dot{\varphi}\cos\varphi$, as described in Sect. 5.1. We use the pre-set parameters of the pendulum. In addition, it may be recalled that we still restrict ourselves to small amplitude oscillations. Activating the presentation mode *'Poincaré Section'* and changing the value of the total energy from the pre-set value of ten to the smaller value of one, and the pre-set accuracy to 10^{-4} to speed up the computation, we then select the initial conditions *'manually'* and colors *'cyclically'*. The Poincaré section is then shown, i.e. a $(\varphi, \dot{\varphi})$-plane extending over the energetically allowed region, where values of $\dot{\psi}$ exist satisfying

$$E_{\rm tot} = \dot{\varphi}^2 + \tfrac{1}{2}\dot{\psi}^2 + \dot{\varphi}\,\dot{\psi} + \tfrac{1}{2}g\,\varphi^2 \tag{5.30}$$

For an initial point at $\varphi = \psi = 0$, this is

$$E_{\rm tot} = \tfrac{1}{2}\dot{\varphi}^2\left\{1 + \left(1 + \frac{\dot{\psi}}{\dot{\varphi}}\right)\right\}, \tag{5.31}$$

yielding

$$\dot{\varphi}_\pm = \pm\sqrt{\frac{E_{\rm tot}}{2 \mp \sqrt{2}}} \tag{5.32}$$

for the initial angular velocities of the two eigenmodes, i.e. $\dot{\varphi}_- = 0.543$ and $\dot{\varphi}_+ = -1.307$ in the present case. Selecting these initial conditions, we find the

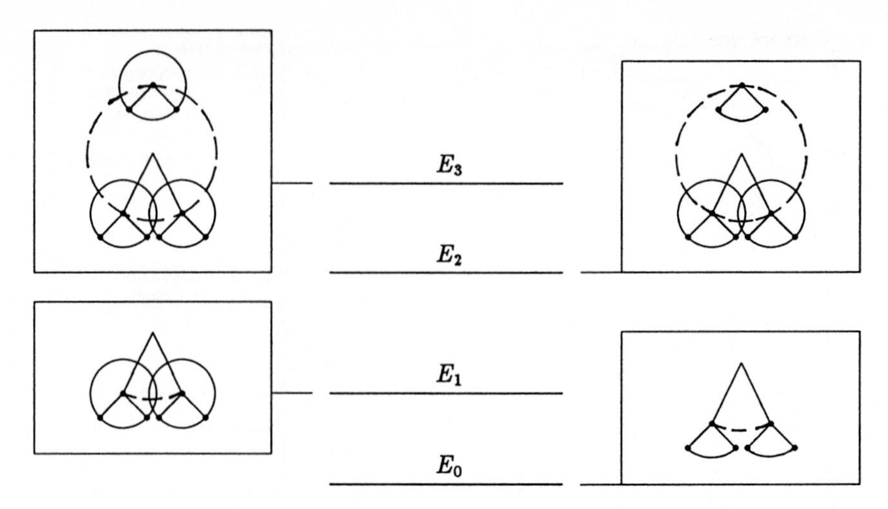

Fig. 5.5. Threshold energies E_k of different types of combined librational (∧) and rotational (○) modes.

two eigenmodes as fixed points of the Poincaré map. One can now select further initial conditions, draw the boundary determined by energy conservation, and produce a phase space plot similar to that shown in Fig. 5.4 (this diagram is pre-computed and can be loaded by pressing ⟨1⟩ and selecting E1.POI). We observe that the whole plane is filled with invariant curves, which signals the integrability of the double pendulum at low energies.

In the following experiments, the dynamics of the double pendulum will be studied at higher energies. Depending on the value of E_{tot}, different types of motion are possible. For low energy, both pendula can only oscillate. At a threshold energy of

$$E_1 = 2m_2 g\, l_2\,, \tag{5.33}$$

the outer pendulum can reach an upright position with zero kinetic energy, and for energies $E_{\text{tot}} > E_1$ the outer pendulum can rotate. In the same way, a threshold energy

$$E_2 = 2Mg\, l_1 \tag{5.34}$$

is obtained for the rotation of the inner pendulum. Finally, for energies exceeding

$$E_3 = E_1 + E_2\,, \tag{5.35}$$

the full rotational motion of both pendula is possible. This is illustrated in Fig. 5.5.

For the standard parameters of equal mass and equal length, the three thresholds are equidistant

$$E_1 = 2mgl < E_2 = 4mgl < E_3 = 6\,mgl\,. \tag{5.36}$$

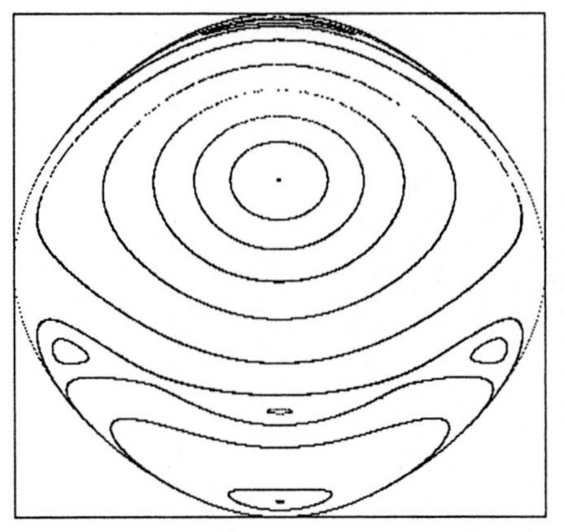

Fig. 5.6. Poincaré section for the double pendulum at energy $E_{\text{tot}} = 4$.

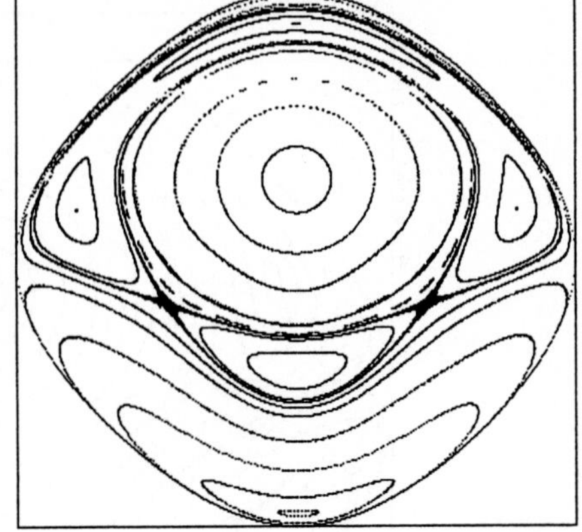

Fig. 5.7. Poincaré section for the double pendulum at energy $E_{\text{tot}} = 8$.

5.4.2 Dynamics of the Double Pendulum

In this numerical experiment, we will explore the dynamics of the pendulum for increasing values of the energy. The pre-set parameters of the system are used. A first orientation are the thresholds for the different types of motion (5.36), which are given by

$$E_1 = 19.62\,,\ E_2 = 39.24\ ,\ E_3 = 58.86 \tag{5.37}$$

for $g = 9.81$.

Figures 5.6 and 5.7 show computed Poincaré sections for energies $E_{\text{tot}} = 4$ and $E_{\text{tot}} = 8$. Each plot shows results from several trajectories. On the screen, these different trajectories can be more easily distinguished by their different colors. The boundary of the energetically accessible phase space region is also shown. In comparison with Fig. 5.4 for $E_{\text{tot}} = 1$, the amplitudes of the motion

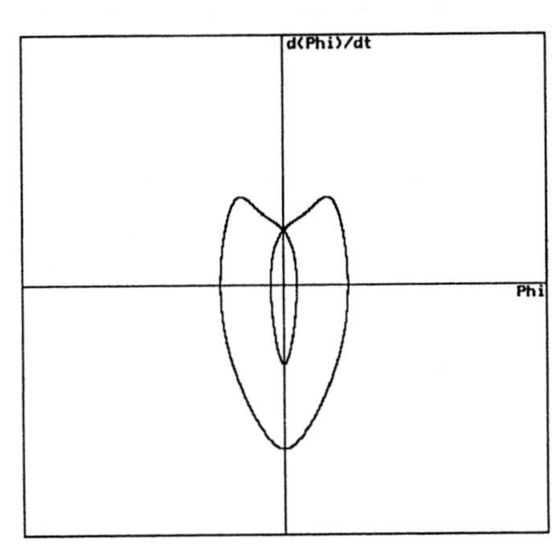

Fig. 5.8. Phase space plot of the periodic orbit in the central fixed point of the Poincaré section in Fig. 5.7.

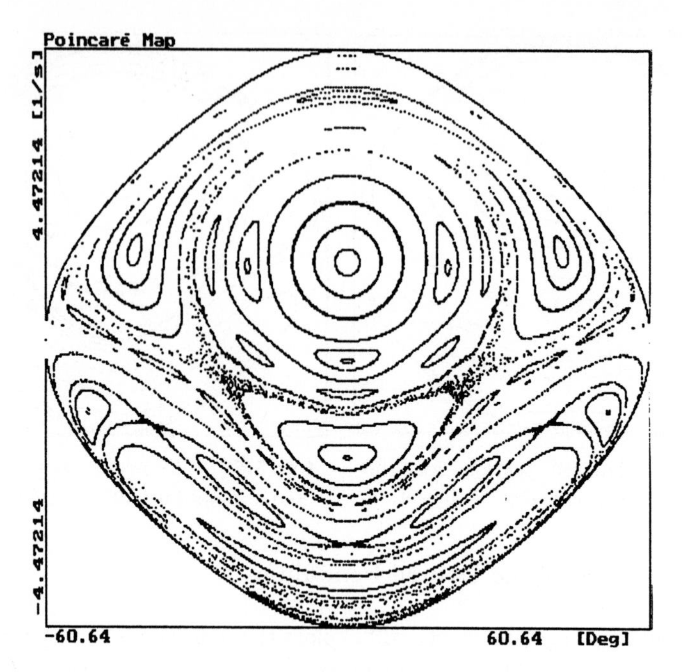

Fig. 5.9. Poincaré section for the double pendulum at energy $E_{\text{tot}} = 10$. This figure is stored (files PEND10.POI, –.POP).

have increased and the nonlinearity of the forces must be taken into account, so that the linear analysis is no longer applicable. For $E_{\text{tot}} = 4$, however, the picture is qualitatively still very similar to Fig. 5.4 for $E_{\text{tot}} = 1$. The overall organization is dominated by two periodic orbits, where the two pendula are co- or counter-oscillating. These orbits appear at fixed points on the $\varphi = 0$ axis of the Poincaré map. For $E_{\text{tot}} = 8$, another fixed point has appeared on the line $\varphi = 0$. The corresponding phase space orbit in Fig. 5.8 shows that this orbit, in fact, performs two rotations in phase space, despite the single point appearing in the Poincaré section due to the symmetry of the orbit. The stability island

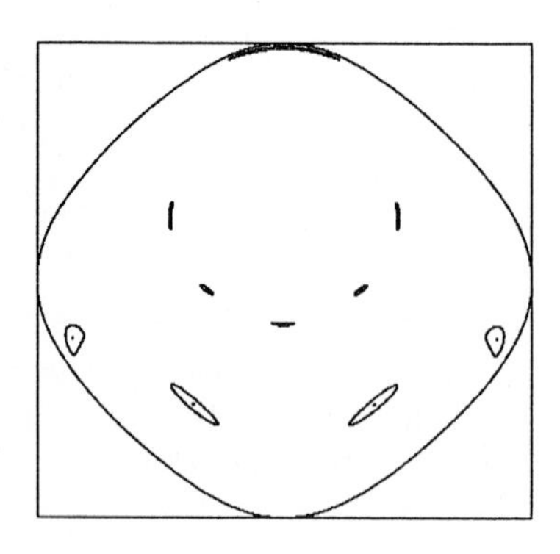

Fig. 5.10. Two selected island chains of Fig. 5.9.

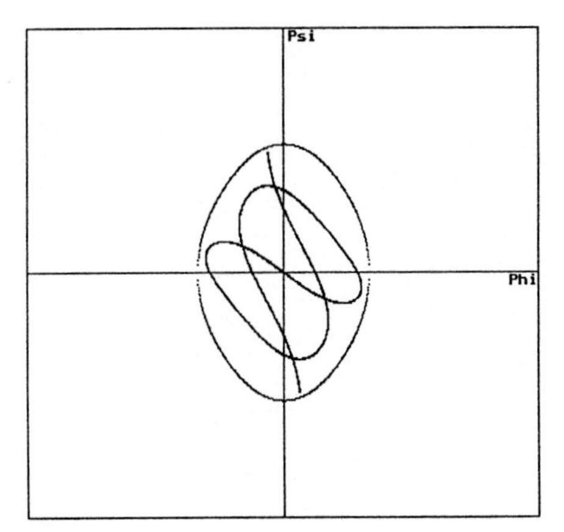

Fig. 5.11. Coordinate space trajectories of the outer periodic orbit shown in 5.10.

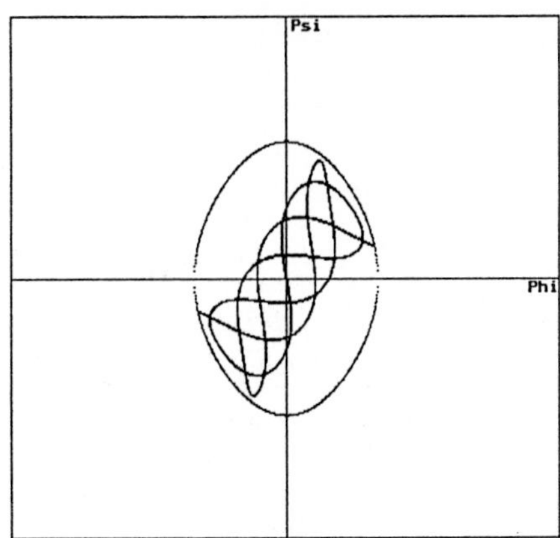

Fig. 5.12. Coordinate space trajectories of the inner periodic orbit shown in 5.10.

surrounding this orbit is bounded by a separatrix, and close to this separatrix one observes the first signs of chaotic motion. With increasing energy the process of destruction of invariant curves continues.

At an energy of about $E_{\text{tot}} = 10$, we have a fully developed KAM scenario as shown in Fig. 5.9. Inside a chain of stability islands we find periodic orbits, which can be also viewed in coordinate space. This can be most simply done by moving the cursor to the center of such an island and starting a phase space trajectory there. Returning to the presentation mode *'Pendulum'* the program displays the same orbit in coordinate space. As an example, the selected phase space orbits of Fig. 5.10 are shown in the (φ, ψ)-plane in Figs. 5.11 and 5.12. The

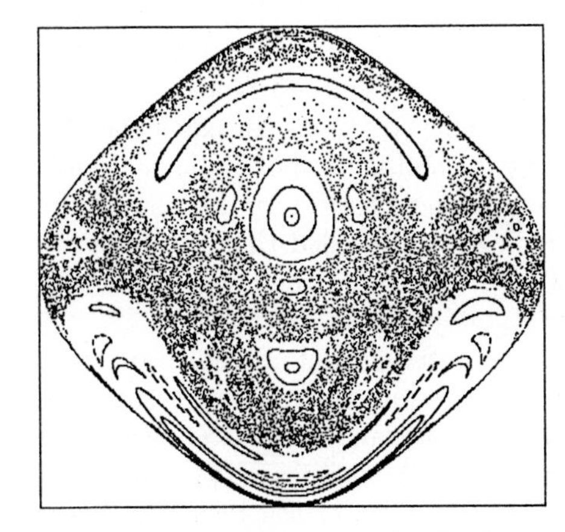

Fig. 5.13. Poincaré section for the double pendulum at energy $E_{\text{tot}} = 15$. This figure is stored on the disk.

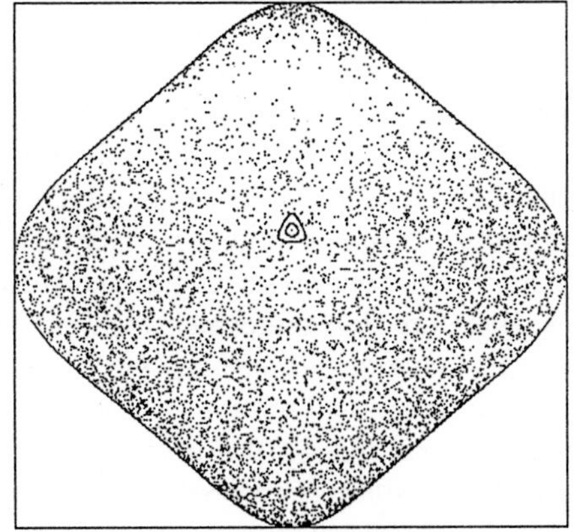

Fig. 5.14. Poincaré section for the double pendulum at energy $E_{\text{tot}} = 25$.

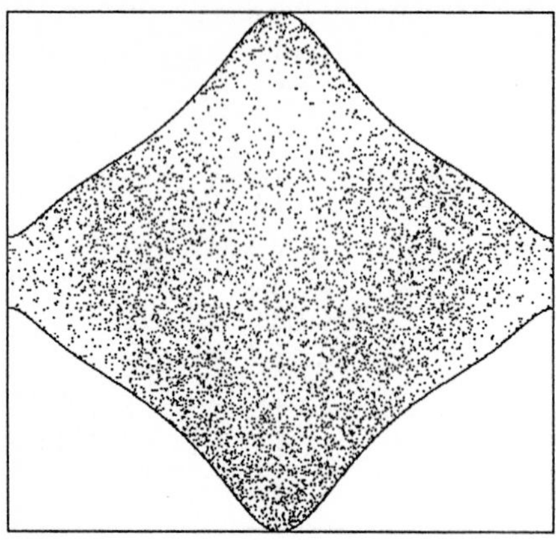

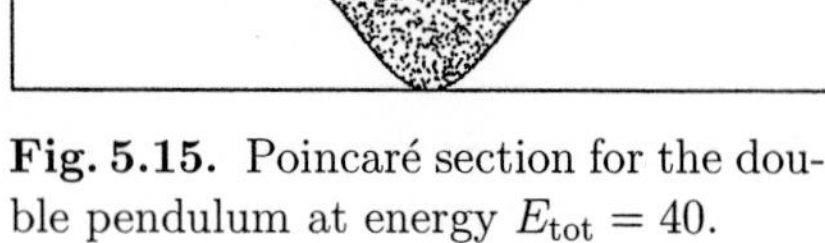

Fig. 5.15. Poincaré section for the double pendulum at energy $E_{\rm tot} = 40$.

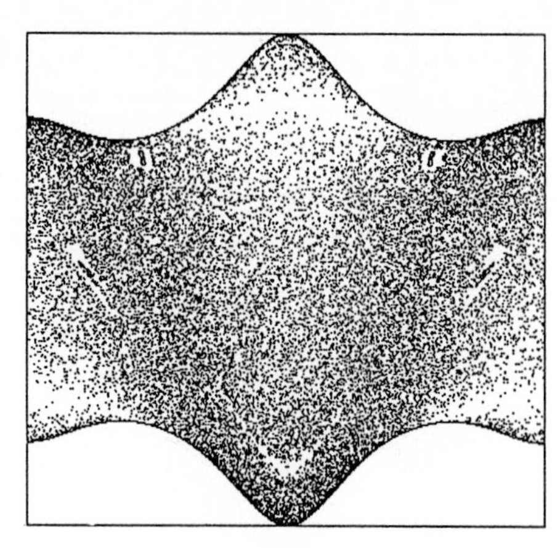

Fig. 5.16. Poincaré section for the double pendulum at energy $E_{\rm tot} = 70$.

initial conditions for these orbits are $\varphi = 52.1°$, $\dot{\varphi} = -1.10$, $\psi = 0$, $\dot{\psi} = 2.38$ and $\varphi = 27.8°$, $\dot{\varphi} = 1.22$, $\psi = 0$, $\dot{\psi} = 2.62$, respectively. Both orbits have five intersections with the $\psi = 0$ axis corresponding to the five points appearing in the Poincaré map, and both meet the energy boundary

$$E_{\rm tot} = V(\varphi, \psi) = M\,g\,l_1\,(1 - \cos\varphi) + m_2\,g\,l_2\,(1 - \cos\psi) \tag{5.38}$$

in coordinate space, where the kinetic energy vanishes. The boundary (5.38) is also plotted in Figs. 5.11 and 5.12. In addition, it can be observed that both orbits are approximately a superposition of two oscillations in the direction of the two eigenmodes $\psi = \pm\sqrt{2}\varphi$.

The chaotic region in Fig. 5.9 grows when the energy is further increased, as illustrated in Fig. 5.13 for $E_{\rm tot} = 15$. When the energy exceeds the threshold $E_1 = 19.68$, the outer pendulum can rotate, but the motion of the inner one is still constrained. Above the threshold $E_2 = 39.24$, the inner one can also rotate, but a full rotation of both pendula is still forbidden. Figures 5.14 and 5.15 show two examples of the chaotic motion in these energy regions. In particular, for $E_{\rm tot} = 40$ the motion seems to be fully ergodic and no sign of stability islands can be seen. All points are computed from a single trajectory. In view of the varying density of points in this plot, it might be recalled that the Poincaré map is *not* area-preserving in the variables φ and $\dot{\varphi}$ and the invariant density is, therefore, *not* uniform.

For energies exceeding $E_3 = 58.86$, the outer pendulum can also rotate and all dynamical motions are energetically accessible. An example is shown in Fig. 5.16 for $E_{\rm tot} = 70$. The motion is mainly chaotic and the greater part of the Poincaré section is covered by a single trajectory. A phase space plot of such a chaotic trajectory is shown in Fig. 5.18 and monitors the erratic behavior. There is, however, also a small stability island in the lower part of the Poincaré map. The orbit in its center is quite interesting and passes through the straight line configuration $\varphi = \psi = 0$ and the configuration $\varphi = 0$, $\psi = 180°$, where the

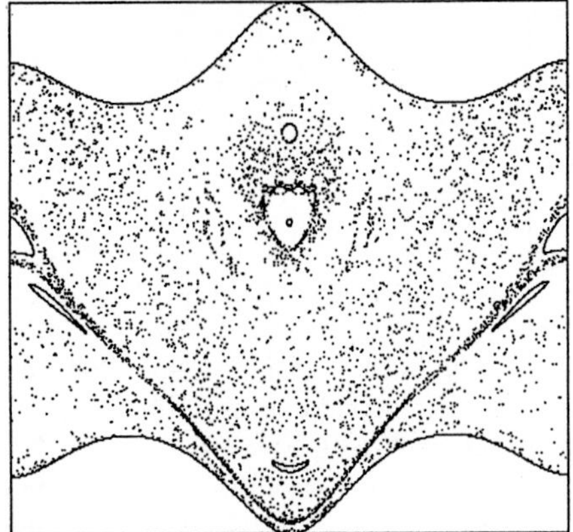

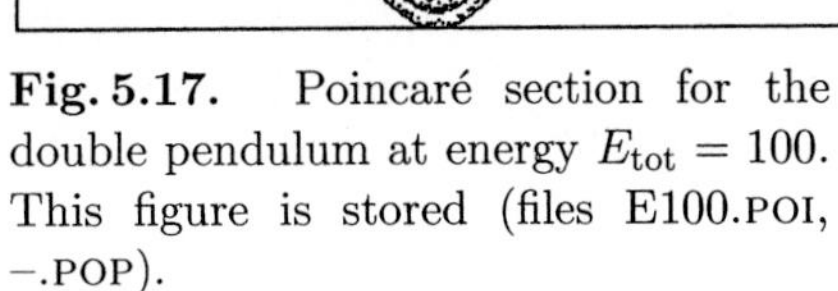

Fig. 5.17. Poincaré section for the double pendulum at energy $E_{\text{tot}} = 100$. This figure is stored (files E100.POI, –.POP).

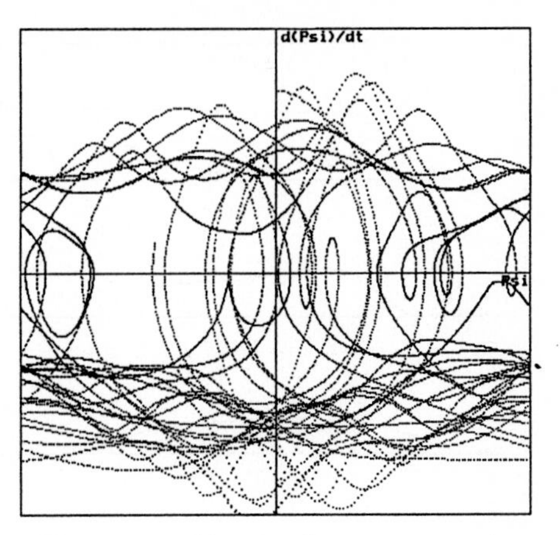

Fig. 5.18. Chaotic phase space trajectory for an energy $E_{\text{tot}} = 70$.

outer pendulum is inverted. The energy flow for this orbit in Fig. 5.19 shows very little potential energy in the inner pendulum, which is always close to its minimum configuration at $\varphi = 0$, whereas its kinetic energy changes drastically. The potential energy of the outer pendulum passes through a maximum in the inverted position of the pendulum.

For an energy of $E_{\text{tot}} = 100$, ordered motion reappears. In the two islands in Fig. 5.17, the double pendulum rotates in space. At even higher energies the dynamics becomes organized again and approaches the integrable limit of an unforced double pendulum, where the angular momentum L_φ is a constant of motion, as discussed in Sect. 5.4.1. More and more stability islands and invariant curves appear, as shown in Figs. 5.20 and 5.21 for energies of 200 and 500. The dominant island in Fig. 5.21 is centered on a stable periodic orbit, which is simply a rotation of the straight-line configuration $\varphi = \psi$ of the pendulum.

5.4.3 Destruction of Invariant Curves

The KAM theorem (compare Sect. 2.2.3) states that for small perturbation of an integrable system, most of the invariant tori survive. Excluded are tori carrying periodic trajectories, i.e. tori with rational values of the winding number w.

The double pendulum is integrable in both the limits of low and high energies and the value of the energy (or its inverse) can be considered as a perturbation parameter. Let us start with the integrable dynamics at high energies, where the influence of the gravitational potential is very small. When the energy is decreased, the integrability is destroyed and the KAM scenario develops. Poincaré plots for energies $E_{\text{tot}} = 500$, 200 and 100 are shown in Figs. 5.21, 5.20, and 5.17. Following the computer experiment shown in the film *The Pla-*

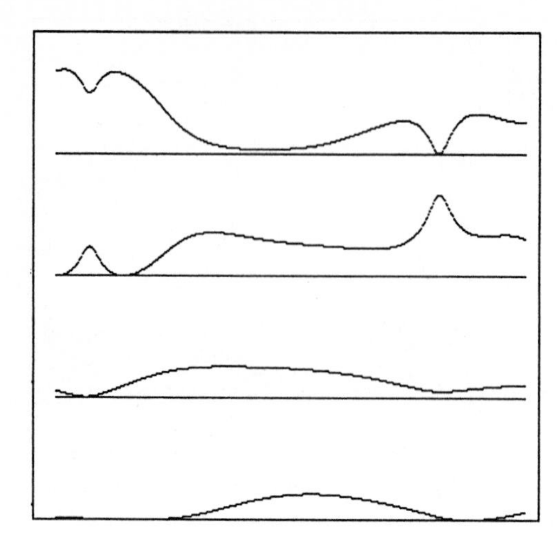

Fig. 5.19. Energy flow for the stable periodic orbit in the Poincaré section 5.16 for $E_{\text{tot}} = 70$. Shown are (from top to bottom) $E_{\text{kin}\,2}$, $E_{\text{kin}\,1}$, $E_{\text{pot}\,2}$, and $E_{\text{pot}\,1}$ for the outer (2) and the inner (1) pendulum.

nar Double Pendulum by Richter and Scholz [5.4], we now study the destruction of the invariant tori in more detail. In Fig. 5.17, we observe two different chaotic regions separated by a boundary. This separation is more evident in the colored Poincaré section displayed on the screen (a pre-computed picture can be loaded from the file E100.POI). Let us study the boundary of the two regions in more detail.

First, we decrease the perturbation by increasing the energy to $E_{\text{tot}} = 120$, where the system is closer to the integrable limit $E_{\text{tot}} = \infty$. The Poincaré plot in Fig. 5.22 shows that the central stability island is more extended. The boundary now appears to be more structured, and in its neighborhood we find several island chains. For initial conditions in the center of these islands, the pendulum swings periodically. Displaying the motion of the double pendulum

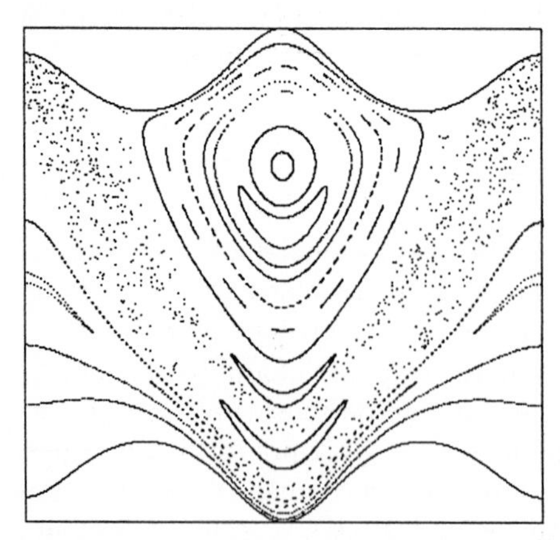

Fig. 5.20. Poincaré section for the double pendulum at energy $E_{\text{tot}} = 200$.

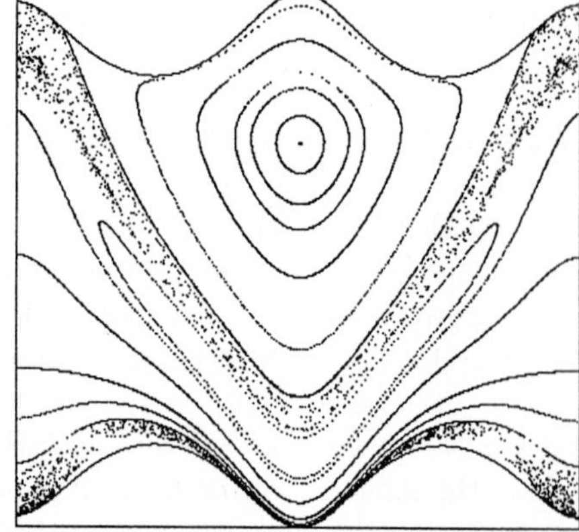

Fig. 5.21. Poincaré section for the double pendulum at energy $E_{\text{tot}} = 500$.

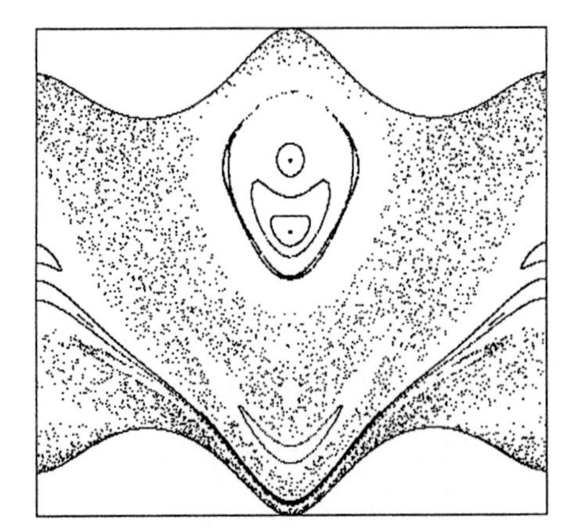

Fig. 5.22. Poincaré section for the double pendulum at energy $E_{\text{tot}} = 120$. This figure is stored on the disk.

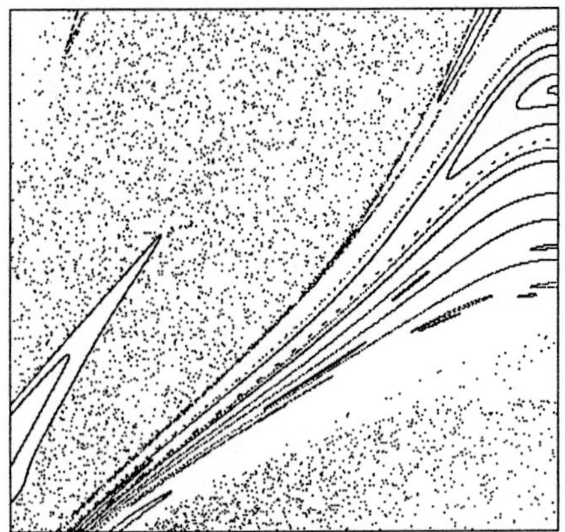

Fig. 5.23. Magnification of the Poincaré section shown in Fig. 5.22 at Energy $E_{\text{tot}} = 120$. This figure is stored on the disk.

on the screen, we can count the number of rotations of both pendula during one period, i.e. the winding number. Close to the boundary, there is a stable period-one orbit (winding number $w_1 = 1$) at $(\varphi, \dot{\varphi}) = (0°, -1.51)$ (colored light blue). For the chain of two islands (colored yellow) centered at $(\varphi, \dot{\varphi}) = (180°, 1.47)$ and $(0°, -1.46)$, we find a periodic orbit with winding number $w_2 = 1/2$, i.e. the outer pendulum rotates twice as fast as the inner one. The two pendula are at resonance. In addition, we find a chain of three islands (colored green) at $(\varphi, \dot{\varphi}) = (146°, -2.92), \ldots$. Counting the rotations of the pendula, we determine a winding number $w_3 = 2/3$. These island chains are separated by invariant curves. For a closer inspection, one can use the *'zoom'* option to blow up the phase space, as shown in Fig. 5.23.

Between these two chains of periodic orbits with winding numbers 1/2 and 2/3 there is another resonance island chain, where the outer pendulum rotates five times, while the inner one performs three rotations, i.e. the winding number is $w_4 = 3/5$. More of these chains may be detected and it should become clear that one is observing the scenario described in Sect. 2.4.2. Between each of the two resonances with winding numbers p/q and r/s, there is another with $w = (p+r)/(q+s)$, which is the rational number with the smallest denominator between p/q and r/s. The series of winding numbers w_n constructed in this way is

$$\frac{1}{1}, \frac{1}{2}, \frac{2}{3}, \frac{3}{5}, \frac{5}{8}, \ldots, \frac{F_n}{F_{n+1}}, \ldots, \tag{5.39}$$

where the F_n are the Fibonacci numbers. This sequence converges to the irrational number $g^* \approx 0.61803$, the *'golden-ratio'* or *'golden-mean'*, i.e. the winding number equals a so-called *'noble number'* (compare Sect. 2.4.2).

The limiting curve – an invariant curve with a noble winding number – supports quasiperiodic trajectories. In view of the KAM theorem, such curves

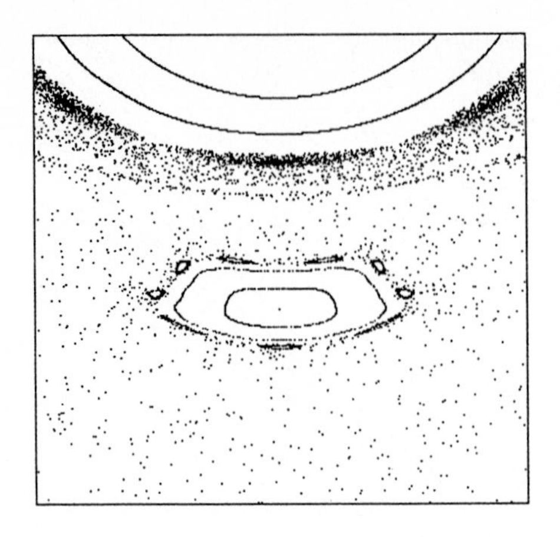

Fig. 5.24. Magnification of the Poincaré section shown in Fig. 5.13 ($E_{\text{tot}} = 15$). This figure is stored on the disk.

are the most stable invariant curves because the winding numbers are the most irrational numbers between zero and one. In phase space, this invariant set is, of course, a torus, a so-called *'noble torus'*.

When the perturbation is increased (i.e. the energy decreased), these noble tori are in many cases the last tori that survive. Looking again at Fig. 5.17 for $E = 100$, and decreasing the energy even more, the noble KAM curve constitutes the last boundary between the two different chaotic regimes. When the energy is further reduced, this invariant curve is destroyed and global chaotic motion sets in.

5.4.4 Suggestions for Additional Experiments

Testing the Numerical Integration. It is recommended that the performance of the four integration methods described in Sect. 5.2 be investigated in order to avoid unacceptable numerical errors or unnecessarily long computation times. This can be done, for instance, by means of the program mode *'Comparison of Numerical Algorithms'*. In any case, it is useful to check the conservation of energy during the integration.

The reader may explore the properties of the algorithms himself, guided by the remarks in Sect. 5.2 . As a hint, one should compare the accuracy of the four methods using the pre-set time step $\Delta t = 3 \cdot 10^{-3}$ for the first three algorithms with the pre-set energy tolerance of 10^{-8} of the adapted Runge-Kutta method. As expected, the Euler and Leapfrog methods will be found to be faster, but more inaccurate. The speed of computation can be measured by displaying the swinging double pendulum on the screen. It is now strongly recommended to increase the time step by, say, a factor of ten and to compare the accuracy and computation time again. This will show some advantages of the pre-set Runge-Kutta integration.

Zooming In. As in the study of billiard dynamics in Chap. 3, the subtleties of the fine structure of chaotic dynamics can be explored by means of the *'zoom'*

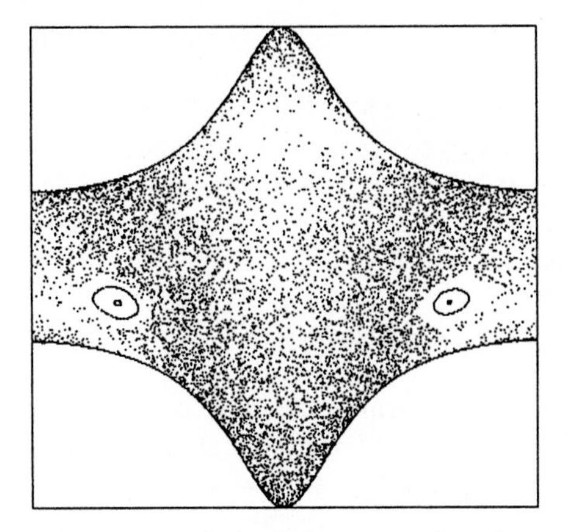

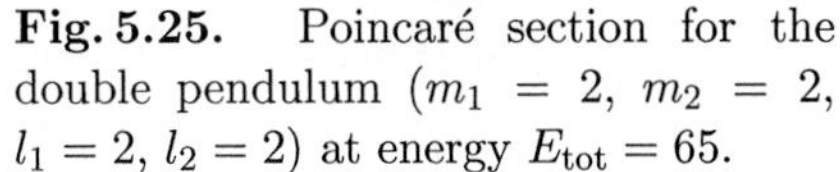

Fig. 5.25. Poincaré section for the double pendulum ($m_1 = 2$, $m_2 = 2$, $l_1 = 2$, $l_2 = 2$) at energy $E_{\text{tot}} = 65$.

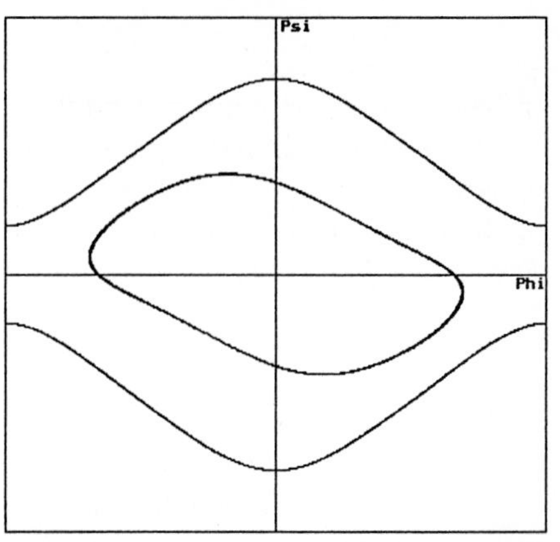

Fig. 5.26. Phase space trajectory for the periodic orbits in the centers of the stability islands in Fig. 5.25.

feature in the presentation mode *'Poincaré Map'*. Such a magnification shows the satellite islands accompanying the bigger ones, the satellites of the satellites and so on. The reader should, however, be aware of the fact that such studies are much more time consuming than for the case of billiard systems, where no differential equations need to be integrated.

To take an example, Fig. 5.24 shows a magnification of the Poincaré section shown in Fig. 5.13 for an energy of $E_{\text{tot}} = 15$. The neighborhood of the small island close to the center is magnified. The stability island of the central orbit, which is similar to that shown in Fig. 5.8 for $E_{\text{tot}} = 8$, is surrounded by a chain of five satellite orbits, which are all connected, i.e. they contain a common periodic orbit. In fact, more such island chains can be found, which may show up in further magnifications.

Different Pendulum Parameters. In the above studies, the pre-set parameters of the pendulum have been used. New effects may be discovered by changing the mass or length of the pendula.

For example, the small angle motion, which is typically quasiperiodic, can be forced to be periodic. This happens for a rational ratio of the eigenfrequencies ω_+ and ω_- given in (5.21). In such a case, the two pendula are at resonance for all initial conditions. As an exercise, one can study the small-amplitude dynamics of the double pendulum by changing the mass ratio from one to $m_1/m_2 = 35$ with $\omega_+/\omega_- = 2$. It may also be of interest to explore this case (and similar ones) in more detail by computing Poincaré sections at low and higher values of the energy.

Another point of interest could be the dynamics of a pendulum where the thresholds E_1 and E_2 in (5.33) and (5.34) are interchanged. This happens for pendula where the length of the inner pendulum is smaller than that of the outer one, more precisely for

$$l_1 < l_2 m_2/M \,. \tag{5.40}$$

In such a case, the inner pendulum can rotate, while the outer one is still constrained to a librational motion. This is expected to lead to differences in the dynamics. As an example, the Poincaré section in Fig. 5.25 shows the dynamics of the double pendulum for $m_1 = 2$, $m_2 = 2$, $l_1 = 2$, and $l_2 = 2$. The energy $E = 65$ is chosen between the thresholds

$$E_2 = 6g = 58.9 < E < E_1 = 78.5 \,. \tag{5.41}$$

A chaotic trajectory fills most of the phase space, where the inner pendulum is allowed to rotate. Two islands of stability are observed, centered at $\varphi = \pm 118°$, $\dot{\varphi} = -1.73$ (the initial conditions for the outer pendulum are $\phi = 0$ and $\dot{\psi} = 1.71$ in both cases). These islands are disconnected, i.e. a trajectory started in the right islands does not appear in the opposite islands. The orbits traced out in coordinate space (Fig. 5.26 are identical, however, one motion ($\varphi = -118°$) is clockwise, whereas the other one is counterclockwise.

5.5 Real Experiments and Empirical Evidence

Modifications of the simple pendulum are prototype setups whose main purpose is to demonstrate qualitatively chaotic vibrations by using different kinds of coupling, geometric nonlinearities, damping, etc.

During the last two decades, from about the time when chaos studies began, these setups have been reinvented several times, slightly modified or so configured as to demonstrate specific aspects. They have been most widely used for educational purposes, to show qualitatively chaotic behavior; only very few examples are such that one can quantitatively determine characteristic dependencies. A selection of such pendula or vibrating systems, attempting to point out the type, the nonlinear term, and the result achieved, is presented in Table 5.1.

The book by Moon [5.5] reviews, among other topics, chaos in a pendulum; viz: the one hinged arch, the parametrically forced pendulum, the rotor with viscous damping and periodically excited torque, the kicked rotor, one- and two-dimensional oscillators in a one- or two-dimensional double well potential, the buckling of an elastic beam due to magnetic body forces, the kicked double rotor. The buckled beam, qualitatively and quantitatively, and chaotic toys (double pendula) are described in detail for desktop lecture experiments.

Table 5.1. Simple experimental setups

System	Nonlinearity	Results achieved
double pendulum [5.4, 5.6]	nonlinear coupling	phase space
double pendulum [5.3]	nonlinear coupling	Lyapunov exponent, stroboscopic monitoring
double pendulum [5.7]	nonlinear coupling	small and large angle motion, Lyapunov exponent
driven spherical pendulum [5.8]	nonlinear damping	periodic/nonperiodic motion in 2D as a function of frequency excitation
2D elastic pendulum [5.6]	small/large angle approximation	coupled modes
two-dimensional elastic pendulum [5.9]	nonlinear coupling of vertical and horizontal motion	Poinçaré section, Lyapunov exponent, power spectrum, KAM theorem
horizontally driven inverted pendulum [5.10]	forced oscillator in a double well potential	frequency doubling
vertically driven inverted pendulum [5.11]	effective double well potential, including damping	stability of inverted state, Hopf bifurcation
vertically driven inverted pendulum [5.12]	double well potential	Hopf bifurcation
vertically driven inverted pendulum [5.13]	double well potential	nonlinear resonance curve
one-dimensional simple pendulum [5.14]	linear damping	transition from rotation to libration
two-dimensional elastic pendulum [5.15]	nonlinear coupling to vibration	stroboscopic picture of motion, nonlinear resonance
torsion pendulum in a static magnetic field [5.16]	double well potential	symmetry breaking critical exponent
torsion pendulum in an alternating magnetic field [5.17]	double well potential	bifurcation, Fourier spectrum

References

[5.1] W. H. Press, B. P. Flannery, S. A. Teukolsky, and W. T. Vetterling, *Numerical Recipes* (Cambridge University Press, Cambridge 1986)
[5.2] R. W. Stanley, *Numerical methods in mechanics*, Am. J. Phys. **52** (1984) 499
[5.3] T. Shinbrot, C. Grebogi, J. Wisdom, and J. A. Yorke, *Chaos in a double pendulum*, Am. J. Phys. **60** (1992) 491
[5.4] P. H. Richter and H.-J. Scholz, *Das ebene Doppelpendel — The planar double pendulum*, Film C1574, Publ. Wiss. Film., Sekt. Techn. Wiss./Naturw., Ser.9 (1986) Nr. 7/C1574
[5.5] F. C. Moon, *Chaotic Vibrations* (John Wiley, New York 1987)
[5.6] D. R. Stump, *Solving classical mechanics problems by numerical integration of Hamilton's equations*, Am. J. Phys. **54** (1986) 1096
[5.7] R. B. Levien and S. M. Tan, *Double pendulum: An experiment in chaos*, Am. J. Phys. **61** (1993) 1038
[5.8] D. J. Tritton, *Ordered and chaotic motion of a forced spherical pendulum*, Eur. J. Phys. **7** (1986) 162
[5.9] R. Cuerno, A. F. Rañada, and J. J. Ruiz-Lorenzo, *Deterministic chaos in the elastic pendulum: A simple laboratory for nonlinear dynamics*, Am. J. Phys. **60** (1992) 73
[5.10] K. Briggs, *Simple experiments in chaotic dynamics*, Am. J. Phys. **55** (1987) 1083
[5.11] J. A. Blackburn, H. J. T. Smith, and N. Grønbech-Jensen, *Stability and Hopf bifurcations in an inverted pendulum*, Am. J. Phys. **60** (1992) 903
[5.12] H. J. T. Smith and J. A. Blackburn, *Experimental study of an inverted pendulum*, Am. J. Phys. **60** (1992) 909
[5.13] N. Alessi, C. W. Fischer, and C. G. Gray, *Measurement of amplitude jumps and hysteresis in a driven inverted pendulum*, Am. J. Phys. **60** (1992) 755
[5.14] D. Permann and I. Hamilton, *Self-similar and erratic transient dynamics for the linearly damped simple pendulum*, Am. J. Phys. **60** (1992) 442
[5.15] Y. Cohen, S. Katz, A. Perez, E. Santo, and R. Yitzak, *Stroboscopic views of regular and chaotic orbits*, Am. J. Phys. **56** (1988) 1042
[5.16] E. Marega, Jr., S. C. Zilio, and L. Ioriatti, *Electromechanical analog for Landau's theory of second-order symmetry-breaking transitions*, Am. J. Phys. **58** (1990) 655
[5.17] E. Marega, Jr., L. Ioriatti, and S. C. Zilio, *Harmonic generation and chaos in an electromechanical pendulum*, Am. J. Phys. **59** (1991) 858

6. Chaotic Scattering

Chaotic dynamics in low dimensional conservative systems is a well established phenomenon and has attracted a great deal of interest during the last decade. Most of this work has been devoted to bounded systems. More recently, however, irregular chaotic phenomena have also been observed and studied for open (scattering) systems. For recent reviews of chaotic scattering, see the articles by Eckhardt [6.1], Smilansky [6.2], and Blümel [6.3]. Chaotic dynamics in such scattering systems is important for many collision processes, e.g. the formation and decay of intermediate collision complexes, the redistribution of energy during a collision and, last but not least, for subsequent processes (such as reactive collision with a third collision partner, the absorption or emission of radiation, etc.), which become more probable the longer the lifetime of the intermediate colliding system.

Scattering systems are, by definition, open systems: the trajectories are not confined to a bounded region. With increasing time, trajectories will leave the interaction region and eventually escape to infinity, where the dynamics can be regular. Therefore, the long time limit, which is an essential concept of many rigorous mathematical results of chaotic dynamics, must be treated differently. Let us assume, for simplicity, that we can clearly distinguish between an

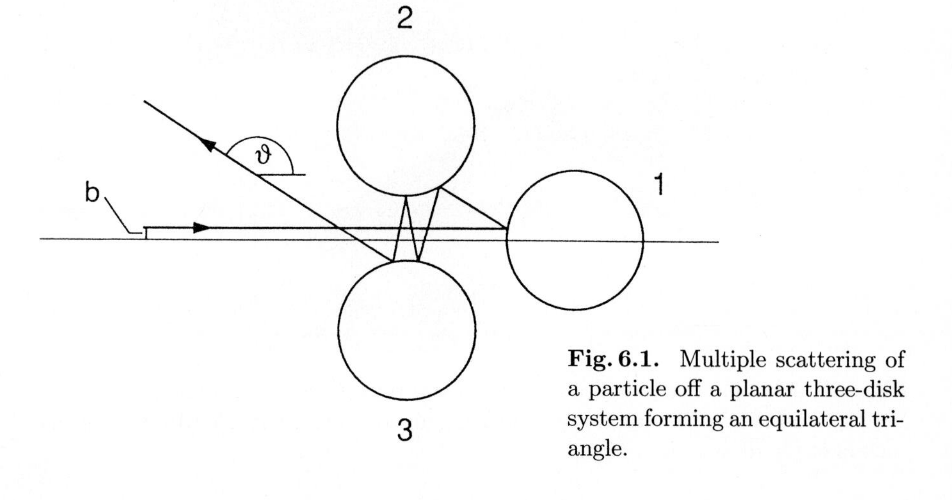

Fig. 6.1. Multiple scattering of a particle off a planar three-disk system forming an equilateral triangle.

asymptotic region and a finite scattering zone. Then, most of the scattering trajectories have a finite collision time. To be more precise, for typical systems the set of all scattering trajectories with infinite collision time ('sticking trajectories') is of measure zero. At first sight, this leaves very little room for manifestly chaotic phenomena. Another important set consists of all bounded periodic trajectories. The so-called 'chaotic repellor' [6.4] is the set of all trajectories which stay trapped for $t \to \infty$ and contains the two sets mentioned above.

Various systems showing chaotic scattering have been analyzed in detail [6.1]–[6.3]. One of the best studied models is the collision of a point mass with three fixed circular disks in a plane, arranged in most cases at the vertices of an equilateral triangle [6.2, 6.4, 6.5]. This system has the advantage of clearly demonstrating several features of classical chaotic scattering without the need for integrating differential equations. The dynamics is of the billiard type, consisting of straight line segments plus elastic reflection from hard walls. In addition, the dynamics only trivially depends on the energy; in particular, the classical repellor undergoes no structural bifurcations when the energy is varied. This also makes the three-disk system well suited for investigating classical–quantum interrelations [6.6, 6.7]. The system has also been used to demonstrate classical chaotic scattering experimentally [6.8].

The classical disk scattering system in a plane has also been extended to three space dimensions (i.e. to scattering off a cluster of three (and more) hard spheres) by Chen et.al. [6.9] and Korsch and Wagner [6.10]. Here, we will confine ourselves to the two-dimensional case.

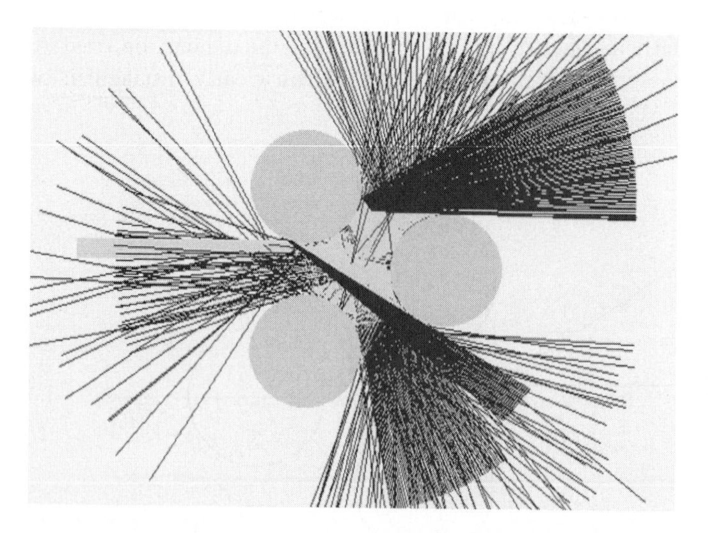

Fig. 6.2. Multiple scattering of a beam of 400 trajectories with impact parameter $1 \leq b \leq 3$ off three disks (radii $r = 6$) forming an equilateral triangle (distance of the centers of the disks $= R = 10$).

6.1 Scattering off Three Disks

This planar model for chaotic scattering consists of a cluster of three hard circular disks $i = 1, 2, 3$ with radii r_i at distances R_i from a center at angular positions φ_i measured with respect to the horizontal scattering axis. A point mass (particle $i = 0$) impinging from the left with impact parameter b undergoes multiple elastic reflections from the disks and, after a number of n collisions, escapes finally with scattering angle ϑ. It is useful to note that the same system can also be interpreted as the reflection of parallel light rays from an arrangement of three perfectly reflecting disks.

Figure 6.1 illustrates the geometry of the system for the case of three disks with the same radii $r_i = r$ and having the same distances $R_i = R$ $(i = 1, 2, 3)$ and forming an equilateral triangle oriented symmetrically with respect to the scattering axis. In this configuration, the distance between the centers of the disks is $L = R\sqrt{3}$.

In Fig. 6.2, the complex dynamics of the three-disk system is illustrated. A beam of 400 trajectories in the impact parameter interval $1.0 \leq b \leq 3.0$ is displayed for a disk of radius $r = 6$ $(R = 10)$, which is analyzed in more detail below. Figure 6.3 shows the collision number, i.e. the number n of impacts which occur before the scattered particle escapes. This is a measure of the lifetime of the collision complex. Also shown is the deflection function, i.e. the scattering angle ϑ (modulo π), as a function of the impact parameter b. A magnification of the figure for the interval $1.95 \leq b \leq 1.98$ is shown in Fig. 6.4, and we see a picture almost identical to Fig. 6.3.

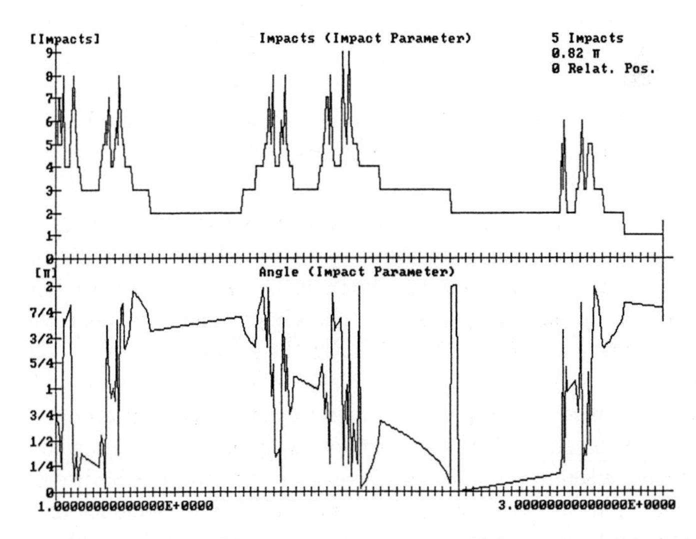

Fig. 6.3. Collision number $n(b)$ and scattering angle $\vartheta(b)$ as a function of the impact parameter b for $r = 6$ and $R = 10$.

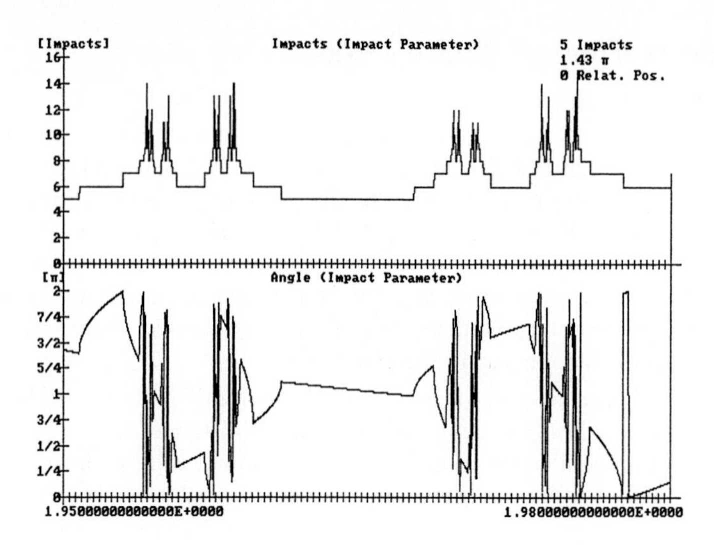

Fig. 6.4. Magnification of the impact parameter interval $1.95 \leq b \leq 1.98$ for $r = 6$ and $R = 10$.

By increasing the value of r, the dynamics becomes more and more complicated. Fig. 6.5 shows the collision number and scattering angle in the impact parameter interval $0.5 \leq b \leq 1.0$ for $r = 8$. Again, magnification of the interval $0.673 \leq b \leq 0.698$ shown in Fig. 6.6 shows no smoothing of these functions. The apparent self-similarity demonstrated in these figures continues under repeated magnification, revealing a fractal structure of the dynamics which can be measured by its fractal dimension d (see computer experiment 6.4.4). The maximum value of the collision number observed in an equidistant scan of an impact parameter interval increases under magnification, and approaches infinity as one gets closer and closer to the chaotic repellor. The dynamics is extremely sensitive to the initial conditions, a feature characteristic of chaotic dynamics.

The collision number shows discontinuities at certain impact parameters; the deflection function is continuous and its first derivative is discontinuous at the same set of impact parameters. The trajectories in the interval between these discontinuities can be characterized by the same collision sequence, for example the orbit shown in Fig. 6.1 is described by the sequence $1 \to 2 \to 3 \to 2 \to 3$, where $1, 2, 3$ are the numbers of the disks. A trajectory cannot hit the same disk twice in direct succession and, therefore, the collision sequence can even be coded binarily, suggesting a treatment by symbolic dynamics [6.5, 6.4]. We associate with a sequence of two collisions in a positive cyclic direction ($1 \to 2$, $2 \to 3$, or $3 \to 1$) the symbol '1', otherwise a '0'. The sequence of collisions of Fig. 6.1 is then given by '1101'. At the boundary of an interval of continuity, the trajectory becomes tangential to one of the disks and changes the dynamics discontinuously by hitting another disk. This set of tangential trajectories

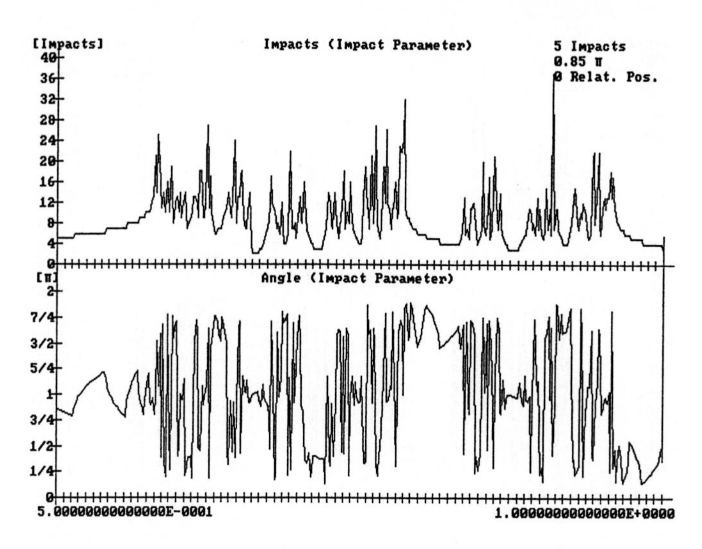

Fig. 6.5. Collision number $n(b)$ and scattering angle $\vartheta(b)$ as a function of the impact parameter b for $r = 8$, $R = 10$.

organizes the structure of the trajectory set. For small values of r, there is a one-to-one relationship between all binary sequences and the topologically different trajectories [6.5]. If r is large, the disks get more and more hidden by each other and some collision sequences become dynamically forbidden. A simple geometrical consideration shows that, for

$$0 \le r/R \le 3/4\,, \tag{6.1}$$

no path between two disks can be affected by the third one, i.e. these screening effects are unimportant for trajectories which collide with more than two disks. In addition, there are outer screening effects, because a disk can hide a part of another disk, i.e. it is partly invisible from the direction of impact. This hidden part can be made visible by changing the impact direction, in contrast to the inner screening effects discussed above.

The dynamics shows an almost self-similar structure, as illustrated by Figs. 6.3 to 6.6. The set of singularities — the set of all impact parameters leading to trapped trajectories — is a fractal. In one dimension, the fractal dimension d of a set measures the asymptotic scaling of the number N of intervals of size ϵ needed to cover the set:

$$N(\epsilon) \sim \epsilon^{-d} \quad \text{for} \quad \epsilon \to 0\,. \tag{6.2}$$

For a self-similar set, invariant under a transformation 'restriction onto $1/m$ of the set and magnification by a scale factor s', (6.2) simplifies to

$$d = \frac{\ln m}{\ln s}\,. \tag{6.3}$$

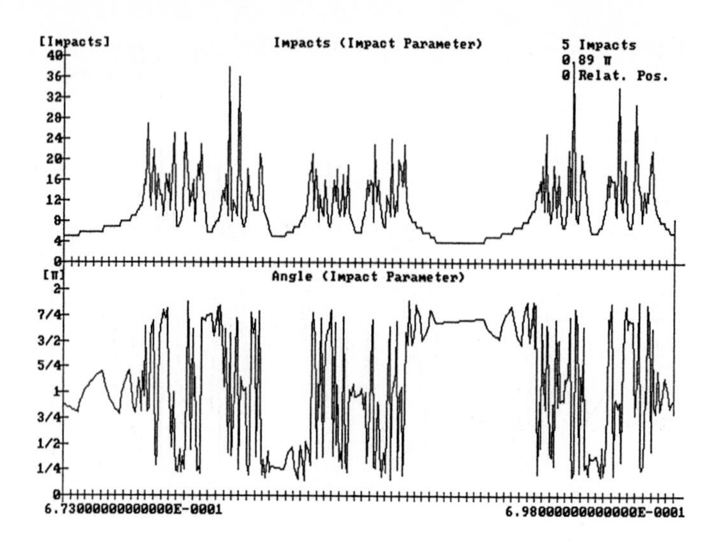

Fig. 6.6. Magnification of the impact parameter interval $0.673 \le b \le 0.698$ for $r = 8$, $R = 10$.

The most famous fractal set is the Cantor set, more precisely the '1/3 Cantor set', which is constructed by erasing the middle 1/3 of a unit interval and iterating this procedure for the subintervals. After n such erasing processes, we have 2^n intervals of length $(1/3)^n$ and the total measure of the remaining set is $(2/3)^n$, which goes to zero for $n \to \infty$, i.e. the limiting Cantor set is of measure zero. Alternatively, one can label two intervals of generation $n = 1$ by '0' and '1', the daughter generation of '0' as an additional '0' or '1', i.e., as '00' and '01', and those of '1' as '10' and '11'. Continuing this process shows that the Cantor set is uniquely labeled by the sequences of '1's and '0's in the form '0110101...', which are the binarily coded real numbers between zero and one. This correspondence is one-to-one, i.e. the '1/3 Cantor set' is not countable. Furthermore, it is nowhere dense. The last two attributes are used to define more general Cantor sets.

The '1/3' Cantor set is self-similar under reduction to 1/2 of the set and magnification by a factor $s = 3$. Its fractal dimension is therefore

$$d = \frac{\ln 2}{\ln 3} = 0.6309\,. \tag{6.4}$$

The fractal dimension of sets observed in the dynamics can be used as a quantitative measure of the chaoticity. A fractal of the three-disk scattering function will be analyzed in the computer experiments described in Sec. 6.4.4.

6.2 Numerical Techniques

Numerically, the problem is very simple and, of course, similar to the closed billiards studied in Chap. 3. The projectile moves with constant velocity in a straight line. The program computes the intersections of this line with the three fixed disks. We obtain six or fewer intersection points. The one at minimum distance to the position of the projectile is the position of the next impact. A line is drawn between the last position of the particle and the new one. Let us assume that a collision with disk i will occur. Then, the angle between the particle trajectory and the radius vector from the position of the center of the disk i to the collision point is computed, and the reflection law determines the direction of the particle after collision. This procedure is repeated until no collision is found and the projectile escapes to infinity. Then, the scattering angle and the collision number are determined and stored. If desired, the program also displays the collision sequence, coded as described above.

In the program, the projectile (particle number 0) is treated as a finite particle (a disk with radius r_0). The pre-set value $r_0 = 10^{-5}$ approximates a mass point.

The program also allows computation of the fractal dimension (6.2). Instead of using the definition directly, we adopt a technique which can be used for fractal sets given as a limiting set of 'generations' n for $n \to \infty$, where each generation is the reduction of the set to $1/m$th of the previous set. A scaling factor s^{-1} is, however, not sharply defined. When A_n is the total measure of generation n, it can be covered by $N(\epsilon_n) \sim \epsilon^{-d}$ intervals of size $\epsilon = A_n m^{-n}$ and we find

$$A_n = \epsilon_n N(\epsilon_n) \sim \epsilon^{1-d} = (A_n m^{-n})^{1-d} \tag{6.5}$$

and, hence, the A_n will decrease exponentially for large n

$$A_n \sim m^{-n(1/d-1)} = \mathrm{e}^{-\gamma n} \tag{6.6}$$

with rate

$$\gamma = (1/d - 1) \ln m \,. \tag{6.7}$$

We end up with

$$d = \frac{\ln m}{\ln m + \gamma} = \frac{\ln m}{\ln(m\delta)} \,, \tag{6.8}$$

where

$$\delta = \mathrm{e}^{\gamma} \approx \delta_n = \frac{A_{n-1}}{A_n} \tag{6.9}$$

is approximately the ratio between the measure of two successive generations. For a self-similar set with scaling factor s we have $A_n = mA_{n-1}/s$, and we recover the result (6.3).

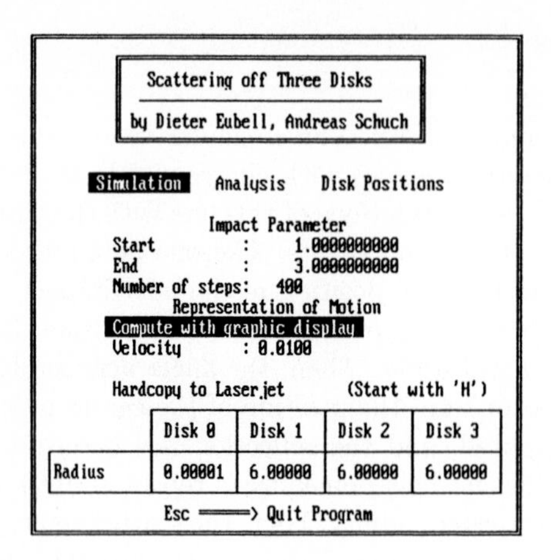

Fig. 6.7. Main menu of program 3DISK.

6.3 Interacting with the Program

The *Main Menu* of the program 3DISK is shown in Fig. 6.7. Here, the user can select the range of impact parameters, choose the mode of representation, change the radii of the disks and activate the simulation.

Main Menu

- **Simulation** — starts the numerical simulation. The trajectory is computed in the selected impact parameter range (see below) in the specified number of steps. The scattering angle ϑ and collision number n are computed. The simulation can be stopped by pressing any key.

  When the menu item *'Compute with graphic display'* is active (i.e. it appears on a grey background), the scattering system and the trajectories are displayed on the screen. When the particle has left the interaction region, the outgoing part of the trajectory is shown in yellow. For trajectories which miss the target, only a part of the incoming path is shown in yellow. The user can control the timescale for this graphical presentation by changing the *'velocity'* (slow mode is pre-set). On pressing the ⟨H⟩ key, a hardcopy of the present graphic image is printed. By means of the ⟨A⟩ key, a hardcopy is printed at the end of the simulation.

- **Analysis** — The computed collision data are displayed in two diagrams. The upper one shows the collision number n and the lower one the deflection function, i.e. the scattering angle ϑ, as a function of the impact parameter b in the selected impact parameter interval. Both diagrams can be printed by pressing the ⟨H⟩ key.

  A new impact parameter interval can be selected by means of two vertical lines marking the lower and upper limits of this interval. The lines can be moved using the cursor keys ⟨→⟩ and ⟨←⟩. Pressing the ⟨TAB⟩ key switches between slow and fast motion, and pressing the ⟨SPACE BAR⟩ toggles the left and right boundary lines. Additionally, ⟨HOME⟩ or ⟨END⟩ switches to the left or right boundary line, respectively.

The value of the number of impacts (i.e. the collision number), the scattering angle at this impact parameter and the relative position ν of the active boundary, i.e. the number of the selected impact parameter within the number of impact parameter steps (see (6.10) below), are displayed on the screen.

The key ⟨S⟩ activates the computation of the sequence of collisions for the trajectory at the active cursor position. This sequence is displayed on the screen (note that the collision number is equal to the length of the sequence of collisions plus one). By pressing the ⟨H⟩ key, it is possible to make a hardcopy of the screen when the collision sequence is displayed .

On pressing key ⟨D⟩, the fractal dimension is computed. The program scans the computed collision function and counts the number of computed impact parameters of collisions with at least n collisions ('generation n'). The ratio δ_n of these numbers is displayed as a function of n, as well as the fractal dimension computed from (6.8). In the limit of large n, these data should converge. It should, however, be noted that the values for larger n can be unreliable when the number of computed trajectories is too small. One should, therefore, use a large number of trajectories in the computation. By pressing the ⟨H⟩ key, it is possible to make a hardcopy of the screen when the collision sequence is displayed.

The ⟨ENTER⟩ key fixes the positions of the cursor keys as a new impact parameter interval and returns to the *Main Menu*, whereas the ⟨ESC⟩ key returns without changes.

- **Disk Positions** — switches to an input mask, where the positions of the three fixed disks can be changed. The locations of the centers of the disks $i = 1, 2, 3$ are given in polar coordinates r_i and θ_i in a 'body-fixed' coordinate system. The whole coordinate system can be vertically shifted (δy) and rotated ($\delta\theta$). An equilateral triangle with $R_i = 10$ is pre-set.

- **Impact Parameter** — specifies impact parameters for trajectory computation.

  - *Start* — lower boundary $b_{\rm min}$ of the impact parameter interval (the pre-set value is 1.00).

  - *End* — upper boundary $b_{\rm max}$ of the impact parameter interval (the pre-set value is 3.00).

  - *Number of steps* — number N of equidistant subdivisions of the impact parameter interval (the pre-set value is 400). Trajectories are computed at impact parameters

$$b_k = b_{\rm min} + (b_{\rm max} - b_{\rm min})k \ , \ k = 0, \ldots, N \, . \tag{6.10}$$

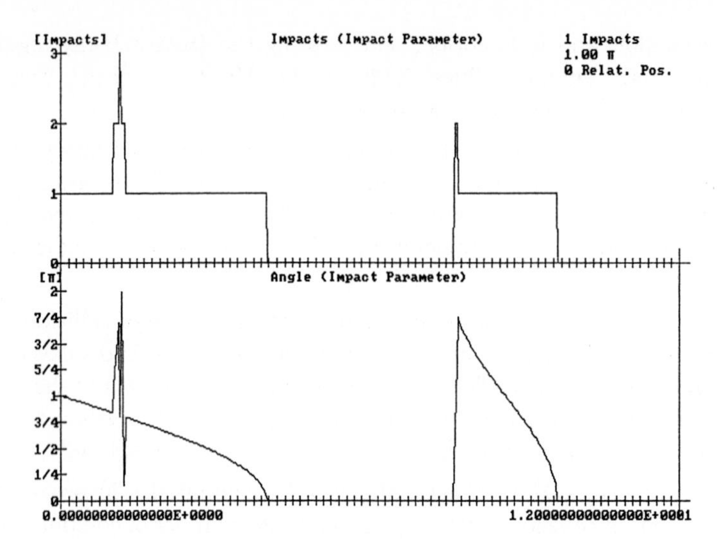

Fig. 6.8. Collision number $n(b)$ and scattering angle $\vartheta(b)$ as a function of the impact parameter b for two disks with radii $r_1 = 4$ and $r_2 = 1$ $(R = 10)$.

The relative position ν at the active cursor position is displayed on the screen.

- **Compute with graphic display** — when this menu item is active (i.e. it appears on a grey background), the scattering system and the trajectories are displayed on the screen during the numerical simulation. Deactivating this item speeds up the computation.

- **Velocity** — controls the time scale for the graphical presentation during the numerical simulation (slow mode is pre-set).

- **Disk radii** — allows one to change the radii r_i of the scattered particle $i = 0$ and the three fixed disks $i = 1, 2, 3$. A 'point-mass' 0 (radius 0.00001) and equal-sized target disks of radii $r_i = 6$ are pre-set.

Pressing the ⟨Esc⟩ key terminates the program.

6.4 Computer Experiments

6.4.1 Scattering Functions and Two-Disk Collisions

For the case where a point particle collides with a single disk (radius r), the deflection function is given by

$$\vartheta = 2\cos^{-1}(b/r)\,, \tag{6.11}$$

where the impact parameter b is measured with respect to the center of the disk. It is instructive to study first the deflection function for a single disk numerically (e.g. by changing the radii in the pre-set set-up to $r_1 = 4$ and $r_2 = r_3 = 0.00001$ in the impact parameter region $0 < b < 12$), which shows the functional dependence (6.11). The collision number n is equal to one if the particle collides with the disk, and zero otherwise. Increasing now the radius of disk 2 to $r_2 = 1$, we observe various changes in the scattering functions $\vartheta(b)$ and $n(b)$ (see Fig. 6.8):

1. In the large impact parameter region, a second scattering structure, which is due to a first collision with disk 2, appears.

2. In the small impact parameter region, the deflection function is distorted due to trajectories hitting first disk 1 and then disk 2. This new region is located at the angular position of the center of disk 2 seen from disk 1, i.e. at $\vartheta = 150^0$ with angular width $\Delta\vartheta \approx 2r_2/L$. The region is bounded by rays tangential to disk 2. A central ray in this region is a radial ray of disk 2, which is therefore reflected backwards with an overall scattering angle $\vartheta = 180^0$ (see Fig. 6.9). One can numerically locate this ray at an impact parameter $b = 1.16468$. The collision number increases by at least one in this region.

3. The additional deflection angle caused by the second collision with disk 2 varies over 360^0. Therefore, a subinterval of these rays must hit disk 1 once again, giving rise to a new substructure where the collision number again increases by one. This procedure can be continued. In this region, there exists an unique ray showing infinitely many reflections (a so-called 'sticking trajectory'), which can be found at $b = 1.16915105067871$.

4. The same behavior is found in the large impact parameter region.

Let us interpret these findings in the language of optics, considering the reflection of parallel light rays from two circular mirrors (the two disks). We will observe a mirror image of disk 2 in mirror 1, and vice versa. The image

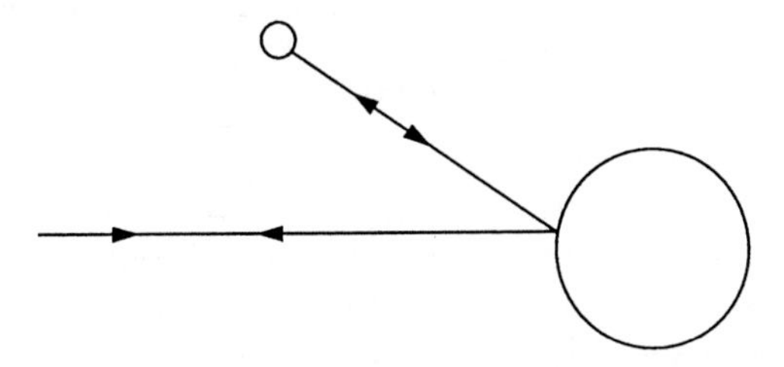

Fig. 6.9. Radial ray with backward reflection in the center of the double collision region (radii $r_1 = 4$ and $r_2 = 1$; $R = 10$).

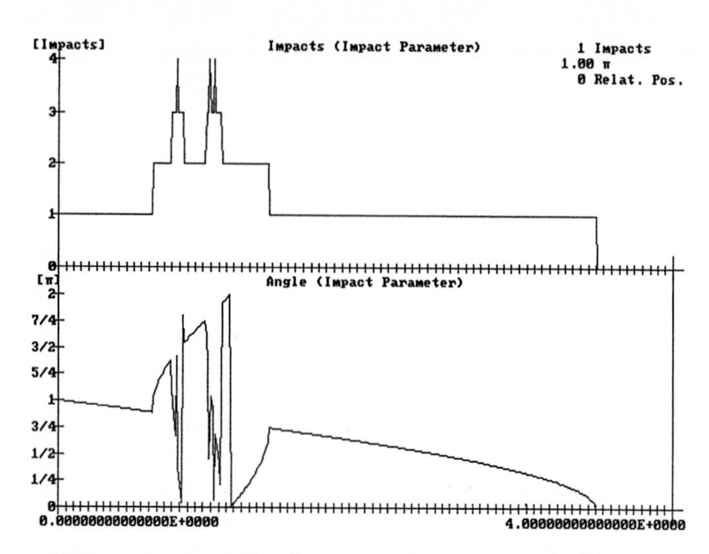

Fig. 6.10. Collision number $n(b)$ and scattering angle $\vartheta(b)$ as a function of the impact parameter b for three disks with radii $r = 3.5$ and $R = 10$ in the impact parameter range $0 \leq b \leq 4$.

of 2 therefore contains an image of disk 1, which carries an image of disk 2, and so on. We find in each disk a nested sequence of repeated mirror images, which converges to a point. The two rays centered at these points are the two long-lived sticking trajectories.

Alternatively, one can start from a periodic orbit consisting of a radial trajectory (or ray) along the line between the centers of the two disks. This trajectory is a periodic two-bounce orbit which is unstable, i.e. any small perturbation will lead to a trajectory which moves away from this orbit, and eventually escapes to infinity. The sticking trajectory is such a path in the reverse direction, where the ray comes arbitrarily close to the unstable periodic orbit after infinite time. More precisely, it is an element of the stable manifold of an unstable fixed point, the periodic orbit.

More details regarding two-disk scattering, e.g. explicit formulae for minimum and maximum collision angles, collision mappings, and the location of the trapped orbits can be found in Ref. [6.11].

In the two-disk case, there is only one unstable periodic orbit and no chaotic scattering, since the repellor consists only of this orbit and, hence, has no fractal structure. The two-disk organization of the collision dynamics will, however, reappear as a substructure in the three-disk case, where the collision dynamics will turn out to be chaotic. This is studied below.

6.4.2 Tree Organization of Three-Disk Collisions

Here, we will explore the organization of three-disk scattering in more detail. Let us start with relatively small disks with the same radius $r = 3.5$ placed again on an equilateral triangle at distance $R = 10$ from the center, where disk 1 is placed at $\theta = 0^0$ (pre-set configuration).

Figure 6.10 shows the collision function in the impact parameter range $0 \leq b \leq 4$, which is sampled by 400 trajectories with equidistant impact parameters (pre-set case). We note that all trajectories in the range $0 \leq b \leq 3.5$ collide first with disk 1. On top of the $n = 1$ plateau, we see two 'chimneys' of the first daughter generation with $n \geq 2$. A magnification of the interval $0.7 \leq b \leq 1.1$ in Fig. 6.11 — again with 400 trajectories — reveals further chimneys on top of the previous ones and so on, as discussed above. Let us now go into more detail by looking at the sequence of collisions in the different regions. This sequence is coded by a sequence of '1's or '0's, where '1' stands for a collision with disk $i+1$, 0 for a collision with disk $i-1$, when the last collision occured with disk i ($i = 1, 2, 3$ modulo 3).

After recomputing the results shown in Fig. 6.11, one can move one of the vertical lines to a point of interest by means of the cursor. Key ⟨S⟩ activates the display of the sequence of collisions. One can easily check that the whole plateau with $n = 2$ yields a sequence of collisions consisting only of a '1', i.e. a collision with disk 2 after the first collision with disk 1.

The two subplateaus with $n = 2$ yield a sequence '11' for the left, and '10' for the right subplateau. Translating to the 'real' trajectory, this is a third collision with disk 3 or 1, respectively. The two subplateaus have different sizes

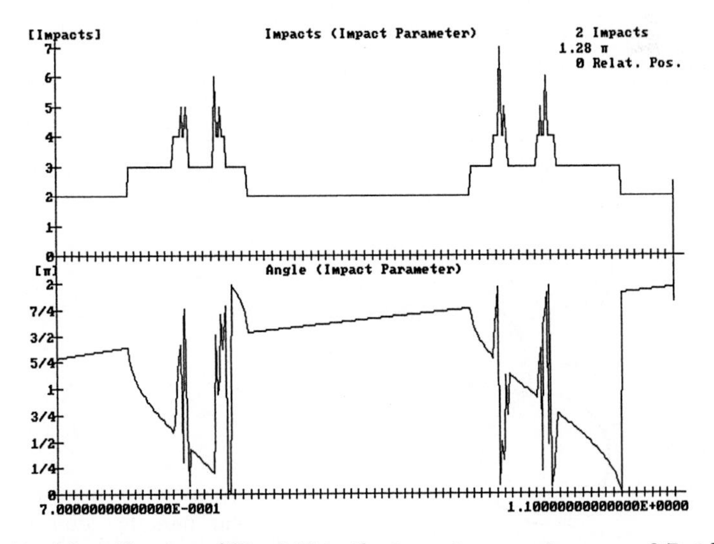

Fig. 6.11. Magnification of Fig. 6.10 in the impact parameter range $0.7 \leq b \leq 1.1$.

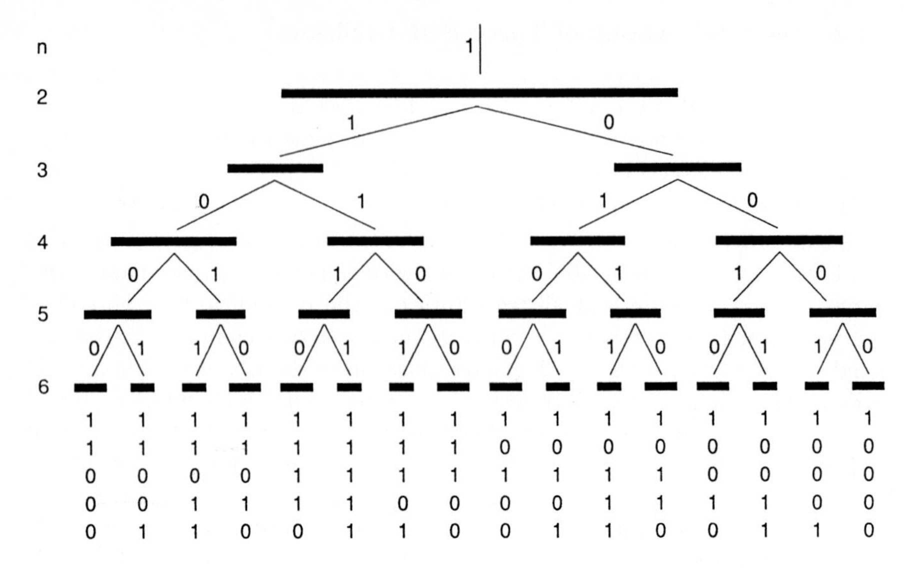

Fig. 6.12. Tree structure of the organization of the sequences of collisions.

Δb; the smaller one appears to the left. The angular width of the plateaus is, however, the same, namely $\Delta\vartheta \approx 0.15\pi$.

Exploring the $n = 4$ generation, we find two subintervals of the left $n = 3$ plateau (symbolic code '110' and '111' from left to right), and two of the right one (symbolic code '100' and '101'). In both cases, the outer interval (the one closer to the boundary of the parent generation $n = 3$) is the larger one.

This scheme is repeated in the following generations. Figure 6.12 shows the organization schematically as a tree diagram. The different plateaus are shown as a function of the impact parameter b (not properly scaled!). The binary code

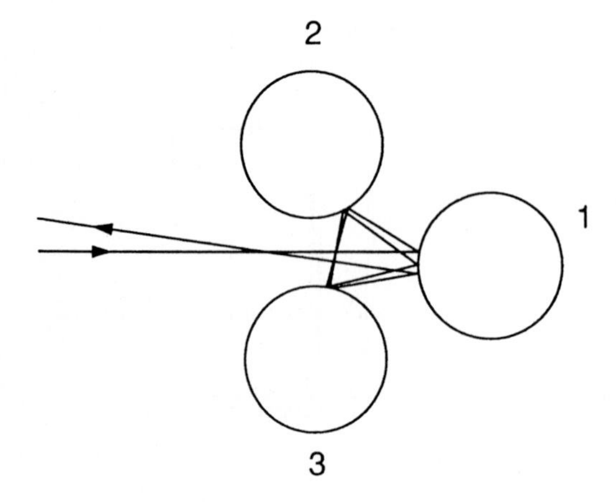

Fig. 6.13. Long-lived trajectory approximately following a triangular periodic orbit for several periods ($r = 0.35$, $R = 10$).

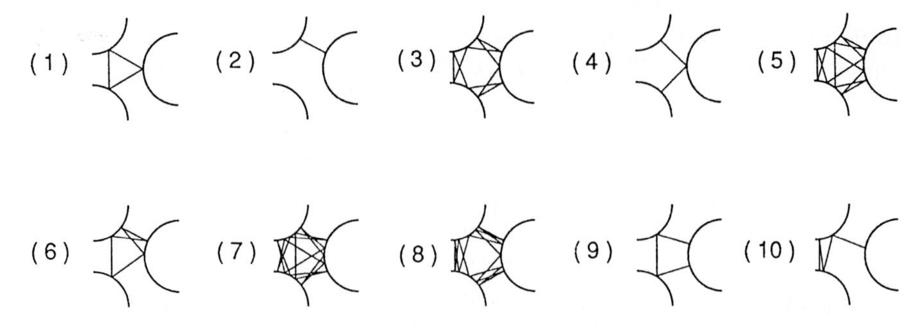

Fig. 6.14. Unstable periodic orbits.

for each interval can be read directly from the diagram by collecting the 1s and 0s following the different branches. The codes for generation $n = 6$ are, e.g., '11010', '11011', '11000', etc.

The 2^{n-1} plateaus for generation n (note that Fig. 6.12 only gives the tree for the 2^{n-2} sequences starting with '1') can be uniquely labeled by the collision sequences with length $n - 1$, and vice versa. There is a one-to-one mapping between the sequences of collisions and the plateaus of the collision function $n(b)$, and the whole scheme of construction closely resembles the Cantor set structure discussed above. In order to localize numerically a trajectory interval with a prescribed sequence of collisions one can start by simply localizing the plateau which reproduces the first few digits of the sequence. Then, by repeated magnification, one can find the subintervals which satisfy the next digits. A selection between two different subintervals is required in each step.

6.4.3 Unstable Periodic Orbits

A sticking trajectory never escapes from the scattering region and the collision sequence for such a trajectory will never terminate. Thus, these trajectories are exactly those with an infinite collision sequence. The set of all sticking trajectories forms a Cantor set of measure zero. A countable subset of these sticking trajectories are those coming infinitely close to an unstable periodic orbit. In the binary code, they appear as sequences ending up in a periodic tail. Trajectories with impact parameters close to such a sticking trajectory are attracted by the periodic orbit, and follow it for a while until they are finally repelled. Figure 6.13 shows an orbit attracted and repelled by the period-three orbit forming an equilateral triangle (binary code '11111...').

Further periodic trajectories — more precisely, impact parameters of trajectories approaching an unstable periodic orbit — are numerically localized by the method described above and shown in Fig. 6.14 for the pre-set configuration (disk radii $r = 6$ and $R = 10$). The data of these orbits are listed in Table 6.1. Theoretically, such an orbit can be localized with arbitrary precision. In

Table 6.1. Trajectories approaching unstable periodic orbits with given collision sequence and period p_s. The period p of the true orbit is equal to p or $3p$.

	b	$n(b)$		p_s	p
(1)	1.17754799767000	28	$\ldots\dot{1}111111111\ldots$	1	3
(2)	1.97534632642241	31	$\ldots\dot{1}\dot{0}10101010\ldots$	2	2
(3)	1.02459608527742	30	$\ldots\dot{1}1\dot{0}1101101\ldots$	3	9
(4)	1.72455696933076	27	$\ldots\dot{1}00\dot{1}100110\ldots$	4	4
(5)	1.21335412735471	30	$\ldots\dot{1}11\dot{0}111011\ldots$	4	12
(6)	1.16957287517956	28	$\ldots\dot{1}111\dot{0}11111\ldots$	5	5
(7)	1.20434893507372	26	$\ldots\dot{1}110\dot{0}11100\ldots$	5	15
(8)	1.01164038132856	28	$\ldots\dot{1}101\dot{0}11010\ldots$	5	15
(9)	1.20627203056642	26	$\ldots\dot{1}1100\dot{0}1110\ldots$	6	6
(10)	1.01080193668915	29	$\ldots\dot{1}101010\dot{0}11\ldots$	8	8

practice, however, one is limited by the finite number of digits and the collision sequence is finite. The maximum collision number $n(b)$ is also listed in the table. In the symbolic code of the unstable orbit, a primitive period is marked by a dot. Periodicity of the trajectory requires that the number of zeros and ones in the symbolic code be equal (modulo three). If this is not the case, the primitive period must be repeated until this condition is met, and the true period is three times longer than the symbolic period.

The periodic trajectories in Table 6.1 are ordered with respect to the length of their symbolic period. It should be noted that the table is *not* complete. As an exercise, one can find the missing orbits with symbolic code '111110', '111100', and '111010'. It is also possible to extend the list and locate orbits with higher period.

Let us point out some features characteristic of periodic orbits (compare the related discussion for the gravitational wedge billiard in Sect. 4.3.1):

- There are orbits with a threefold symmetry in space (examples (1), (3), (5), (7), and (8) in Fig. 6.14 and Table 6.1). These orbits perform a full rotation in space with respect to the center in the triangular disk configuration. It should be noted that there also exist duplicates of these orbits, transversed in the inverse direction, i.e. the '1's and '0's are interchanged in the binary code.

- If an orbit contains a central ray orthogonal to a disk (examples (2), (4), and (9)), it is simply reflected onto itself. This implies that the binary code is invariant against a reflection, interchanging '1's and '0's at some point. It is also evident that there must be exactly two such orthogonal points, or none.

- The orbits which are *not* three-fold symmetric appear three times, because the two rotated orbits are also allowed trajectories. They have the same collision sequence, starting, however, on a disk different from disk number one.

- Orbits (1)–(9) show a reflection symmetry with respect to a symmetry line. This is, of course, not necessarily the case and orbit (10) does *not* show such a symmetry.

The number of periodic orbits shows an exponential proliferation with increasing period. As an exercise, one can try to work out an analytic formula for this number.

6.4.4 Fractal Singularity Structure

In the examples shown above, collision functions have been computed for equilateral triangle configurations with $R = 10$ and disk radii $r = 3.5$, 6.0, and 8.0. The character of the observed structures in the collision functions was found to be strongly dependent on the radii of the disks. For small disks we have rather clumsy structures ('chimneys' on flat roof tops for $r = 3.5$). For larger disks the structures become finer ('cathedral towers' for $r = 6$) and more and more ragged in the limit of disks touching each other (compare Figs. 6.5 and 6.6 for $r = 8$). This different impressions can be quantitatively expressed in terms of the fractal dimension, as discussed above.

Let us first look at the scaling behavior of the width of the Δb plateaus. The following technique can be used: we compute a series of magnifications following a certain structure, e.g. the larger plateaus shown in Figs. 6.10 and 6.11 for $r = 3.5$ and $R = 10$. Using always the same resolution of, e.g., 400 computed impact parameters in each plot, we move the two vertical lines to the lower and upper boundary of the Δb plateau, read the relative coordinates of the boundaries displayed on the screen, and compute the difference. By pressing the ⟨ENTER⟩ key and activating the item *'simulation'*, a magnification of this plateau is computed. Again we move the cursor to the next plateau, measure its boundary coordinates once more, and so on. Table 6.2 lists the results when the large plateau is traced in generations $n = 2$ to $n = 6$. Both the length and coordinate of the interval apparently converge and we find a scaling factor between subsequent generations of $s = 400/53 = 7.547$. Assuming self-similarity with this scaling factor formula (6.3) yields a fractal dimension of $d = \ln 2/\ln s = 0.343$ (evidently we have $m = 2$). We can, however, also trace the smaller plateau. From Fig. 6.12 we see that, in this case, we follow the '1111...' orbit instead of the '101010...' in the case of the larger interval. The results are also listed in Table 6.2. The scaling factor is $s = 400/43 = 9.302$ and, therefore, a self-similar fractal dimension would be equal to $d = 0.311$. Assuming a multiscale fractal [6.12] with two scales $s_1 = 7.547$ and $s_1 = 9.302$, one can derive a fractal dimension from the implicit equation

Table 6.2. Width Δb of $n(b)$ plateaus for a sequence of magnifications, following the larger or smaller interval ($r = 3.5$, $R = 10$,). Values of b are given as relative coordinates, where the width of the parent interval is always set to 400.

	large plateau			small plateau		
n	$b_{\max}$	$b_{\min}$	Δb	$b_{\max}$	$b_{\min}$	Δb
2	184	235	51	66	107	41
3	176	227	51	333	288	45
4	176	228	52	117	74	43
5	174	227	53	328	285	43
6	174	227	53	118	75	43

$$s_1^{-d} + s_2^{-d} = 1 \tag{6.12}$$

which yields $d = 0.326$, which is close to the fractal dimension $d = 0.325$ computed from the average of the two scaling factors $\bar{s} = (s_1 + s_2)/2 = 8.425$.

It has already been noted that the hard disk collisions can also be interpreted as mirror images in a perfectly reflecting circular mirror, or a sphere in three dimensions. The imaging properties of such spheres have been analyzed by Berry [6.13]. Here, we need the lateral reduction of an image, which is approximately

$$s = \left[2(\sqrt{3}R/r - 1)\right]^{-1} \tag{6.13}$$

for large angle scattering [6.10]. Using this as an approximate average scaling ratio in expression (6.3), for the fractal dimension we obtain [6.10]

$$d = \frac{\ln 2}{\ln 2(\sqrt{3}R/r - 1)}, \tag{6.14}$$

which reduces to zero in the limit $r \to 0$ and to unity for $r = \sqrt{3}R/2$, i.e. for disks in contact with each other. For the case of $r/R = 0.35$ considered above, (6.1) yields $d = 0.335$.

Let us now use directly the implemented program feature for calculating the fractal dimension. Computing a (magnified) collision function and pressing the ⟨D⟩ key generates a listing of the ratio δ_n between the measure A_{n-1} and A_n versus n, as well as the approximate fractal dimension (6.8). Table 6.3 lists the results of such a computation for $r = 6$ and $R = 10$. Data from 16 000 trajectories in an impact parameter interval $0.5 \le b \le 2.5$ have been used. The data for the largest values of n are unreliable, since they are based on a small number of trajectories. From the results shown in the table, we can assign a fractal dimension of $d \approx 0.321$ for $r/R = 0.35$ and $d = 0.489$ for $r/R = 6$. This can be checked by computing data from magnified intervals and higher collision numbers. The fractal dimension for $r/R = 0.35$ is in good agreement

Table 6.3. Ratio $\delta_n = A_{n-1}/A_n$ of total measure of successive collision number generations and fractal dimension d.

	$r/R = 0.35$		$r/R = 0.6$	
n	δ_n	d	δ_n	d
2	2.6426	0.4163	1.0654	0.9163
3	4.3405	0.3207	1.6136	0.5916
4	4.3323	0.3210	2.0694	0.4880
5	4.3514	0.3204	2.0633	0.4890
6	4.1111	0.3290	2.0664	0.4885
7	6.0000	0.2789	2.0727	0.4874
8			2.0360	0.4936
9			2.0000	0.5000

with the numerical values in Ref. [6.5]; fractal dimensions for larger radii are not reported there. The simple estimate (6.14) yields $d = 0.52$, which is somewhat larger than the numerical value.

Figure 6.15 compares finally the fractal dimensions calculated using the program with those of the approximate formula (6.14) and shows good overall agreement, remarkable in view of the simplicity of the approximation.

6.4.5 Suggestions for Additional Experiments

Long-Lived Trajectories. For larger disk radii, more and more trajectories get trapped for longer times. This happens, in particular, in the case of disks which almost touch each other, i.e. $r \lesssim \sqrt{3}R/2 \approx 0.866R$ for an equilateral

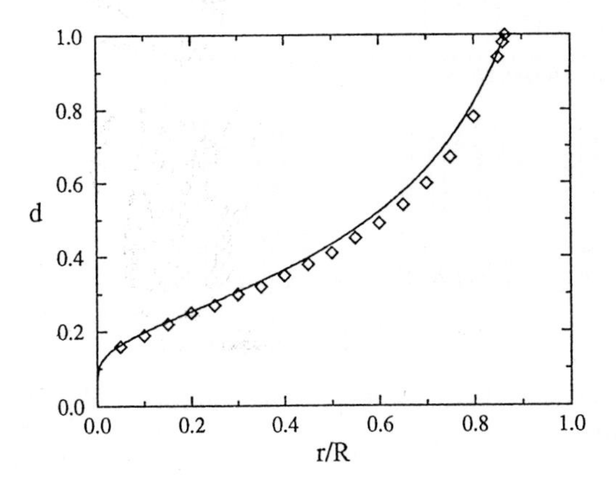

Fig. 6.15. Fractal dimension d as a function of the disk radius. Numerical results ($\diamond$) are compared with the estimate (6.14) (—).

triangular configuration. Figure 6.16 shows the scattering functions for $r = 8.5$ and $R = 10$ and a small impact parameter interval $0.22296 \le b \le 0.22297$ scanned by 400 trajectories. The computation time becomes longer, reflecting the occurrence of long-lived trajectories. The fractal dimension is found to be not much below unity.

One can locate an orbit with 155 impacts at an impact parameter b_ν with relative coordinate $\nu = 352$ (see (6.10)). The collision sequence for this orbit is

$$\text{`}1^4 0^3 1 0^7 1 0^2 1^2 0 1 0^2 1^6 (01)^3 0^4 1^2 0 1 0^2 1^4 0^2 1 0^3 1^2 (01)^6 0^5$$
$$1 (01)^9 1 0^2 1 (01)^2 0^3 1^4 0^2 1^2 0 1^2 (01)^5 0^7 1 0 1^8 0 1^3 0^4 1 0 1\text{'}.$$

Several sub-sequences occur in this collision sequence, where the trajectory gets close to a periodic orbit. The simple triangular orbit (1) in Fig. 6.14 appears several times (' 1^4 ', ' 0^7 ', ' 1^6 ', ' 0^7 ', ' 1^8 ', ...) and the two-bounce orbit ' 10 ', in particular, appears quite often. Here, the particle enters the small exit channel between two disks and is reflected on both sides, while energy is converted from transverse to vibrational motion until a reflection back into the interior region occurs.

Looking at Fig. 6.16, one observes characteristic smooth structures between erratic looking regions. Here, the collision sequences are also quite long, albeit highly organized. Magnifying such an interval reveals a clear sawtooth structure of the deflection function, accompanied by a staircase in the collision number. This can be understood by looking at the collision sequence, which ends up with a tail '...$(01)^k$'. This tail is an oscillation between two disks until the particle finally escapes. When the impact parameter is varied, the scattering

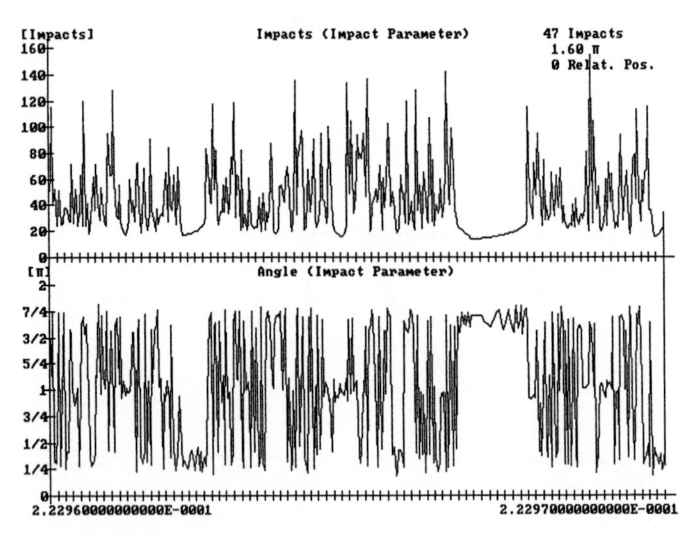

Fig. 6.16. Collision number $n(b)$ and scattering angle $\vartheta(b)$ as a function of the impact parameter b for $r = 8.5$ and $R = 10$.

angle changes over a large range until the trajectory becomes tangential to the other disk and an additional impact occurs. This effect is also visible in a two-disk system, where the projectile is trapped in the small channel between the two disks.

Incomplete Symbolic Dynamics. When the disk radii become larger, the disks are more and more screened by each other. This results in the fact that the one-to-one correspondence between periodic orbits and symbolic binary collision sequences is destroyed. Some of the sequences no longer appear as periodic orbits; they are 'forbidden', because otherwise the projectile would have to transverse one of the disks. One can investigate this process, find forbidden collision sequences, discuss the organization of the binary coding tree and study the fractal properties of such a case.

Multiscale Fractals. As discussed above, the fractals appearing in the three-disk system are not single-scale fractals. They are multi-scale fractals, as discussed, e.g., in the review article by Tél [6.12] . In the above examples, the scaling differences are, however, quite small. One can try to emphasize these differences by varying the disk sizes, the distances and positions of the disks and studying the fractal properties.

6.5 Suggestions for Further Studies

Chaotic scattering for smooth potentials has been studied by Jung for a potential with three maxima [6.14, 6.15, 6.16]. Periodic orbit organization, Cantor set singularity structure, and the semiclassical limit of quantum scattering have been investigated.

Quantum scattering: three-disk scattering also serves as a model for exploring chaotic phenomena in quantum scattering [6.4, 6.6, 6.7]. In the semiclassical limit of small wavenumbers, the scattering cross sections can be described using the Gutzwiller trace formula, which is based on the classical periodic orbits. In particular, the symbolic dynamics described above has been successfully used in such an analysis.

6.6 Real Experiments and Empirical Evidence

The chaotic reflections from a system of three shiny (Christmas-tree) balls, which is a three-dimensional version of the planar model in this case, can be studied in an everyday experiment (see Walker [6.17], as well as the numerical simulation by Korsch and Wagner [6.10]).

Berkovich et al. [6.8] monitored the scattered laser light from highly reflecting cylinders. Their set-up is such that a laser is positioned on a traveling microscope ($\Delta x \sim 0.1$mm) and the cylinders positioned on a plane parallel to each other; the scattering objects, mounted on this plane, can be rotated and this angle recorded. Ray-tracing is achieved using smoke, the laserspot then being scattered onto a screen. It is possible to demonstrate qualitatively the scattering pattern, measure the Lyapunov exponent and introduce the Cantor set and fractal dimensions.

References

[6.1] B. Eckhardt, *Irregular scattering*, Physica D **33** (1988) 89

[6.2] U. Smilansky, *The classical and quantum theory of chaotic scattering*, in: *Les–Houches Summer School on Quantum Chaos 1989*, page 371, Elsevier, Amsterdam 1990

[6.3] R. Blümel, *Quantum chaotic scattering*, in: H. Bai-Lin, D. H. Feng, and J.-M. Yuan, editors, *Directions in Chaos*, World Scientific, Singapore 1991

[6.4] P. Gaspard and S. A. Rice, *Scattering from a classically chaotic repellor*, J. Chem. Phys. **90** (1989) 2225

[6.5] B. Eckhardt, *Fractal properties of scattering singularities*, J. Phys. A **20** (1987) 5971

[6.6] P. Gaspard and S. A. Rice, *Semiclassical quantization of the scattering from a classically chaotic repellor*, J. Chem. Phys. **90** (1989) 2242

[6.7] P. Gaspard and S. A. Rice, *Exact quantization of the scattering from a classically chaotic repellor*, J. Chem. Phys. **90** (1989) 2255

[6.8] C. Bercovich, U. Smilansky, and G. P. Farmelo, *Demonstration of classical chaotic scattering*, Eur. J. Phys. **12** (1991) 122

[6.9] Q. Chen, M. Ding, and E. Ott, *Chaotic scattering in several dimensions*, Phys. Lett. A **145** (1990) 93

[6.10] H. J. Korsch and A. Wagner, *Fractal mirror images and chaotic scattering*, Computers in Physics Sept./Oct. (1991) 497

[6.11] J. V. José, C. Rojas, and E. J. Saletan, *Elastic particle scattering from two hard disks*, Am. J. Phys. **60** (1992) 587

[6.12] T. Tél, *Fractals, multifractals and thermodynamics*, Z. Naturforsch. 43a (1988) 1154

[6.13] M. V. Berry, *Reflections on a Christmas–tree baule*, Phys. Edu. **7** (1972) 1

[6.14] C. Jung and H. J. Scholz, *Cantor set structure in the singularities of classical potential scattering*, J. Phys. A **20** (1987) 3607

[6.15] C. Jung and S. Pott, *Classical cross section for chaotic potential scattering*, J. Phys. A **22** (1989) 2925

[6.16] C. Jung and S. Pott, *Semiclassical cross section for a classically chaotic scattering system*, J. Phys. A **23** (1990) 3727

[6.17] J. Walker, *The distorted images seen in Christmas-tree ornaments and other reflecting balls*, Scientific American **259** (December 1984) 84

7. Fermi Acceleration

The so-called *Fermi acceleration* – the acceleration of a particle through collision with an oscillating wall – is one of the most famous model systems for understanding nonlinear Hamiltonian dynamics. The problem was introduced by Fermi [7.1] in connection with studies of the acceleration mechanism of cosmic particles through fluctuating magnetic fields. Similar mechanisms have been studied for accelerating cosmic rockets by planetary or stellar gravitational fields. One of the most interesting aspects of such models is the determination of criteria for stochastic (statistical) behavior, despite the strictly deterministic dynamics.

Here, we study the simplest example of a Fermi acceleration model, which was originally studied by Ulam [7.2]: a point mass moving between a fixed and oscillating wall (see Fig. 7.1). Since then, this model system has been investigated by many authors, e.g. Zaslavskii and Chirikov [7.3], Brahic [7.4], Lichtenberg, Lieberman, and their coworkers [7.5]–[7.8], and it is certainly one of the first examples of the variety of kicked systems which appear in numerous studies of nonlinear dynamics.

In contrast to the previous examples, this Hamiltonian system has only one degree of freedom. From several aspects, it is, therefore, simpler than the examples studied in the preceding chapters. It is, however, explicitly time-dependent (otherwise it would be trivially integrable). It should be noted that a time-dependent Hamiltonian system with N degrees of freedom can be rewritten as a $(N+1)$-degree system, without explicit time dependence, by introducing the

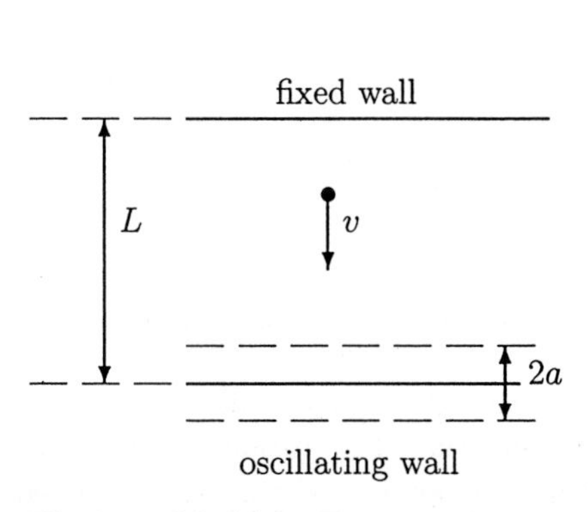

Fig. 7.1. Model for Fermi acceleration.

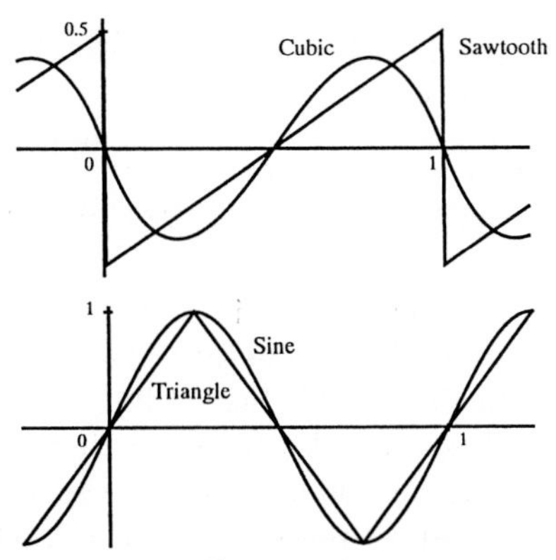

Fig. 7.2. Different wall oscillations.

time t as an additional canonical coordinate $q_t = t$ with canonical momentum $p_t = -E$, where E is the energy. The system is conservative in this extended phase space.

7.1 Fermi Mapping

We simplify the setup of the model even further by assuming that the amplitude, a, of the wall oscillation is very small compared to the distance L of the walls. In addition, we consider the limit of a massive wall compared to a small particle mass, i.e. the vibrating wall does not move in coordinate space. This leads to the 'simplified Ulam map' [7.6]. For particle velocities much larger than the wall velocity, the simplified map reproduces most features of the exact system.

When the particle with velocity v is reflected elastically from the massive wall moving with velocity V, it rebounds with velocity $v' = -v+2V$. This follows directly from energy and momentum conservation. In the case considered here, we register the particle velocity just before the nth impact as v_n (we assume v_n to be positive, see below). The particle hits the oscillating wall at a phase ϕ_n and is reflected with velocity $-v_n + 2V(\phi_n)$. It then moves to the fixed wall, is reflected and returns with velocity

$$v_{n+1} = v_n - 2V(\phi_n) \,. \tag{7.1}$$

after a time $2L/|v_{n+1}|$. If the wall velocity oscillates with period T, the phase of the wall oscillation at the next impact is given by

$$\phi_{n+1} = \phi_n + \frac{2L}{T\, v_{n+1}} \,. \tag{7.2}$$

Measuring the velocities in terms of a velocity unit V_0 and defining

$$u_n = \frac{v_n}{V_0} \quad , \quad U(\phi) = -2V(\phi)/V_0 \,, \tag{7.3}$$

we finally obtain as our working equations ($M = 2L/TV_0$)

$$\begin{aligned} u_{n+1} &= |u_n + U(\phi_n)| \\ \phi_{n+1} &= \phi_n + \frac{M}{u_{n+1}} \quad \text{modulo } 1 \,. \end{aligned} \tag{7.4}$$

The absolute value in (7.4) takes care of the region of low particle velocities, where unphysical negative recoil velocities may occur (for more details see the discussion in [7.5, Sect. 3.4]).

It should also be noted that systems with walls which oscillate with non-zero amplitude in coordinate space have been studied by various authors (see [7.5] and references given there). The simple mapping (7.4) can be considered as a limit for $a \ll L$.

It can easily be checked that the mapping (7.4) is area-preserving, and that the mapping in terms of the variables u_n and ϕ_n can be considered as Poincaré sections of a Hamiltonian system. We can expect to observe structural changes, familiar from the KAM theorem (see Sect. 2.2.3), where the wall oscillations can be considered as a perturbation of an integrable system with a non-oscillating wall. A condition for an application of the KAM theorem is, however, the smoothness of the perturbation (a sufficient number of continuous derivatives must exist). It is especially interesting to study differences in the dynamics for different types of perturbation, e.g. discontinuous, continuous, non-differentiable or differentiable wall velocities. The different forms of wall oscillations used in the program are shown in Fig. 7.2.

The program computes the iteration of an initial point in phase space and shows the density $\varrho(u, \phi)$ of these points in phase space. In addition, the probability of exciting the particle to a velocity u is calculated. This velocity distribution function is determined by integration over the phase angle ϕ

$$P(u) = \int_0^1 \varrho(u, \phi) \mathrm{d}\phi \,, \tag{7.5}$$

i.e. by projection of the phase space distribution onto the u-axis.

7.2 Interacting with the Program

In the *main menu* of the program FERMI, the user can select a wall oscillation, modify parameters, and start the iteration. The available items can be activated by means of the cursor keys and by pressing ⟨ENTER⟩. The ⟨F1⟩ key displays information about the program.

Selection of wall oscillation: Pressing a key toggles the type of motion of the wall. The following 1-periodic wall velocity functions $U(\phi)$ $(0 \leq \phi < 1)$ can be chosen:

1. *Sawtooth* — Linearly increasing wall velocity:

$$U(\phi) = \phi - 1/2 \,. \tag{7.6}$$

2. *Sine* — Harmonic wall velocity:

$$U(\phi) = \sin(2\pi\phi) \,. \tag{7.7}$$

3. *Cubic* — Wall velocity is a cubic polynomial:

$$U(\phi) = (2\phi - 1)\left(1 - (2\phi - 1)^2\right) \,. \tag{7.8}$$

4. *Triangle* — Triangular straight line approximation of a Sine function:

$$U(\phi) = 4 \begin{cases} \phi & : \quad 0.0 \leq \phi < 0.25 \\ 0.5 - \phi & : \quad 0.25 \leq \phi < 0.75 \\ -1 + \phi & : \quad 0.75 \leq \phi < 1.0 \,. \end{cases} \tag{7.9}$$

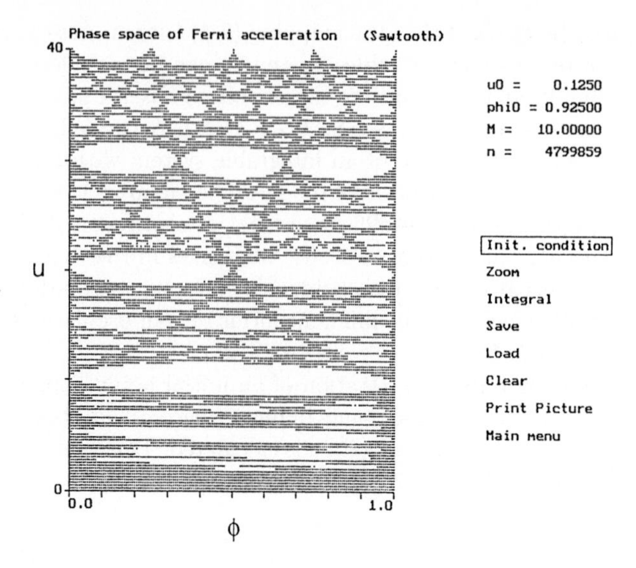

Fig. 7.3. Phase space menu of program FERMI. A phase space plot for a sawtooth wall oscillation ($M = 10$) is shown, which is generated by a single trajectory starting with low velocities. This figure is stored (file SAW10.FMI).

The user can then enter numbers for the initial particle velocity u_0 and initial phase ϕ_0, the system parameter $M = 2L/TV_0$, and the maximum particle velocity $u_{\max}$ displayed on the screen. The pre-set values are $u_0 = 0.1$, $\phi_0 = 0.25$, $M = 10$, $u_{\max} = 30$. The permitted range of these parameters is given by $0 \leq u_0 \leq 1000$, $0 \leq \phi_0 \leq 1$, $0 \leq M$, $0 \leq u_{\max}$. One should choose $u_{\max} > M$. Pressing ⟨ENTER⟩ switches to the phase space menu and starts the numerical iteration.

The phase space diagram (velocity u_n versus angle ϕ_n) is shown, and the succeeding impact values of the trajectory are marked. The different colors reflect the number of times a point has been reached, i.e. the time integrated phase space density. The selected phase space section is divided into 161×161 cells and the number of times $N(u,\phi)$ the trajectory enters a cell centered at (u,ϕ) is counted. The accumulated phase space density $N(u,\phi)$ is displayed on the screen, coded by different colors. The following coloring scheme is used: blue for $0 < N \leq 4$, green for $4 < N \leq 20$, red for $20 < N \leq 60$ and yellow for $N > 60$. The number n counts the iterations of the trajectory.

The following menu items can be activated in the *phase space menu*:

- **Init. condition** — In the phase space, a cross, which can be positioned by means of cursor keys, appears. Pressing the ⟨TAB⟩ key switches between slow and fast motion. The numerical values of u_0 and ϕ_0 at the cursor position are displayed. If the cross is in the desired position, pressing the ⟨ENTER⟩ key fixes the corresponding phase space coordinate as the new initial condition, and the computation is started. ⟨Esc⟩ cancels the operation. It should be noted that this feature can also be used to read numerical data from the screen simply by moving the cursor to the point of interest of the phase space.

- **Zoom** — Allows magnification of the phase space sections. A rectangular window appears. The boundary lines of this window can be moved using the cursor keys. One can switch between the upper and lower, or the right and left boundary, by using the ⟨Space Bar⟩. Pressing the ⟨Tab⟩ key switches between fast and slow motion. If the rectangle is in the desired position, pressing the ⟨Enter⟩ key starts the iteration. The chosen phase space section is magnified. On leaving this item and using the ⟨Enter⟩ key the iteration starts. ⟨Esc⟩ cancels the operation.

- **Integral** — Shows the velocity distribution (7.5) in addition to the phase space density.

- **Save** — Upon calling this menu item, one is asked under which name the picture is to be stored. Suffixes are ignored. After the input of a valid file name, a file under this name with the suffix *.FMI is created on the logged drive. If this file is loaded, simulation can be continued.

- **Load** — Enquires the name under which a picture has been stored and reads the data from file *.FMI. The results of the present simulation are lost. The program is in the same state as when the loaded picture was stored. The phase space distribution for a sawtooth wall oscillation with M=10 shown in Fig. 7.3 has already been computed and can be loaded from file SAW10.FMI .

- **Clear** — Clears the graphic screen and erases the data.

- **Print Picture** — A hardcopy of the entire screen is printed.

- **Main menu** — Returns to the main menu. Data not secured are lost.

7.3 Computer Experiments

7.3.1 Exploring Phase Space for Different Wall Oscillations

Let us begin our numerical experiments with a review of the classic studies published in the literature. First, we will consider a sinusoidal wall velocity. In this case, the mapping traditionally appears in the literature as

$$\begin{aligned} u_{n+1} &= |u_n + \sin\psi_n\,| \\ \psi_{n+1} &= \psi_n + \frac{2\pi M}{u_{n+1}}\,, \end{aligned} \tag{7.10}$$

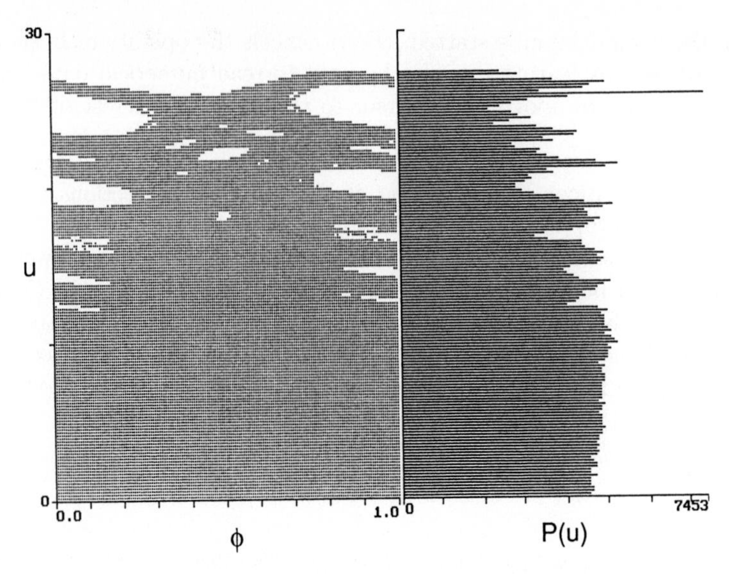

Fig. 7.4. Poincaré phase space section (u_n, ϕ_n) for a sinusoidal wall velocity with $M = 100$. The points are 623.000 iterations of a single trajectory. Also shown is the velocity distribution $P(u)$.

with the phase ψ extending over 2π rather than unity. A simple rescaling $\phi = 2\pi\psi$ immediately yields the standard form (7.4) used in the program. The value of M remains the same.

After activating the *'sine'* wall oscillation, we change the pre-set value of the strength parameter to the much stronger value of $M = 100$. Starting our trajectory with the pre-set initial conditions of ϕ_0 and u_0 and keeping the maximum displayed velocity $u_{\mathrm{max}} = 30$, we observe rapid spreading of the iterated points in phase space, showing an almost erratic behavior. The velocity oscillates up and down, i.e. the particle is accelerated and decelerated in a seemingly stochastic manner, though the dynamics is rigorously deterministic. Some phase space cells are approached again (visualized by the coloring) and the underlying structure slowly develops. The phase space distribution and the velocity distribution $P(u)$ found after some considerable time (about 623 000 iterations) is shown in Fig. 7.4 (compare [7.5, Fig. 1.14]). We can list three main features of the results of this numerical experiment:

- There exists an upper bound $u_{\mathrm{b}} \approx 27.5$ for the particle velocity, i.e. $u_n \leq u_{\mathrm{b}}$, for all n and all trajectories starting with low initial velocity.
- At low velocities ($u \leq u_{\mathrm{s}} \approx 12.0$), the iterated points seem to uniformly fill the phase space.
- In the intermediate region, a sequence of embedded islands centered at $\phi = 0$ is observed, which cannot be entered by the trajectory from outside.

The centers of the most prominent ones are located at velocities $u = 25.0$, 20.0, 16.7, 14.3, and 12.6.

These features are also visible in the velocity distribution.

It should be noted that it is possible to read numerical data from the screen by activating the menu item *'Init. condition'* and moving the cursor to the point of interest of the phase space.

In Fig. 7.5, the results of a second classic experiment for a sawtooth wall oscillation are shown for $M = 10$, $u_{\text{max}} = 16$, and 500 000 iterations of a single trajectory with low initial velocity (compare [7.5, Fig. 3.12]). Here, the visual impression of the evolution of the phase space pattern is different:

- The uniform region at low velocities is reduced to $u_n \leq u_s \approx 1.5$.
- A highly structured web of unoccupied islands appears, some of which are arranged in chains with the same area. The single islands are centered at $\phi = 0.5$ and $u = 10.0$, 5.0, 3.3, 2.5, 2.0, and 1.7. The trajectory injected into the stochastic region at low velocity diffuses slowly through this web of islands.
- The existence of an upper boundary u_b for the particle velocity is not evident from the plot.

This structured organization of phase space continues for higher velocities. As discussed in Ref. [7.6], an upper barrier u_b for the velocity does not exist, due to

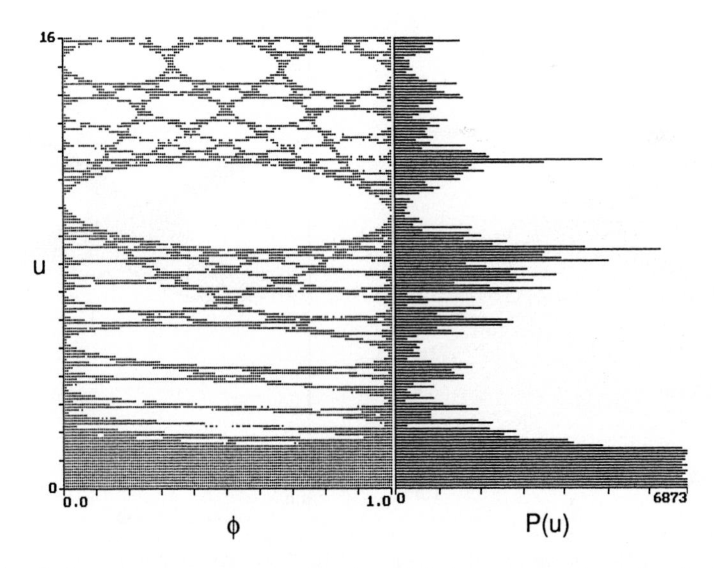

Fig. 7.5. Poincaré phase space section (u_n, ϕ_n) for a sawtooth wall velocity with $M = 10$. The points are iterations of a single trajectory. Also shown is the velocity distribution.

the discontinuous nature of the wall oscillation. The reader is invited to explore this regime. As an example, the phase space plot generated by about 4 800 000 iterations of a single trajectory started at $(u_0, \phi_0) = (0.125, 0.925)$ in the region $u < u_s$ is shown in Fig. 7.3 above (this picture can be loaded in the program from file SAW10.FMI).

The difference between sinusoidal and sawtooth wall oscillation is *not* due to the different magnitudes of the strength constant M, as can easily be verified by running the program with different values of M. Results for a sawtooth oscillation with $M = 1000$ can be found in [7.6], and a harmonic oscillation with $M = 10$ is considered below.

7.3.2 KAM Curves and Stochastic Acceleration

In this section, we will start to explore the sinusoidal wall oscillation in more detail. In this case, the force is analytic and the KAM-theorem is applicable (see Sect. 2.2.3). We find typical KAM phase space organization: quite generally, the phase space plots show regular islands embedded in a chaotic sea with chains of alternating k-periodic elliptic and hyperbolic fixed points. The stable ones appear in the center of the regular islands, which are separated from the outside by invariant curves, the so-called KAM curves. The motion inside the islands is mainly governed by the rotation number σ (see Sect. 2.4.5). Higher order island

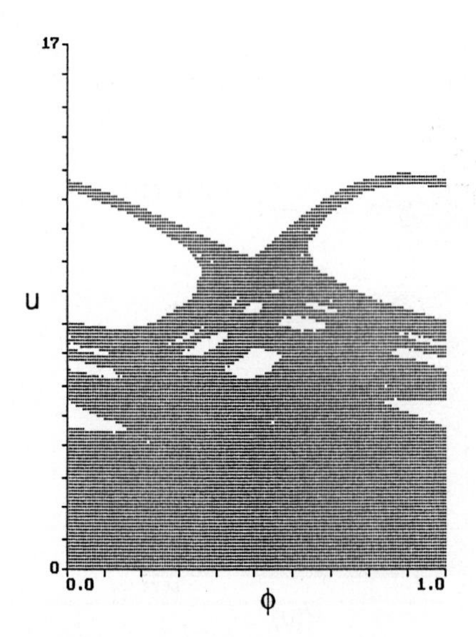

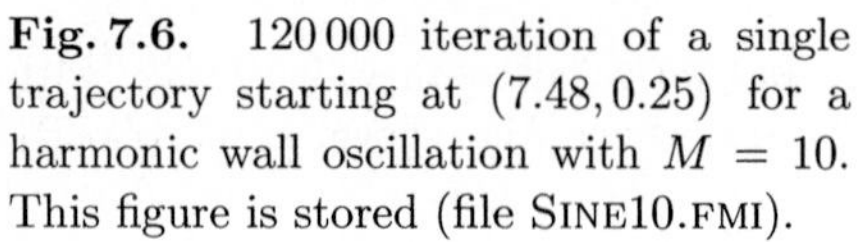

Fig. 7.6. 120 000 iteration of a single trajectory starting at (7.48, 0.25) for a harmonic wall oscillation with $M = 10$. This figure is stored (file SINE10.FMI).

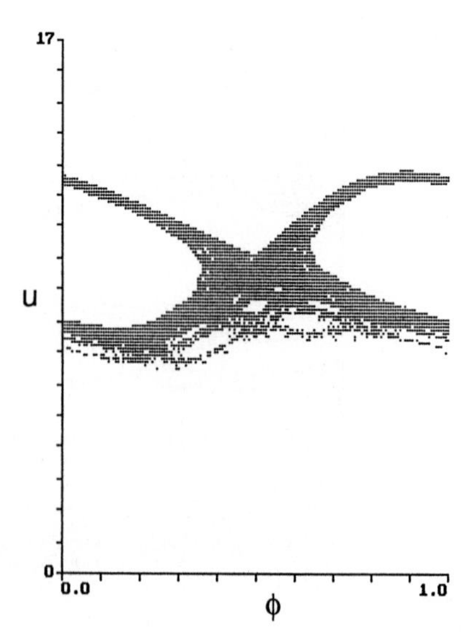

Fig. 7.7. As in Fig. 7.6, the first 20 000 iteration of a single trajectory starting at (7.48, 0.25) are shown.

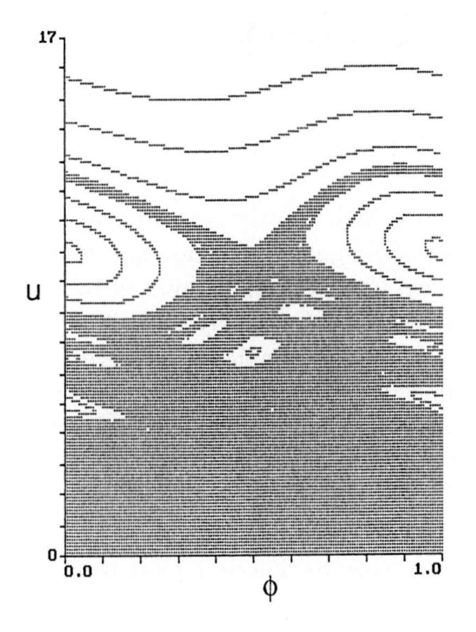

Fig. 7.8. Poincaré phase space section for a harmonic wall oscillation with $M=10$. This figure is stored on the disk.

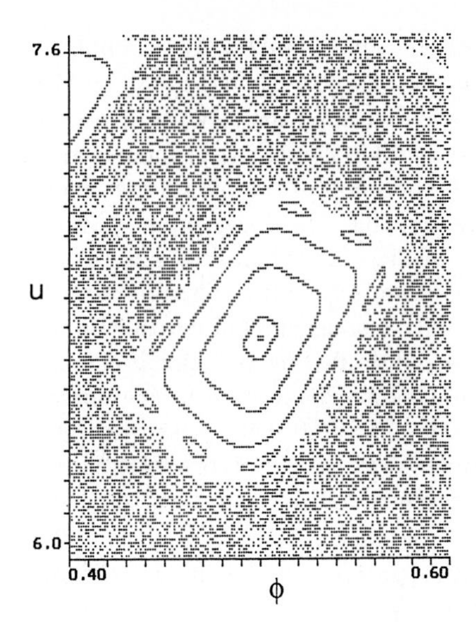

Fig. 7.9. Magnification of the period-two stability island at $(u, \phi) = (6.67, 0.5)$ in Fig. 7.8. This figure is stored on the disk.

structures are observable on closer inspection, the most pronounced being found in the vicinity of the island boundary.

The role of the invariant curves is very important to the acceleration process since they divide the phase space into disconnected regions, thus limiting the flow in the chaotic sea. In particular, invariant curves extending over the whole angular interval act as impenetrable barriers to the phase space evolution.

The case $M = 100$ was explored in Fig. 7.4 above. The chaotic phase space region for $M = 10$ is shown in Fig. 7.6. The distribution is independent of the starting condition in the long-term time limit, provided that the starting point is inside this connected chaotic region. This condition can most easily be met by starting at very low energies. In Fig. 7.6, however, we choose the initial point $(u_0, \phi_0) = (7.48, 0.25)$ to observe dynamical behavior at higher velocities. The iterated points spend a considerable time in the high velocity region, being almost captured in a thin layer close to the boundary surrounding a large island at $(u, \phi) = (10, 0)$. This can be seen from Fig. 7.7, which is restricted to the first 20 000 iterations. The trajectory suddenly finds its way through the lower island chain and also approaches the low velocity regime.

The motion inside the islands can be explored by choosing starting points for the iteration from inside these regions. Figure 7.8 shows a synoptic phase space plot of the iterations for several selected trajectories. In the middle of the large central island at $(u, \phi) = (10, 0.5)$, we have a one-periodic trajectory, i.e. a fixed point of the mapping (7.4). More period-one fixed points appear at $(u, \phi) = (10/m, 0.5)$, for $m = 1, 2, \ldots$. In addition, there are chains of k-

periodic fixed points. An example is the two-periodic trajectory alternating between $(u, \phi) = (6.67, 0)$ and $(u, \phi) = (6.67, 0.5)$.

The motion around the fixed points is determined by the rotation number σ. This is studied in more detail in the following section. Magnifying the neighborhood of the period-two fixed point at $(6.67, 0.5)$, as shown in Fig. 7.9, we see that the fixed point is surrounded by higher order periodic orbits, i.e. chains of stability islands. This process of magnification can be continued down to arbitrary small scales. In practice, however, one is limited by the rapidly increasing computation time needed to explore strongly magnified regions. Two further magnifications are shown in Figs. 7.10 and 7.11, where first the neighborhood close to the unstable region in the upper right corner close to $(u, \phi) = (6.9, 0.56)$ is shown, revealing a sub-structure. Finally, the island centered at $(u, \phi) = (6.9894, 0.5638)$ — the central orbit has period 46 — is magnified, and is again seen to be surrounded by satellite islands.

At lower velocities $u < u_\mathrm{s} \approx 4$, no stability islands are observed. As discussed in more detail in the following section, this stochastic region is given by

$$u < u_\mathrm{s} \approx \sqrt{\pi M/2}\,, \tag{7.11}$$

where all fixed points are unstable. Here, the dynamics can be approximated by a stochastic acceleration model based on a Markov process in u, as discussed in Ref. [7.5, 7.6, 7.7].

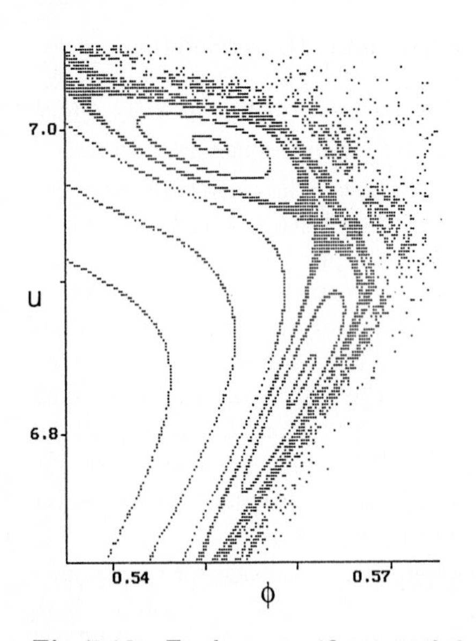

Fig. 7.10. Further magnification of the unstable region close to $(u, \phi) = (6.9, 0.56)$ in Fig. 7.9. This figure is stored on the disk.

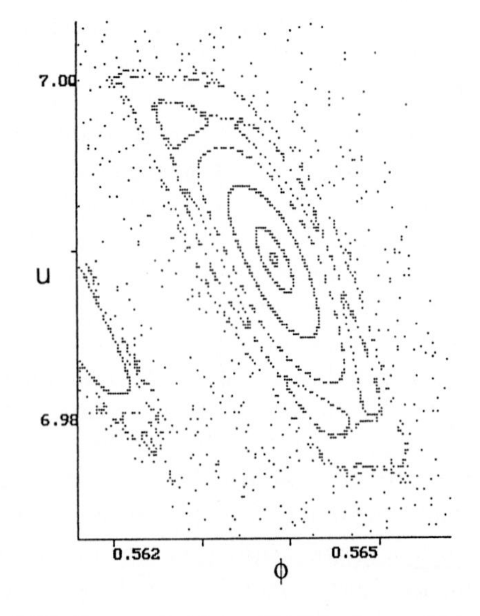

Fig. 7.11. Further magnification of the island at $(u, \phi) = (6.99, 0.564)$ in the upper right of Fig. 7.10. This figure is stored on the disk.

Table 7.1. Stable period-one fixed points $(u_m, \phi), m = 1, 2, \ldots$ and stability boundaries u_s.

wall oscillation	(u_m, ϕ)	u_s
sawtooth	$(M/m, 0.5)$	$\sqrt{M}/2$
harmonic	$(M/m, 0)$	$\sqrt{\pi M/2}$
cubic	$(M/m, 0.5)$	$\sqrt{M/2}$
triangle	$(M/m, 0)$	$\sqrt{M}$

At higher velocities, we have invariant KAM curves extending over the whole interval $0 \leq \phi \leq 1$, which act as a barrier to the acceleration process, as discussed in more detail in Sect. 7.3.4.

7.3.3 Fixed Points and Linear Stability

The dynamics in the vicinity of the fixed points can be understood by linearization of the mapping (7.4) at the fixed point. In the following, we assume for simplicity $u > U(\phi)$.

The period-one fixed points are determined by $u_{n+1} = u_n$ and $\phi_{n+1} = \phi_n$ modulo 1, which yields

$$U(\phi) = 0 \quad \text{and} \quad u = M/m\,, \quad m = 1,\, 2, \ldots\,. \tag{7.12}$$

The condition for ϕ is satisfied for all wall oscillations considered here at $\phi = 0.5$, and also at $\phi = 0$, with the exception of the sawtooth oscillation. Linearization of (7.4) yields

$$\begin{pmatrix} \Delta u_{n+1} \\ \Delta \phi_{n+1} \end{pmatrix} = \mathbf{A}_{n+1,n} \begin{pmatrix} \Delta u_n \\ \Delta \phi_n \end{pmatrix}, \tag{7.13}$$

with

$$\mathbf{A} = \mathbf{A}_{n+1,n} = \begin{pmatrix} 1 & U'(\phi_n) \\ -M/u_n^2 & 1 - MU'(\phi_n)/u_n^2 \end{pmatrix}, \tag{7.14}$$

where prime denotes differentiation with respect to ϕ, and

$$\det \mathbf{A} = 1 \quad \text{and} \quad \operatorname{Tr} \mathbf{A} = 2 - MU'(\phi_n)/u_n^2\,. \tag{7.15}$$

The condition for stability is $|\operatorname{Tr} \mathbf{A}| < 2$, which can only be met for $U' > 0$ and

$$u > u_s = \tfrac{1}{2}\sqrt{MU'(\phi)}\,. \tag{7.16}$$

Table 7.1 lists the stable period-one fixed points (u, ϕ), and the stability boundaries u_s for the various types of wall oscillation.

For $u < u_s$ the fixed points are hyperbolic with reflection. The transition from stability to instability at $\mathrm{Tr}\,\mathbf{A} = -2$ appears at rotation number $\sigma = \pi$, as is evident from

$$\cos\sigma = \tfrac{1}{2}\,\mathrm{Tr}\,\mathbf{A}\,. \tag{7.17}$$

At the transition point, the excitation is in resonance with the half-period of the oscillation and the periodic oscillation is destroyed. This phenomenon is well-known [7.5, Sect. 3.4c], e.g. for the resonant excitation of a swing.

These considerations can be numerically tested using the program. In the computer experiments discussed in Sects. 7.3.1 and 7.3.2, several period-one fixed points were located: for the sawtooth map with $M = 10$, they were computationally found at $\phi = 0.5$ and $u = 10.0$, 5.0, 3.3, 2.5, 2.0, and 1.7 with stability boundary $u_s = 1.5$. For the sinusoidal case with $M = 100$, fixed points appear at $\phi = 0$ and $u = 25.0$, 20.0, 16.7, 14.3, and 12.6 with $u_s = 12$ and similarly for $M = 10$. All these numerical results are in agreement with theoretical analysis: the fixed points were found at $u = M/m$ with $m = 1, 2, \ldots$ for the sawtooth and $m = 4, 5, \ldots$ for the sinusoidal case with $M = 100$.

7.3.4 Absolute Barriers

For a sinusoidal wall oscillation, the phase space at higher velocities is filled with invariant KAM curves extending over the whole interval $0 \le \phi \le 1$. These KAM curves act as impenetrable barriers to the acceleration process. The lowest of these curves determines the maximum value u_b of the velocity when the acceleration is started at low velocity.

For the absolute barrier, one finds the simple approximate expression [7.6]

$$u_b \approx 2.8\sqrt{M}\,, \tag{7.18}$$

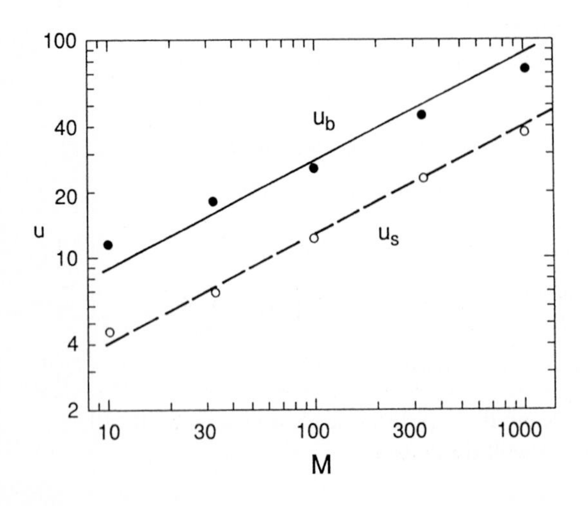

Fig. 7.12. Absolute barrier u_b (•) and stochastic transition velocity u_s (∘) (adapted from Lichtenberg and Liebermann [5]).

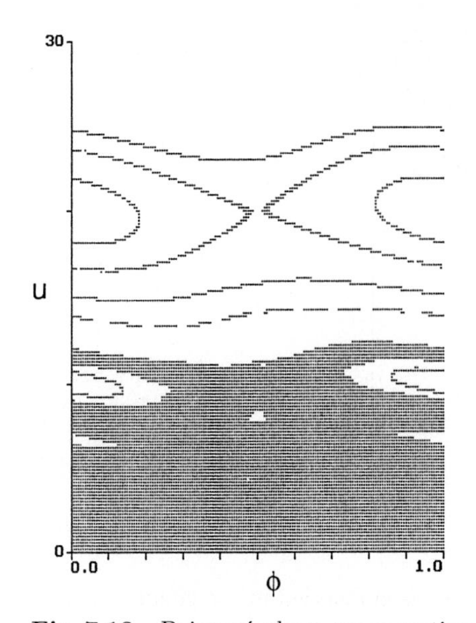

Fig. 7.13. Poincaré phase space section for a harmonic wall oscillation with $M = 20$. Iterations of several selected trajectories.

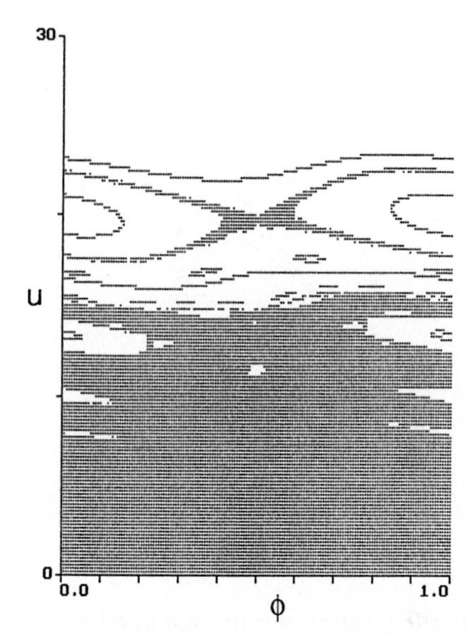

Fig. 7.14. Poincaré phase space section for a harmonic wall oscillation with $M = 40$. Iterations of several selected trajectories.

which is supported by the double logarithmic plot in Fig. 7.12 (taken from Ref. [7.5]). These data agree, of course, with the values $u_b = 12.9$ and 27.5 obtained numerically in the above experiments.

Equation 7.18, however, describes only the overall features of the growth of the stochastic sea with increasing M and does not account for the finer details. When we compare the Poincaré phase space sections for $M = 10$ (Fig. 7.5) and $M = 100$ (Fig. 7.4), we observe the highest period-one fixed point at $u = M/m$, with $m = 1$ in the first case and $m = 4$ in the second. Therefore, the three fixed points with $m = 1$, 2, and 3 disappear from the chaotic sea when M is increased from 10 to 100. Here, we study this mechanism in some more detail.

Figures 7.13 and 7.14 show Poincaré sections for $M = 20$ and 40. Locating the fixed points in the centers of the large stability islands embedded in the chaotic sea at $u = 10$ or 13.3, respectively, we readily identify them as $m = 2$ or $m = 3$ fixed points. The maximum values of the velocity accessible for acceleration from small velocities is given by $u_b = 12.5$ and 15.8, respectively.

The fixed points with lower m are located above these u_b values and a KAM curve separates these higher fixed points from the lower ones. It should also be noted that the value of $u_b = 12.5$ for $M = 20$ is *smaller* than that that of $u_b = 12.9$ for $M = 10$, despite the overall increase predicted in (7.18) and shown in Fig. 7.12. This justifies a more detailed numerical study of the M-dependence of u_b. Figure 7.15 summarizes some numerical values computed using the program. Also shown is the approximation (7.18). At low values of

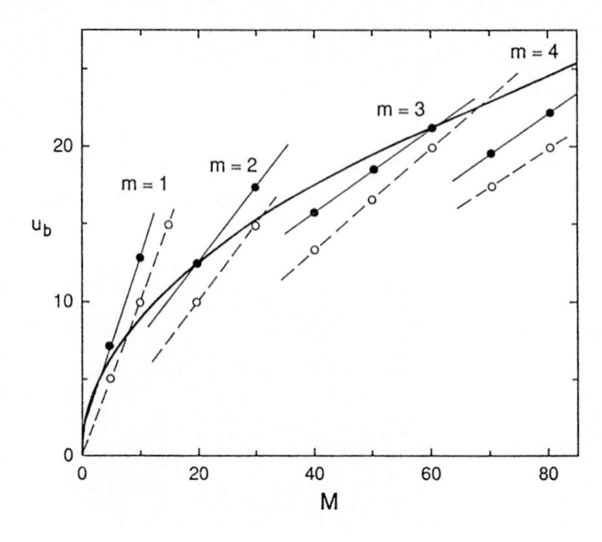

Fig. 7.15. Absolute barrier u_{b} (•) and period-one fixed points (∘) for $m = 1, 2, 3, 4$. Also shown is approximation (7.18) (full curve).

M, the (stable) period-one fixed point $m = 1$ is embedded in the chaotic regime ($u_{\mathrm{b}} > M/m = 10$). Both the absolute barrier and the fixed point increase almost linearly with M and the phase space distribution remains structurally similar. Then — in a critical region close to $M \approx 12$ — a structural transition occurs, where the fixed point $m = 1$ is ejected from the chaotic region and a new KAM barrier is formed which separates the fixed point $m = 1$ from the fixed point $m = 2$ at lower velocity, thus reducing the absolute barrier for the velocity to a value below the $m = 1$ fixed point.

When M is increased further, u_{b} again grows linearly, albeit with a reduced slope, until the next fixed point $m = 2$ is ejected. This scheme is repeated for the fixed points $m = 3, 4, \ldots$.

7.3.5 Suggestions for Additional Experiments

Higher Order Fixed Points. The investigation of first order fixed points in Sect. 7.3.3 can be extended to k-periodic ones. Here, the analysis becomes progressively more difficult with increasing k. Analytic formulae for the two-periodic orbits $(u_1, \phi_1) \to (u_2, \phi_2) \to (u_1, \phi_1)$ can quite easily be derived. For the case of a sawtooth oscillation, they are located at

$$\left.\begin{aligned} (u_1, \phi_1) &= (2M/m + 1/4\,,\, 1/4) \\ (u_2, \phi_2) &= (2M/m - 1/4\,,\, 3/4) \end{aligned}\right\}\ m = 1, 3, 5, \ldots \tag{7.19}$$

with stability for $u > \sqrt{M}$. In the sinusoidal case, some higher order fixed points are found at

$$\left.\begin{aligned} (u_1, \phi_1) &= (2M/m, \alpha_{m,h}) \\ (u_2, \phi_2) &= (2M/m, \alpha_{m,h} + m/2) \end{aligned}\right\}\ m = 1, 3, 5, \ldots \tag{7.20}$$

with

$$\alpha_{m,h} = \sin^{-1}\{\,2M(m^{-1} - h^{-1})\,\} \quad h = 1, 3, 5, \dots \,, \tag{7.21}$$

which are stable for $h = m$ if $u_1 = u_2 > \sqrt{\pi M}$ (see Ref. [7.6, Tab.II]). An example is the two-periodic motion alternating between $(6.67, 0)$ and $(6.67, 0.5)$, which has been found for a harmonic oscillation with $M = 10$. This compares well with the theoretical result for $m = h = 3$.

The properties of these higher order fixed points an be explored in more detail both analytically and theoretically.

Standard Mapping. Following Ref. [7.5, Sect. 4.1b], we can derive the so-called standard mapping

$$\begin{aligned} I_{n+1} &= I_n + K \sin\theta_n \\ \theta_{n+1} &= \theta_n + I_{n+1} \end{aligned} \tag{7.22}$$

from the Fermi mapping for a sinusoidally oscillating wall (7.10)

$$\begin{aligned} u_{n+1} &= |u_n + \sin\psi_n| \\ \psi_{n+1} &= \psi_n + \frac{2\pi M}{u_{n+1}} \end{aligned} \tag{7.23}$$

by linearization in action space near a period-one fixed point $u = M/m$ with integer m, defining

$$\theta_n = \psi_n - \pi \quad , \quad I_n = -K(u_n - u) \tag{7.24}$$

and

$$K = 2\pi M u^{-2} = 2\pi m^2/M \,. \tag{7.25}$$

The (nonlinear) standard mapping (7.22) is extensively studied (see, e.g., Ref. [7.5] and references therein) and can be used to model the behavior of a nonlinear system in the vicinity of the fixed point. A noticeable difference between the true map (7.23) and the standard map is the 2π-periodicity of the latter in action space.

The parameter K — known as the stochasticity parameter — determines the global stochasticity of the standard map. For small K, we have local stochasticity confined to a thin layer, which is trapped between invariant curves. With increasing K this layer grows, and at a critical value found as $K_{\text{crit}} \approx 0.9716$ the transition from local to global chaos takes place. For $K > 4$, the central fixed point $(I, \theta) = (0, 0)$ of the standard map is unstable.

The standard map can be useful for developing criteria for the transition from local to global stochasticity (resonance overlap criteria), the location and thickness of the separatrix layers etc. More details can be found in the book by Lichtenberg and Liebermann [7.5].

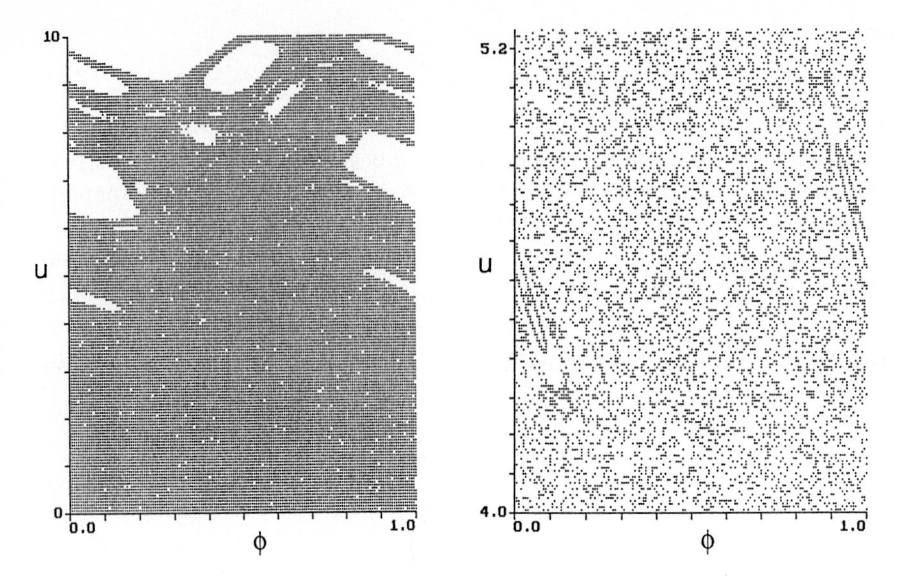

Fig. 7.16. Phase space showing a bifurcated fixed point slightly below u_s for the harmonic oscillation with $M = 14$.

Fig. 7.17. Magnification of Fig. 7.16 showing the period-two fixed points.

Bifurcation Phenomena. The dynamics near a period-one fixed point as it passes from stability to instability (see Sect. 7.3.3) has been studied in great detail in Ref. [7.5, 7.7] for the harmonic case (7.23) with $M = 14$. On examining the phase space plot in Fig. 7.16, we find a stability island structure in the vicinity of the stochastic transition velocity $u_s = 4.68947$. The period-one fixed point at $(u_0, \phi) = (14/3, 0) \approx (4.67, 0)$ is, however, expected to be unstable due to $u_0 < u_s$. Following Ref. [7.5, 7.7], Fig. 7.17 shows a magnified phase space section centered at velocity u_0. We observe an elongated stability island region surrounding the pair of period-two fixed points. The unstable fixed point $(u_0, 0)$ is located on the separatrix layer of these islands. The stable period-two fixed points can be located at $(u_1, \phi_1) = (4.55, 0.037)$ and $(u_2, \phi_2) = (4.79, 0.963)$, in agreement with the analytic result [7.5, 7.7]

$$u_i = \frac{M}{2\phi_i + m}, \quad i = 1, 2, \tag{7.26}$$

where the ϕ_i are the two solutions of

$$\frac{4M\phi}{m^2 - 4\phi^2} = \sin 2\pi\phi. \tag{7.27}$$

For $M = 14$ and $m = 3$, we find $\phi_{1,2} = \pm 0.0372$ and, hence, $u_1 = 4.554$ and $u_2 = 4.785$. It is, furthermore, of interest to explore the behavior of the fixed points when M is decreased, thus driving the fixed points deeper into the unstable region. As suggested by Lichtenberg and Liebermann, this can be

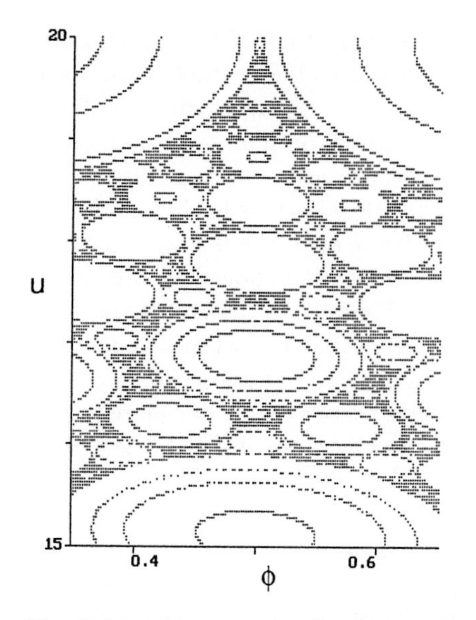

Fig. 7.18. Phase space distribution for sawtooth wall oscillation with $M = 10$. Magnification of Fig. 7.5. This figure is stored on the disk.

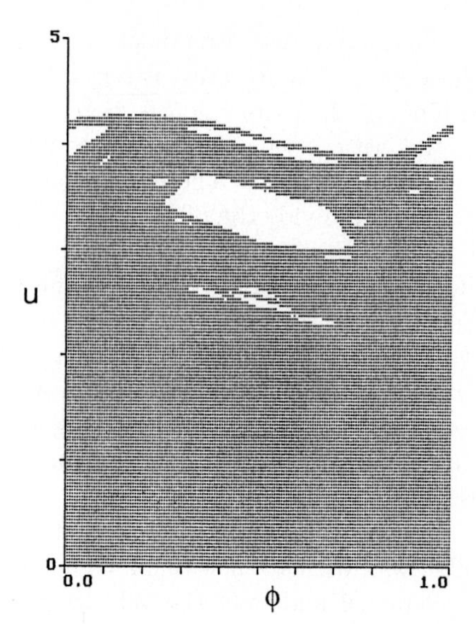

Fig. 7.19. Phase space distribution for cubic wall oscillation with $M = 10$. This figure is stored (file CUBIC10.FMI).

understood by approximating ϕ near a resonance by (7.26), which reduces the two-dimensional to a one-dimensional mapping

$$u_{n+1} = u_n + \sin \pi (M/u_n - m) . \tag{7.28}$$

Mappings of this type are discussed in Chap. 9 . Generally, they show a cascade of successive period-doubling bifurcations. As a numerical example, one can try to locate a period-four bifurcation of the fixed points shown in Figs. 7.16 and 7.17 .

Influence of Different Wall Velocities. As already pointed out in Sect. 7.1, the applicability of the KAM-theorem is based on a sufficiently smooth perturbation. In the present case, we consider four different wall oscillations with different properties: The *'sawtooth'* is discontinuous, the *'triangle'* is continuous with a discontinuity in the slope, the *'cubic'* oscillation is continuously differentiable with a discontinuity in the second derivative ($U''(0) \neq U''(1)$) and, finally, the *'sine'* oscillation is analytic.

It is of interest to study the differences in the dynamics for different types of perturbation, e.g. discontinuous, continuous, non-differentiable or differentiable wall velocities.

In the case of sawtooth wall oscillation, the discontinuities at some fixed points prevent the formation of invariant curves extending over the entire phase interval $0 \leq \phi \leq 1$, which stops the acceleration of the particle at an absolute barrier velocity u_b.

Sinusoidal oscillation displays the characteristics of a KAM organization of phase space. The structure is very different from that of the sawtooth case. Some results are shown in Fig. 7.5. Here, the basic underlying structure is produced by the critical set of those points which hit the discontinuity after a certain number of iterations. These points are organized on curves ('lines of discontinuity'), which divide the phase space plane into various regions having a different dynamics. Initial points inside such a region never meet the discontinuity and the dynamics is strictly linear in the interior of such a region. The organizing discontinuity lines are clearly seen in Fig. 7.5. Figure 7.18 illustrates this kind of behavior by a magnification of a central part of Fig. 7.5, which is again different from the KAM organization. As already pointed out above, for sawtooth mapping no invariant curve forming an absolute barrier exists. For more results see Ref. [7.5, 7.6], as well as the study of an exact Fermi mapping for sawtooth oscillation by Brahic [7.4].

Figure 7.19 finally shows the stochastic low velocity region for the case of cubic wall oscillation with $M = 10$. In this case, an invariant curve limits the acceleration and only the velocity region $u < u_b \approx 4.28$ is accessible. It is recommended that the reader explore more details of the cubic case (internal structure of the island in Fig. 7.19, formation of invariant curves, ...) and study the dynamics of continuous non-differentiable *'triangle'* wall oscillation.

7.4 Suggestions for Further Studies

Exact Fermi mapping. The simplified Fermi mapping considered in this program results from the approximation of a negligible wall amplitude in coordinate space. An exact treatment results in a somewhat more complicated set of equations, whose dynamics has been investigated in some detail (see [7.3, 7.4, 7.5, 7.6]).

The bouncing ball. A system closely related to the Fermi acceleration model is that of a ball bouncing on a vibrating table [7.9]–[7.12]. Here, the reflection from the fixed wall is replaced by 'reflection' from the gravitational field. Assuming vertical motion and neglecting wall motion in coordinate space once more, the ball falls back onto the table after a time of flight $2v/g$, where v is the velocity after impact and g the gravitational constant. It follows that the mapping equations (7.4) take the slightly modified form

$$\begin{aligned} u_{n+1} &= |u_n + U(\phi_n)| \\ \phi_{n+1} &= \phi_n + M u_{n+1} \quad \text{modulo } 1\,. \end{aligned} \tag{7.29}$$

Such bouncing ball mapping has been studied by Holmes [7.13, 7.14], who also included damping effects. (Note the similarity between (7.29) and the standard map (7.22) for sinusoidal wall oscillation!)

Quantum Fermi acceleration. The quantum mechanical version of Fermi acceleration has been considered by various authors [7.15]–[7.18]to explore the quantum to classical correspondence for systems which are classically chaotic. The manifestation of classical chaos in quantum systems (so-called *'Quantum Chaos'*) is currently of considerable interest. Fermi acceleration is one of the few model systems analyzed in such studies.

7.5 Real Experiments and Empirical Evidence

The original idea of a ball bouncing elastically from a vibrating table goes back to Pippard [7.9]. Briggs [7.10] describes a setup using a loudspeaker cone vibrating at about 1 kHz as a table, small steel balls, and an electrete microphone. As the amplitude of the table oscillation, directly monitored on one channel of an oscilloscope, is increased subharmonics of the driving frequency are heard or seen on the second channel, or as harsh noise or chaotic waveforms in the chaotic states.

Tufillaro et al. [7.11] used the same excitation, however a piezoelectric film for monitoring the motion; if the ball hits the film, it generates a voltage which is readily detectable by one channel of the oscilloscope. The authors [7.19] then changed the method of data acquisition and analysis in order to construct an impact map monitoring the strength and phase of each collision with the piezoelectric film and appropriate electronics. When these collisions are stored on an oscilloscope, the impact map shows period-two orbits of the bouncing ball, period-four orbits, strange attractors and plane filling motions.

Zimmerman et al. [7.12] realized an electronic analogue of the bouncing ball: the free fall of the ball when a diode is reverse biased, and impact when it conducts. A dual channel oscilloscope traces the bouncing ball simulator. Periodic and chaotic orbits, Poincaré sections and bifurcation diagrams can easily be shown on the oscilloscope.

Moon [7.20, p. 74] describes another real Fermi accelerator for qualitative demonstrations. A mass slides horizontally and freely on a shaft with viscous damping until it hits springs on either side, with a deadband in the restoring force.

A fascinating extension of this Fermi accelerator is described by Moon [7.20, p. 165], in which, periodically, surface waves in a cylinder filled with water are vertically periodically excited by a loudspeaker cone. Keolian et al. [7.21] studied the amplitude – frequency parameter space using an equivalent setup. They point out that the first observation of subharmonic parametric excitation is due to Faraday (~1830), who studied shallow water waves, when the containing vessel was driven vertically. Rayleigh (~1880) analyzed and confirmed the results of Faraday.

References

[7.1] E. Fermi, *On the origin of cosmic radiation*, Phys. Rev. **75** (1949) 1169 (see also *Collected Works* 2, 978)

[7.2] S. Ulam, *On some statistical properties of dynamical systems*, 4th Berkeley Symp. on Math. Stat. and Probabil. **3** (1961) 315

[7.3] G. M. Zaslavskii and B. V. Chirikov, *Fermi acceleration mechanism in the one–dimensional case*, Sov. Phys. Dokl. **9** (1965) 989

[7.4] A. Brahic, *Numerical study of a simple dynamical system*, Astron. Astrophys. **12** (1971) 98

[7.5] A. J. Lichtenberg and M. A. Lieberman, *Regular and Stochastic Motion* (Springer, New York 1983)

[7.6] M. A. Lieberman and A. J. Lichtenberg, *Stochastic and adiabatic behaviour of particles accelerated by periodic forces*, Phys. Rev. A **5** (1972) 1852

[7.7] A. J. Lichtenberg, M. A. Lieberman, and R. H. Cohen, *Fermi acceleration revisited*, Physica D **1** (1980) 291

[7.8] J. E. Howard, A. J. Lichtenberg, and M. A. Lieberman, *Two–frequency Fermi mapping*, Physica D **5** (1982) 243

[7.9] A. B. Pippard, *Response and Stability* (Cambridge Press, Cambridge 1985)

[7.10] K. Briggs, *Simple experiments in chaotic dynamics*, Am. J. Phys. **55** (1987) 1083

[7.11] N. M. Tufillaro and A. M. Albano, *Chaotic dynamics of a bouncing ball*, Am. J. Phys. **54** (1986) 939

[7.12] R. L. Zimmermann, *The electronic bouncing ball*, Am. J. Phys. **60** (1992) 378

[7.13] P. J. Holmes, *The dynamics of repeated impacts with a sinusoidally vibrating table*, J. Sound Vib. **84** (1982) 173

[7.14] J. Guckenheimer and P. Holmes, *Nonlinear Oscillations, Dynamical Systems, and Bifurcations of Vector Fields. Springer, New York 1983*,

[7.15] J. V. José, *Study of a quantum Fermi-acceleration model*, Phys. Rev. Lett. **56** (1986) 290

[7.16] G. Karner, *On the quantum Fermi accelerator and its relevance to "quantum chaos"*, Lett. Math. Phys. **17** (1989) 329

[7.17] C. Scheininger and M. Kleber, *Quantum to classical correspondence for the Fermi-acceleration model*, Physica D **50** (1991) 391

[7.18] J. V. José, *Quasi energy and eigenfunctions of time-dependent periodic Hamiltonians*, in: H. A. Cerdeira, R. Ramaswamy, M. C. Gutzwiller, and G. Casati, editors, *Quantum chaos*, World Scientific 1991

[7.19] T. M. Mello and N. M. Tufillaro, *Strange attractors of a bouncing ball*, Am. J. Phys. **55** (1987) 316

[7.20] F. C. Moon, *Chaotic Vibrations* (John Wiley, New York 1987)

[7.21] R. Keolian, L. A. Turkevich, S. J. Putterman, I. Rudnick, and J. A. Rudnick, *Subharmonic sequences in the Faraday experiment: Departures from period doubling*, Phys. Rev. Lett. **47** (1981) 1133 (reprinted in: B.-L. Hao, *Chaos*, World Scientific, Singapore) 1984.

8. The Duffing Oscillator

The Duffing oscillator is one of the prototype systems of nonlinear dynamics. It first became popular for studying anharmonic oscillations and, later, chaotic nonlinear dynamics in the wake of early studies by the engineer Georg Duffing [8.1]. The system has been successfully used to model a variety of physical processes such as stiffening springs, beam buckling, nonlinear electronic circuits, superconducting Josephson parametric amplifiers, and ionization waves in plasmas. Despite the simplicity of the Duffing oscillator, the dynamical behavior is extremely rich and research is still going on today.

8.1 The Duffing Equation

The Duffing oscillator is described by the differential equation

$$\ddot{x} + r\dot{x} + {\omega_0}^2 x + \beta x^3 = f \cos \omega t \,, \tag{8.1}$$

which differs from the elementary textbook example of a forced and damped harmonic oscillator only by the nonlinear term βx^3, which changes the dynamics of the system drastically. The *harmonic oscillator*

$$\ddot{x} + r\dot{x} + {\omega_0}^2 x = f \cos \omega t \tag{8.2}$$

offers, however, a convenient starting point for a subsequent discussion of the nonlinear case. For $\beta = 0$, a solution is given by

$$x(t) = A \cos(\omega t - \phi) \,, \tag{8.3}$$

which appears to be an ellipse

$$\left(\frac{x(t)}{A}\right)^2 + \left(\frac{v(t)}{\omega A}\right)^2 = 1 \tag{8.4}$$

in phase space $(x, v) = (x, \dot{x})$. The amplitude A and phase shift ϕ depend on the driving frequency ω, which shows typical resonance behavior when ω is varied:

$$A(\omega) = \frac{f}{\sqrt{({\omega_0}^2 - \omega^2)^2 + (r\omega)^2}} \tag{8.5}$$

$$\tan\phi(\omega) = \frac{r\omega}{{\omega_0}^2 - \omega^2}\,. \tag{8.6}$$

The resonance frequency is

$$\omega_{\text{res}} = \sqrt{{\omega_0}^2 - (r/2)^2}\,. \tag{8.7}$$

The general solution of the equation of motion is a superposition of the solution (8.3) and the general solution of the homogeneous equation, i.e. it can be written as

$$x(t) = A\sin\left(\omega t + \phi\right) + c_+ \mathrm{e}^{\lambda_+ t} + c_- \mathrm{e}^{\lambda_- t}\,, \tag{8.8}$$

where the constants $c_\pm$ are determined by the initial conditions $x(0)\,,\dot{x}(0)$, and

$$\lambda_\pm = -r/2 \pm \sqrt{(r/2)^2 - {\omega_0}^2}\,. \tag{8.9}$$

The solution of the inhomogeneous differential equation (8.3) is the unique limit cycle. For $r > 0$, the contribution of the homogeneous equation vanishes in the long time limit and each solution $x(t)$ approaches (8.3). It should be recalled that the two regions can be distinguished: in the underdamped case $r < 2\omega_0$, the approach to the limit cycle is slow and oscillatory, while in the overdamped case, $r > 2\omega_0$, it is slow and exponential. Both regions are separated by the aperiodic limiting case $r = 2\omega_0$, where the limit cycle is reached in the shortest time. For decreasing values of the force amplitude f, the amplitude A goes to zero and the limit cycle — the ellipse (8.4) — shrinks to a limit point. In a stroboscopic picture (i.e. using a flashlight at times 0, T, $2T$, $3T$, ..., where $T = 2\pi/\omega$ is the period of the excitation frequency) the limit cycle appears as a single point in phase space, which appears at

$$(x, v) = (A\,\cos\phi, -A\omega\,\sin\phi)\,. \tag{8.10}$$

It should be noted that such a stroboscopic view can be considered as a Poincaré section. This discussion exhausts the possible dynamical behavior of the pure harmonic case.

The appearance of the nonlinear term βx^3 in the Duffing equation (8.1) changes the situation completely:

- An analytic solution is no longer available.

- The superposition principle is no longer valid. A linear combination of solutions does not solve the equation of motion.

- The qualitative behavior is infinitely richer:

  There appear coexistent limit cycles of different periods and the long time limit of the motion is sensitively dependent on the initial conditions.

  One observes an entirely different type of long time behavior, where the trajectory approaches a strange attractor.

  Different strange attractors and limit cycles can coexist. Their basins of attraction can show fractal structures in the phase space plane of initial conditions.

When a parameter of the system is varied, interesting bifurcation phenomena can be observed and the regions of qualitatively different dynamics are delicately intertwined in parameter space.

As a first step towards an understanding of the Duffing oscillator, it is useful to look at the potential belonging to the time-independent part of the force ${\omega_0}^2 x + \beta x^3$:

$$V(x) = \tfrac{1}{2}{\omega_0}^2 x^2 + \tfrac{1}{4}\beta x^4 \,. \tag{8.11}$$

Let us first remark that (8.11) can be regarded as the first terms of a Taylor expansion of a general (symmetric) potential. This shows the importance of the Duffing oscillator for approximating more general interactions. Secondly, we should like to point out that the *parameter* ${\omega_0}^2$ can also be negative, in which case ω_0 cannot be considered as a frequency. Figure 8.1 shows $V(x)$ for positive and negative values of ${\omega_0}^2$. Close to $x = 0$ the potential is approximately harmonic, for large x the quartic term dominates, and for ${\omega_0}^2 < 0$ the system is bistable, showing two potential minima at

$$x_\pm = \pm\sqrt{-{\omega_0}^2/\beta} \tag{8.12}$$

with depth $-\omega_0^4/4\beta$.

The differential equation for the Duffing oscillator contains five parameters, r, ${\omega_0}^2$, β, f, and ω. The complete dynamical behavior of the system can, however, be explored by varying only three parameters, since two of them can be removed by a simple scaling transformation $x \longrightarrow ax$, $t \longrightarrow bt$. The choice of $a = f\omega^{-2}$, $b = \omega^{-1}$, for instance, leads to the equation

$$\ddot{x} + \tilde{r}\dot{x} + \tilde{\omega}_0^2 x + \tilde{\beta}x^3 = \cos t \tag{8.13}$$

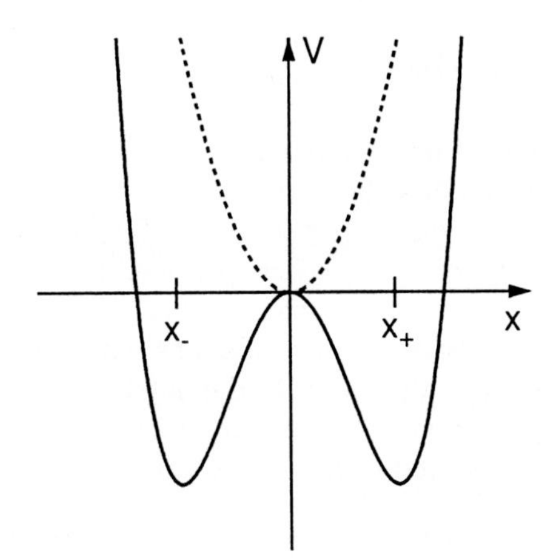

Fig. 8.1. Potential $V(x)$ for positive $(\cdots)$ and negative (—) values of ${\omega_0}^2$.

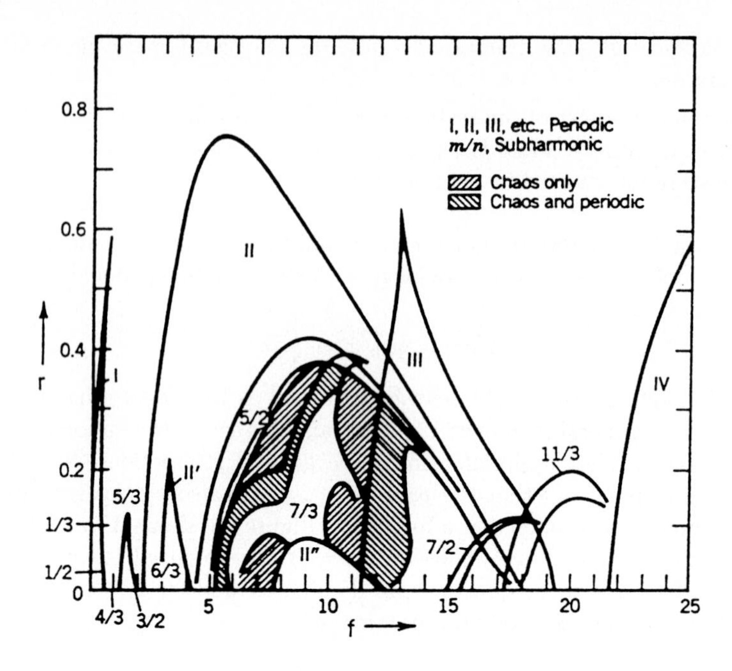

Fig. 8.2. Parameter space diagram for the behavior of the simplified Duffing equation $\ddot{x} + r\dot{x} + x^3 = f\cos t$ (adapted from Moon [4]). Qualitatively different dynamics is found in different regions.

with $\tilde{r} = r\omega^{-1}$, $\tilde{\omega}_0^2 = {\omega_0}^2\omega^{-2}$, $\tilde{\beta} = \beta f^2\omega^{-6}$, which contains only three parameters. For practical purposes it is convenient to use the unscaled equations.

A complete understanding of the dynamics of the system for all values of the parameters has not yet been achieved. The best studied case is the simplified Duffing equation with $\omega_0 = 0$ and $\beta = \omega = 1$ (the so-called *Ueda oscillator*)

$$\ddot{x} + r\dot{x} + x^3 = f\cos t\,, \tag{8.14}$$

which depends on two parameters, r and f, only. Figure 8.2 is based on a study by Ueda [8.2, 8.3] (adapted from Ref. [8.4]) and shows regions of qualitatively different behavior in the (r, f)-plane in the region $0 < r < 0.8$, $0 < f < 25$. In regions marked by I, II, II', II", III, and IV, we find periodic attractors with period one. In regions marked by m/n, subharmonic or ultrasubharmonic solutions of order m/n exist[3]. Fig. 8.2 is restricted to oscillations with $n < 3$. Higher order oscillations are observable, but not shown. In the hatched regions chaotic attractors appear, which may coexist with regular attractors. According to Ueda [8.3], the map (8.2) is not complete and many details have to be filled

[3] An ultrasubharmonic motion of order m/n is a periodic motion whose principal frequency is m/n times the frequency of the external force.

in, in particular in the region $5 < f < 15$. This offers many possibilities for numerical experiments.

8.2 Numerical Techniques

The program solves the second order differential equation (8.1) for the Duffing oscillator numerically, using discretization in equidistant time steps. The number, n, of these time steps per period $T = 2\pi/\omega$ of the driving force must be specified. It should be noted that numerical errors are to be expected when the motion is fast as compared to T, e.g. for large values of the harmonic frequency ω_0. The computation time is reduced by calculating and storing the values of the force $f \cos \omega t$ at the discrete times at the beginning of the computation.

After each cycle T, during the plotting of the phase space trajectory, the phase space point (x, v) is shown in a different color. If only a stroboscopic Poincaré plot is desired, only the sequence of these points is stored.

8.3 Interacting with the Program

The program DUFFING starts up with a screen showing the Duffing equation and the program's authors. On pressing a key, the user accesses the main menu. By entering the appropriate number, an item from among the following can be selected:

(0) Change parameters — The program displays the equation's parameters, the initial conditions and the number of iteration steps. A parameter can be changed by pressing the displayed key and then entering the new value. Inputting of a parameter can be canceled by pressing ⟨ESC⟩.

key	parameter	pre-set value
$\langle n \rangle$	total number of time steps per excitation period	200
$\langle r \rangle$	friction coefficient r	0.2
$\langle 0 \rangle$	harmonic coefficient ${\omega_0}^2$	-1
$\langle b \rangle$	anharmonic coefficient β	0.1
$\langle f \rangle$	amplitude f of external excitation	1
$\langle w \rangle$	frequency ω of external excitation	1
$\langle x \rangle$	initial value of x	0
$\langle v \rangle$	initial value of $v = \mathrm{d}x/\mathrm{d}t$	0
$\langle s \rangle$	current time (in time step units) modulo $n: 0 \leq s < n$	0

A value between 75 and 200 is recommended for n. For higher precision, the value has to be increased. Smaller values may result in abortion of the program due to calculation errors.

It should be noted that in order to reproduce a trajectory, one has to reset the value of the 'time' s. Otherwise, the trajectory is continued at the current value of s.

Any other key calls back the main menu.

(1) Display phase space orbit — The program integrates the Duffing equation and plots the results in phase space (v versus x). Points of the Poincaré map (i.e. at times, which are integer multiples of the excitation period) are shown in bright color. The permissible keys are displayed at the top of the screen. One may change the scaling of the x- and the y- (i.e. v-) axis using the keys ⟨x⟩, ⟨X⟩, ⟨y⟩ and ⟨Y⟩. By entering a small/capital letter, the scale is reduced/expanded.
The ⟨SPACE BAR⟩ clears the current screen and ⟨ESC⟩ exits the phase space display.

(2) Display Poincaré map — The program integrates the Duffing equation and plots the points of the Poincaré map. The active keys are the same as in (1). In addition, the ⟨+⟩ and ⟨−⟩ keys can be used to select the relative phase of the Poincaré section shown with respect to the driving force. Picture number m shows the stroboscopic map at times $t_m = nT + mT/M$, $(0 \leq m < M)$. The value of M depends on the graphics card installed (number of colors, existence of a second graphics page). It should be noted that all M Poincaré sections are stored simultaneously in one picture. The selected section is shown in white, and all others in black. Pressing ⟨SHIFT⟩–⟨A⟩ shows all sections on the active graphics page simultaneously. Stepping forward in time by means of the ⟨+⟩ key displays the attractor at different times. Pressing the key continuously simulates a motion in time, which is especially impressive in the case of strange attractors.

(3) Compute resonance curve — The excitation frequency ω is varied from an initial to a final value (and vice versa) sweeping over the interval in a certain number of steps. All parameters — except the excitation frequency ω — have to be specified before activating this item. The program prompts for the required parameters in turn:

- Number of calculation steps for the resonance curve (2,... ,80)
- The excitation frequency is varied. Enter initial and final value of the frequency
- Transient periods before data are recorded (0,..., 100 periods)
- The calculation will be stopped if the relative change of the amplitude in one time step is smaller than the value chosen for the ratio '(change in amplitude)/amplitude'
- Maximum number of oscillations per frequency step (1,..., 100)

The input of parameters can be interrupted by pressing ⟨ESC⟩, and the program then returns to the main menu. When all required parameters have been specified, the program computes the transient periods. The calculation of the

resonance curve is then performed in two runs. The program displays the current run and step number within the interval. The first run sweeps from the initial to the final frequency, and the second run in the reverse direction. When the calculation is finished, a diagram displays the amplitude as a function of ω. The green curve represents the variation from initial to final frequency, the red that in the reverse direction.
The values of the parameters β, r, f, ${\omega_0}^2$ and the frequency interval are shown on the left. Pressing any key exits the diagram and the results can be stored into a file. The suffix *.RES is added automatically.

(4) Compute Feigenbaum diagram — This procedure works in a similar way to (3), but here the excitation amplitude f is varied. All parameters — except the excitation amplitude f — have to be specified before activating this item. The prompts for entering all required parameters are:

- Number of calculation steps for the Feigenbaum diagram $(2, \ldots, 400)$
- Excitation amplitude f is varied. Enter range:
  - initial amplitude f
  - final amplitude f
- Transient periods before data are recorded $(0, \ldots, 100$ periods$)$
- Transient periods for each step (i.e. for each value of f) $(0, \ldots, 100$ periods$)$
- Calculation time for each step (i.e. for each value of f) $(1, \ldots, 64$ periods$)$

The input of parameters can be interrupted by pressing ⟨Esc⟩, and the program then returns to the main menu. When all parameters have been entered, the program computes the initial transient period. The Feigenbaum diagram is then calculated according to the adjusted parameters. During the calculation, the varying value of f is displayed. The diagram shows f on the vertical axis, while the horizontal axis represents the distance of the points of the Poincaré map from the origin of the phase space. Pressing a key exits the diagram and the data can be stored into a file. The suffix *.FGB is added automatically.

(5) Show resonance curve — A previously stored resonance curve can be loaded and displayed. In the graphics mode, pressing any key returns one to the main menu. Three examples are already given and can be loaded (files RES1.RES, ..., RES3.RES for parameters $r = 0.1$, $\omega_0^2 = \beta = 1$ and $f = 0.05$, 0.1, 0.2, respectively.

(6) Show attractors belonging to a resonance curve — A resonance curve file has to be chosen using the file select box. First, the resonance curve is displayed. After pressing a key, the corresponding attractors are shown for each ω of the resonance curve (a green/red attractor corresponds to the green/red branch of the resonance curve — see item (3)). The keys ⟨x⟩, ⟨X⟩, ⟨y⟩ and ⟨Y⟩

allow one to adjust the scaling of the x- and y- axis. It is possible to step over all calculated values of ω within the interval using the $\langle + \rangle$ and $\langle - \rangle$ keys. With $\langle 1 \rangle, \ldots, \langle 9 \rangle$ the step size can be changed. Pressing ⟨ESC⟩ returns to the main menu. Three examples are already given and can be loaded (files RES1.RES, ..., RES3.RES) for parameters $r = 0.1$, $\omega_0^2 = \beta = 1$ and $f = 0.05, 0.1, 0.2$, respectively.

(7) Show Feigenbaum diagram — A previously stored Feigenbaum diagram can be loaded and displayed. After a file has been chosen (see Appendix B.2), the stored values of r, ${\omega_0}^2$, β, ω, and the interval of f are displayed. Pressing any key shows the Feigenbaum diagram. Again, pressing any key returns to the main menu. The bifurcation diagram shown in Fig. 8.18 below is already stored (file FEIGB.FGB).

(8) Information about the program — The program gives information on the use of the program DUFFING. The Cursor keys $\langle \uparrow \rangle, \langle \downarrow \rangle$ allow one to go through the text line by line, ⟨PGUP⟩ and ⟨PGDN⟩ page by page. ⟨HOME⟩ brings us back to the beginning, and ⟨END⟩ to the end of the text. ⟨ESC⟩ returns to the main menu.

(Esc) Exit — Pressing the ⟨ESC⟩ key in the main menu quits the program DUFFING without further warning. All data are lost. Warning: entering unreasonable values for the parameters may result in abortion of the program.

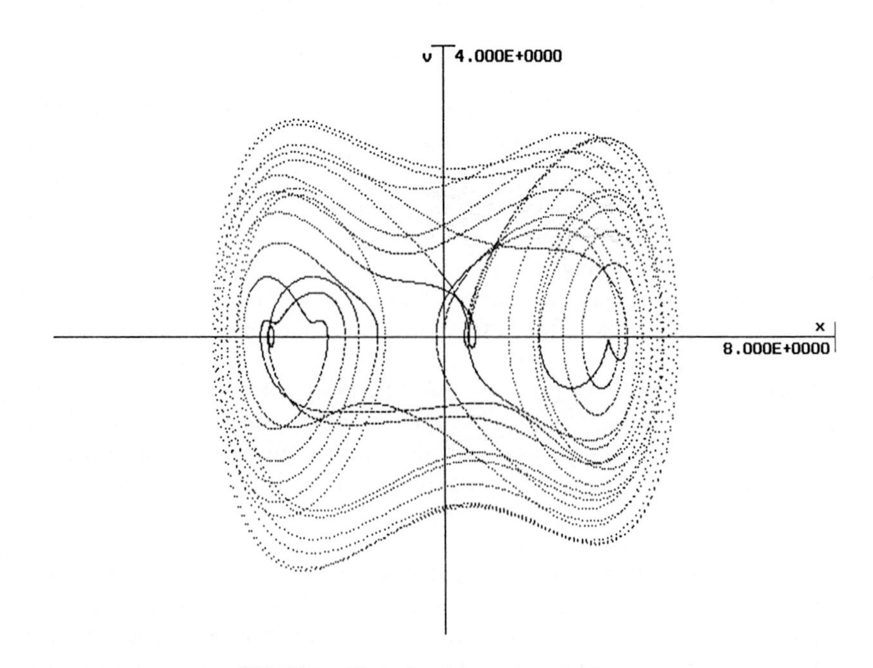

Fig. 8.3. Chaotic phase space trajectory.

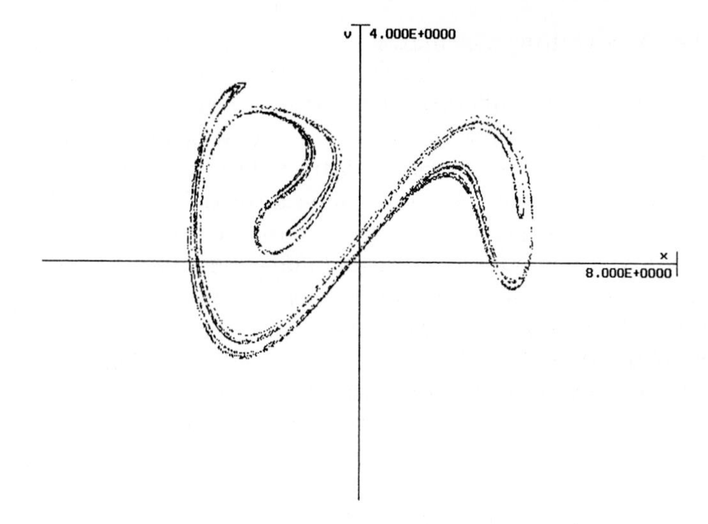

Fig. 8.4. Poincaré map of a strange attractor.

8.4 Computer Experiments

8.4.1 Chaotic and Regular Oscillations

As a first experiment, the program can be started using the pre-set parameters. The phase space orbit can be displayed by activating the item *'Display phase space orbit'*. The orbit shown on the screen will be highly irregular (see Fig. 8.3), closely resembling a moth fluttering around two candles. The motion continues in the same way thereafter (the SPACE BAR can be pressed to clear the screen).

Closer inspection shows that the trajectory approaches a special orbit in the long time limit. This limiting orbit is a strange attractor. The menu item (2) *'Display Poincaré map'* produces a stroboscopic picture of the orbit, showing only phase space points at times which are integer multiples of the period of the driving force. Figure 8.4 shows this strange attractor. The same attractor appears in the long time limit for different initial conditions (x, v).

As described in menu item (2) above, the $\langle + \rangle$ key can be used to show the Poincaré section at different times. Stepping slowly forward in time by pressing the $\langle + \rangle$ key shows how the attractor changes with time. Pressing the key continuously displays the dynamics of the strange attractor, which closely resembles a continuous mixing of a phase space 'fluid'. Further numerical experiments with strange attractors are discussed below.

8.4.2 The Free Duffing Oscillator

Let us first consider the Duffing oscillator without driving force ($f = 0$). Here, for ${\omega_0}^2 > 0$, we find a single limit point — a focal point — in the center of the phase space and all trajectories are attracted to this point. In the bistable case ${\omega_0}^2 < 0$, the dynamics is more complicated: there are two focal points at $(x, v) = (x_\pm, 0)$ (compare (8.12)) and an unstable fixed point at $(0, 0)$. This transition is a typical example of pitchfork bifurcation: when ${\omega_0}^2$ passes through zero, the limit cycle at $x = 0$ becomes unstable and two new stable limit cycles at $x_\pm$ are born. This is illustrated in Fig. 8.5.

In the conservative case with no damping, $r = 0$, the trajectories stay close to one of the two minima at $x_\pm$ for negative energies. The focal points are stable (elliptic) fixed points. For positive energies, the oscillation widens and the orbits encircle both minima. These structurally different regions are separated by the trajectory for zero energy, the *'separatrix'*, which passes through the hyperbolic fixed point at the center. This is illustrated in Fig. 8.6.

In the damped case, the separatrix of the hyperbolic fixed point is transformed into the stable and unstable manifolds of the fixed point. The stable manifold separates the phase space into two entangled regions. All trajectories starting inside one of these regions approach the same focal points, i.e. the stable manifold separates the basins of attraction of the two focal point. This is illustrated in Fig. 8.7 for the same parameters as those used in Fig. 8.6 and $r = 0.2$.

As a numerical exercise, one can try to compute the stable and unstable manifold. (Hint: the stable manifold can easily be found by starting a trajectory close to the unstable fixed point. The unstable manifold can be discovered, e.g., by varying the initial condition on the x-axis until the trajectory hits the fixed point $x = 0$, $v = 0$ with sufficient accuracy. For the case shown in Fig. 8.7, this gives $x = 1.5153$).

8.4.3 Anharmonic Vibrations: Resonances and Bistability

Here, we study the case of a Duffing oscillator where the nonlinear term is still weak and perturbative methods are often applicable. To be more precise, we

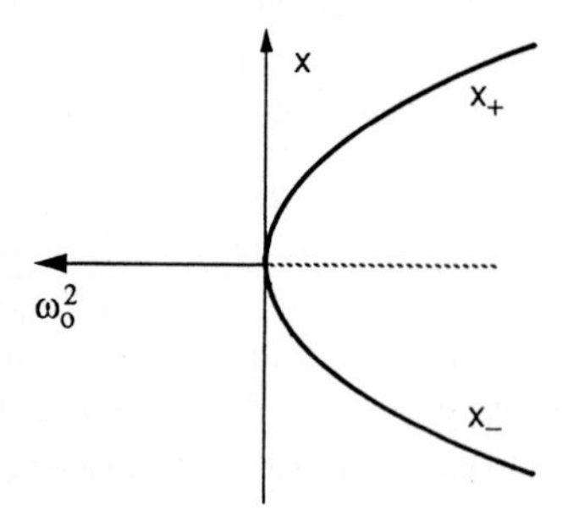

Fig. 8.5. Pitchfork bifurcation of limit cycles for the free Duffing oscillator as a function of ${\omega_0}^2$.

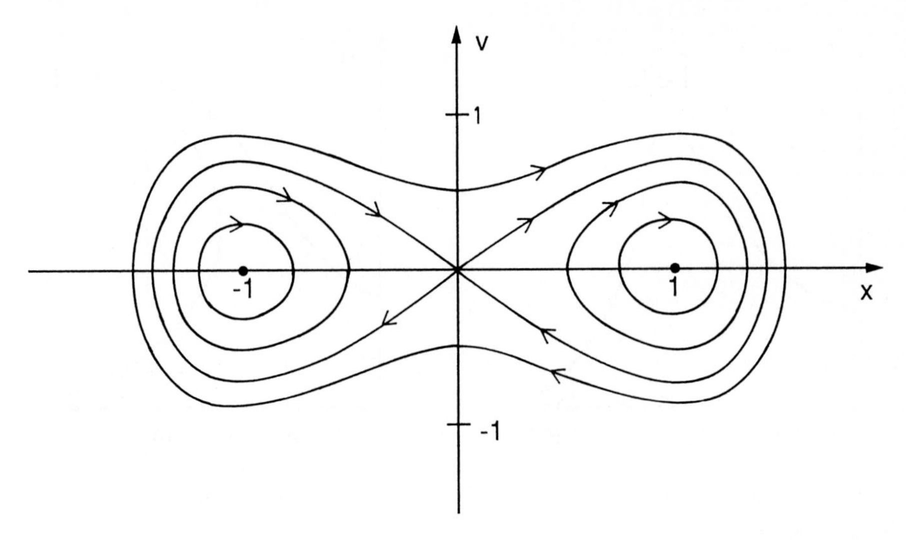

Fig. 8.6. Phase space trajectories for the free Duffing oscillator for the case of no damping ($\omega = \beta = -\omega_0{}^2 = 1$).

assume both β and the amplitude x to be small. It should be noted that this also limits the amplitude of the driving force. In such a situation, one speaks of *'anharmonic vibrations'* or *'nonlinear oscillations'*, which are analyzed in many engineering and applied science textbooks by a variety of highly developed techniques (see, e.g., the classical texts by Minorsky [8.5] or Nayfeh and Mook [8.6]). In the following, we use a simple perturbative approach [8.7], which

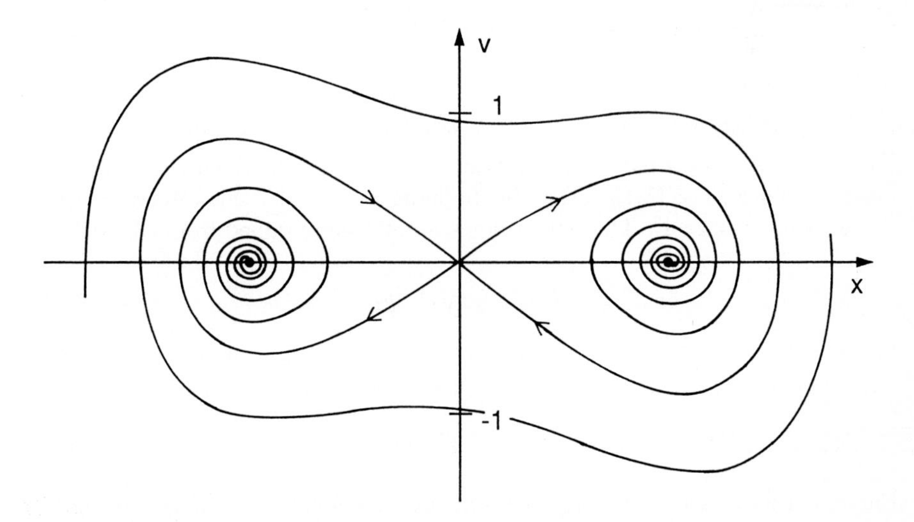

Fig. 8.7. Phase space trajectories for the free Duffing oscillator for the damped case.

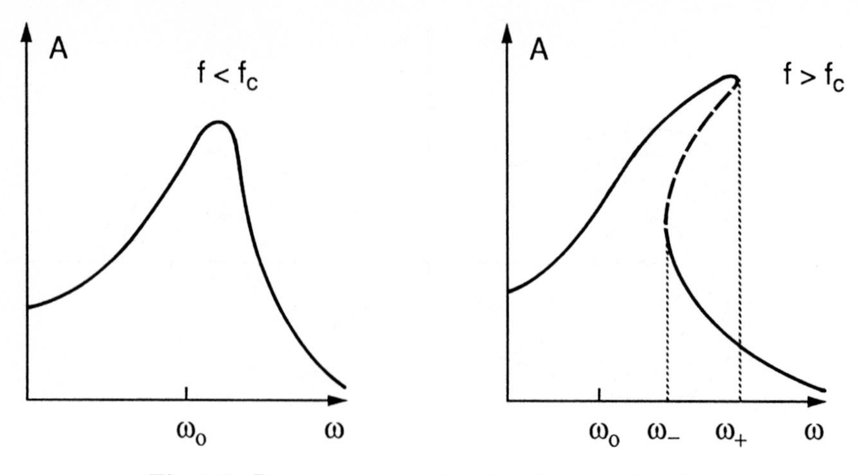

Fig. 8.8. Resonance curves for $f < f_{\text{crit}}$ and $f > f_{\text{crit}}$.

shows that in the case of small damping, the resonance curve (8.5) is modified in a characteristic way: the amplitude $A(\omega)$ appears as the solution of a cubic equation for A^2,

$$A^2\left[(\epsilon - \kappa A^2)^2 + \tfrac{1}{4}r^2\right] = \left(\frac{f}{2\omega_0}\right)^2 \tag{8.15}$$

with $\epsilon = \omega - \omega_0$ and $\kappa = 3\beta/8\omega_0$. The real solutions of (8.15) determine the amplitude of the limit cycle. There can be one or three real solutions. A branching of the solutions occurs at those points, where the derivative of A^2 with respect to ϵ (i.e. the frequency ω) diverges. Differentiation of (8.15) leads to the condition

$$\epsilon^2 - 4\kappa A^2\epsilon + 3\kappa^2 A^4 + \tfrac{1}{4}r^2 = 0 \tag{8.16}$$

with solutions

$$\epsilon_\pm = 2\kappa A^2 \pm \sqrt{\kappa^2 A^4 - \tfrac{1}{4}r^2}\,. \tag{8.17}$$

For $r < 2\kappa A^2$, there exists an interval $\epsilon_- < \epsilon < \epsilon_+$, in which three real and positive solutions of (8.15) exist. In the limit $r \to 2\kappa A^2$, this interval shrinks to the point $\epsilon = 2\kappa A^2$. The critical force amplitude obtained from (8.15) is

$$f_{\text{crit}} = \omega_0\sqrt{r^3/|\kappa|}\,. \tag{8.18}$$

For $f < f_{\text{crit}}$ there is only one solution while for $f > f_{\text{crit}}$ there are three. The frequency at which this bifurcation occurs follows from (8.17) as $\epsilon = r$ or

$$\omega_{\text{crit}} = \omega_0 + r\,. \tag{8.19}$$

Figure 8.8 shows schematically the resonance curves in both regions and Fig. 8.9 shows a chart of the behavior of the system in the (ω, f)-plane for parameter values $r = 0.1$, $\omega_0 = \beta = 1$, and therefore $f_{\text{crit}} = 0.052$, $\omega_{\text{crit}} = 1.1$.

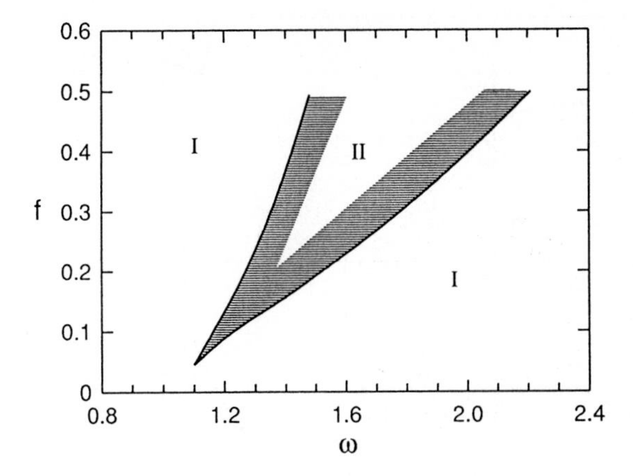

Fig. 8.9. Chart for the behavior of the weakly nonlinear Duffing oscillator. In the hatched region, three limit cycles, two stable and one unstable, exist (for parameters see text).

As an exercise, one can find the limit cycles in the different regions of Fig. 8.9 numerically. It turns out that only two limit cycles, and not three, can easily be found for parameter values in the hatched region of Fig. 8.9. The third (represented by a dashed curve in Fig. 8.8) is unstable. Fig. 8.10 shows, by way of example, two coexisting limit cycles for $r = 0.1$, $\omega_0 = \beta = 1$, $f = 0.2$, $\omega = 1.4$ as well as the unstable limit cycle (dashed curve).

Let us start a trajectory which is attracted by the 'large' limit cycle in Fig. 8.10. Increasing the value of ω in small steps, at a critical value ω_+, we observe a sudden jump in the trajectory, which approaches the 'small' limit cycle after some transient oscillations. When ω is decreased once more, the

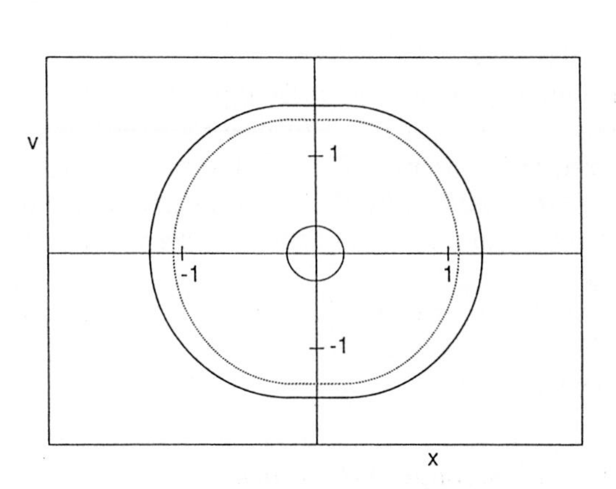

Fig. 8.10. Stable (—) and unstable (- - -) limit cycles.

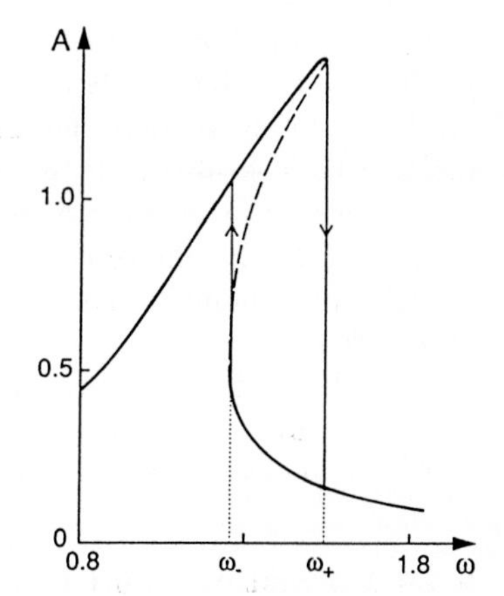

Fig. 8.11. Resonance curve. This figure is stored (file RES3.RES).

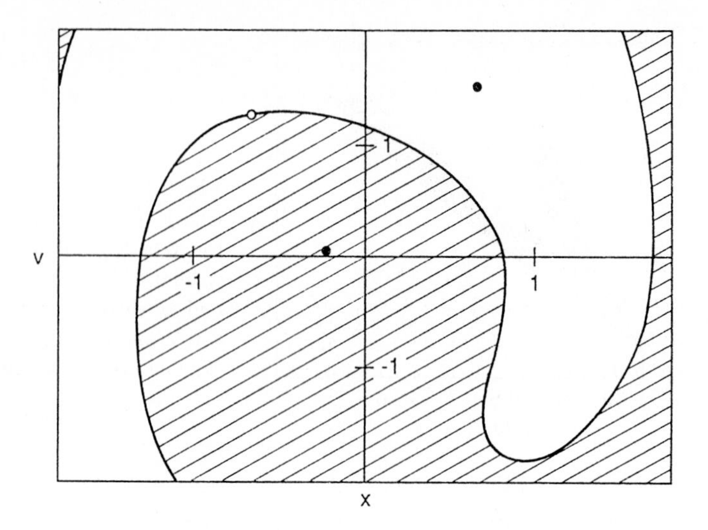

Fig. 8.12. Poincaré map: Basins of attraction of the two stable fixed points (•) and stable manifold of the unstable (∘) fixed point.

trajectory follows the 'small' limit cycle until another jump up to the 'large' cycle occurs at frequency $\omega_- < \omega_+$. In the region $\omega_- < \omega < \omega_+$, the system is bistable. Such jump and hysteresis phenomena are typical of anharmonic oscillations.

A resonance curve can be scanned automatically by the program (menu item *'Compute resonance curve'*). The amplitude $A(\omega)$ of the oscillation is stored for increasing and decreasing variation of the frequency. As an example, Fig. 8.11 shows the computational result for the same parameter values as in Fig. 8.10. Three such resonance curves for this case are already computed and can be displayed by activating menu item *'Show resonance curve'*. The files are RES1.RES, ..., RES3.RES for $f = 0.05$, 0.1, 0.2.

In the Poincaré map, the three limit cycles appear as fixed points, as shown in Fig. 8.12. Two are stable (•) and one is unstable (∘). Also shown are the basins of attraction of the two stable fixed points (hatched and unhatched regions), which are simply connected. They are separated by the stable manifold of the unstable fixed point (to construct this diagram compare the remarks at the end of Sect. 8.4.2).

Finally, it should be noted that the menu item *'Show attractors belonging to a resonance curve'* is very useful for studying the transformation of the attractors along a computed resonance curve.

8.4.4 Coexisting Limit Cycles and Strange Attractors

In the previous experiment, weakly nonlinear oscillations were studied. New phenomena occur for strongly nonlinear systems and the dynamics becomes much more complicated. We find many coexisting limit cycles and strange attractors whose basins of attraction show a highly complex organization with fractal basin boundaries. The following examples can only illustrate a few features of the rich dynamical behavior.

We confine ourselves to the restricted Duffing equation (8.14) (the Ueda oscillator)

$$\ddot{x} + r\dot{x} + x^3 = f \cos t \,, \tag{8.20}$$

which is discussed in the introduction to this chapter (parameter values: ${\omega_0}^2 = 0,\ \beta = \omega = 1$). Ueda [8.2, 8.3] used this equation to model an electronic circuit with a nonlinear inductor, and obtained many beautiful Poincaré maps of strange attractors. The chart in Fig. 8.2 gives an impression of the complex dynamics for different values of the parameters r and f. We give a few examples here:

- Let us first choose parameter values of $r = 0.08$ and $f = 0.2$. We can localize five coexisting stable limit cycles, which are shown in Fig. 8.13. They can be detected numerically by varying the initial conditions until the desired limiting oscillation is found. The five limit cycles shown in Fig. 8.13 can be obtained by starting trajectories at $v = 0$ and various initial coordinates: $x = -0.8$ (a); 1.0 (b); -1.0 (c); -0.9 (d); -0.7 (e) (it may be helpful to recall that the time step s must be reset to zero before starting a new trajectory). The points at times nT are marked (•). These marks show that

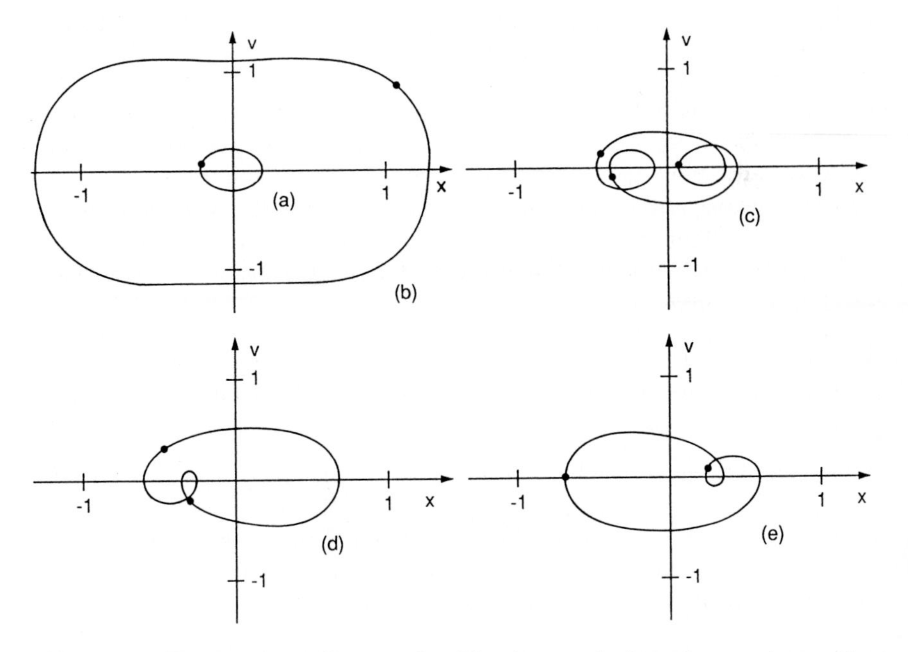

Fig. 8.13. Five coexisting limit cycles. The dots mark the points at times nT.

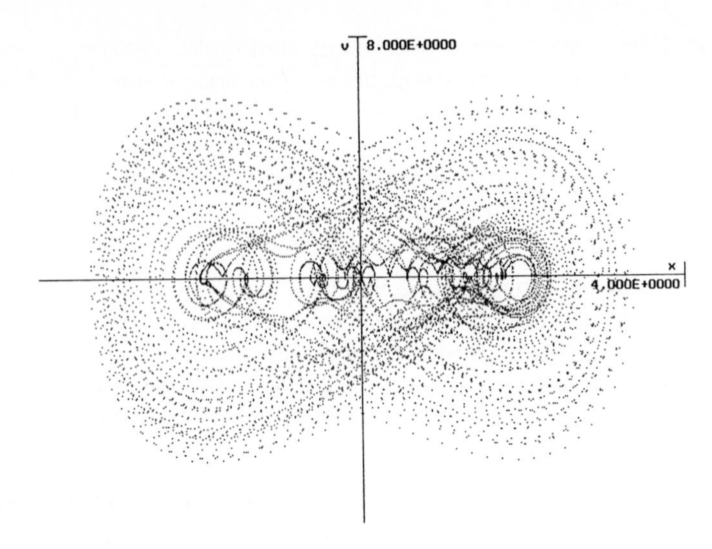

Fig. 8.14. Phase space plot of a trajectory approaching a strange attractor.

cycles (a) and (b) have period T, (d) and (e) period $2T$, and (c) period $3T$, i.e. the corresponding frequency ratios are 1, 1/2, and 1/3. It should be noted that the two unsymmetrical orbits with frequency 1/2 are mirror images of each other.

- The basins of attraction of the five limit cycles shown in Fig. 8.13 can be explored. They show a complex fine structure.

- Looking at the dynamical chart in Fig. 8.2, one observes that there exists a period-one limit cycle for $(r, f) = (0.2, 1.5)$, as well as for $(r, f) = (0.2, 17.5)$. Both parameter points belong to the same region; the orbits are, however, structurally different. One can follow the structural deformation of the limit cycle when the parameters (r, f) are varied from $(0.2, 1.5)$ to $(0.2, 18.5)$ without crossing any boundary. One observes a characteristic transition curve $\rightarrow$ cusp $\rightarrow$ loop.

- In a similar way, the behavior of the two stable limit cycles in region II of Fig. 8.2 where the parameters come close to the boundary of the region marked by '0', can be studied.

- Also of interest is the behavior of the two stable limit cycles in region III of Fig. 8.2, where the parameters come close to the boundary with the region marked by '0', e.g., along the straight line from $(r, f) = (0.2, 16.5)$ to $(0.2, 18.5)$. Comparison with the previous results shows both similarities and differences.

- Strange attractors exist in the hatched regions of Fig. 8.2. Choosing, for example, $(r, f) = (0.05, 7.50)$, the trajectory approaches a complicated orbit,

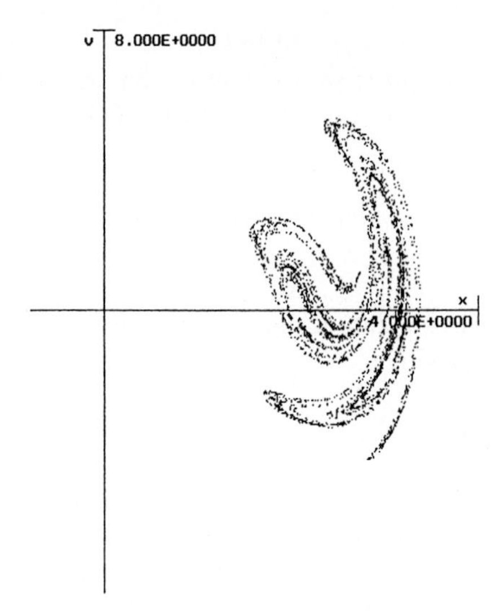

Fig. 8.15. Poincaré section of a strange attractor.

a strange attractor, which is shown in Figs. 8.14 and 8.15. The phase space trajectory closely resembles the example studied in Sect. 8.4.1, but in the present case there is *no* underlying potential with two minima, and the two centers which appear in the phase space plot are generated by the dynamics alone.

8.4.5 Suggestions for Additional Experiments

Harmonic Oscillator. The program DUFFING can also be used to study the simple harmonic oscillator numerically by setting $\beta = 0$. Such computer experiments are useful in a first introduction to numerical experiments using the Duffing oscillator. The concepts of phase space trajectories, focal points, stroboscopic Poincaré maps and resonance curves can be clarified, and the numerical results compared with analytic formulae (see the first section of the present chapter). Some simple numerical exercises are listed below ($\omega_0{}^2 > 0$ in all cases):

- Verify that the phase space trajectories for the undamped and undriven case ($r = f = 0$) are ellipses.

- Explore the behavior of the unforced damped evolution ($f = 0$). Distinguish between the different regions of small damping $r < 2\omega_0$, large damping $r > 2\omega_0$ and critical damping $r = 2\omega_0$ (compare (8.7)).

- Investigate the dynamics of the driven oscillator and, in particular, the limit cycle, the ellipse given in (8.4), and the convergence to the limit cycle for small, large, and critical damping.

- The amplitude A and phase shift ϕ (see (8.5) and (8.6)) can be extracted from the phase space trajectories (the phase shift can easily be seen as the rotation angle of the stroboscopically marked point). Study the behavior for small and large damping.

- Compute numerically the resonance curve and compare this with (8.5) (menu item *'Compute resonance curve'*).

Gravitational Pendulum. The simple pendulum in a constant gravitational field is described by the differential equation

$$\ddot{x} + \frac{g}{l} \sin x = 0 \,, \tag{8.21}$$

where g is the gravitational acceleration and l is the length of the pendulum. In most textbooks, the problem is greatly simplified by linearization of the force, which leads to the harmonic oscillator. For increasing values of the amplitude, one has to keep the next term in the expansion $\sin x = x - x^3/3! + x^5/5! \ldots$, yielding

$$\ddot{x} + \frac{g}{l}(x - x^3/6) = 0 \,, \tag{8.22}$$

i.e. a Duffing equation results with ${\omega_0}^2 = g/l$ and $\beta = -{\omega_0}^{-2}$. In addition, damping and driving terms can be included (see the classroom computer experiments in Ref. [8.8]).

Exact Harmonic Response. It can be shown that in the case of the undamped driven Duffing oscillator

$$\ddot{x} + {\omega_0}^2 x + \beta x^3 = f \cos \omega t \,, \tag{8.23}$$

an exact subharmonic response is possible for special parameter combinations (see Ref. [8.9]). There exist solutions

$$x(t) = A \cos(\Omega t + \Phi) \tag{8.24}$$

with

$$\Omega = \omega/3 \quad , \quad A = 2\,(f/2|\beta|)^{1/3} \tag{8.25}$$

and phase shift

$$\Phi = 0\,, \pm 2\pi/3 \text{ for } \beta > 0 \quad \text{or} \quad \Phi = \pm\pi/3\,\pi \text{ for } \beta < 0 \tag{8.26}$$

provided that the driving frequency is given by

$$\omega = 3\sqrt{{\omega_0}^2 + 3\beta\,(f/2|\beta|)^{2/3}} \tag{8.27}$$

($f > 0$ is assumed, and for negative values of ${\omega_0}^2$ the radicand must be positive). These subharmonic solutions can be stable or unstable, depending on the values of the parameters. In the simplified Ueda case (${\omega_0}^2 = 0$), we have stability for

$\beta > 0$. It is a point of interest that the amplitude of the 3-periodic solutions (8.24) goes to infinity in the limit of *vanishing* anharmonicity, $\beta \to 0$, i.e. the periodic orbits appear in a region where the anharmonic term βx^3 is large enough to support this anharmonic oscillation.

Numerically, the appearance of these analytic solutions can be studied. Of interest are:

- The stability properties dependent on the parameters ${\omega_0}^2$ and β (a short discussion can be found in Ref. [8.9]).
- The behavior of the solutions (8.24) when the stability of the oscillation changes.
- The sensitivity of the solutions with respect to violation of condition (8.27).
- The influence of the damping term in the Duffing equation (8.1) on the harmonic response solutions.

Period-Doubling Bifurcations. The Duffing oscillator with negative parameter ${\omega_0}^2$ shows an extremely rich behavior (compare the examples discussed in the book by Guckenheimer and Holmes [8.10] and references therein). Cascades of period-doubling bifurcations can be observed (compare the more detailed description of computer experiments for the chaos generator in Chap. 10, and the logistic map in Chap. 9). Examples are the frequency region $1.4 \geq \omega \geq 1.0$ (parameter values $r = 0.25$, ${\omega_0}^2 = -1$, $\beta = 1$, $f = 0.3$). For $\omega = 1.4$ we find a period-one, and for $\omega = 1.3$ a period-two oscillation. The bifurcation occurs at $\omega = 1.325$. Fig. 8.16 shows the coordinate $x(t)$ of the limit cycle for $\omega = 1.4$ (below the bifurcation point; period-one) and $\omega = 1.3$ (above the bifurcation point; period-two). The phase space plots of the same cycles are shown in Fig. 8.17. Further bifurcations can be localized for decreasing values of ω (second bifurcation at $\omega = 1.27$). The cascade of period doublings leads to a regime of chaotic motion.

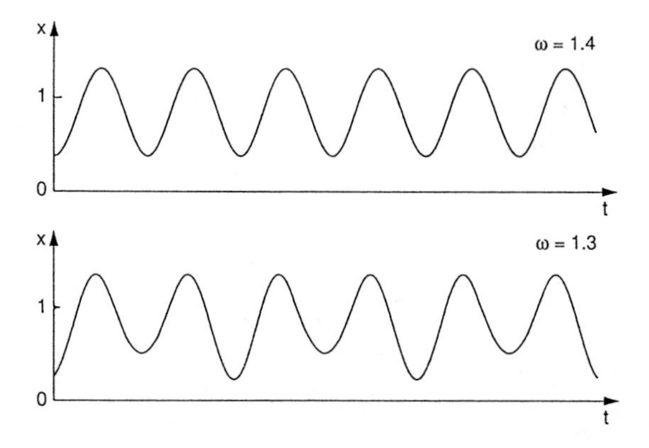

Fig. 8.16. Period-doubling bifurcation. Coordinate $x(t)$ as a function of time.

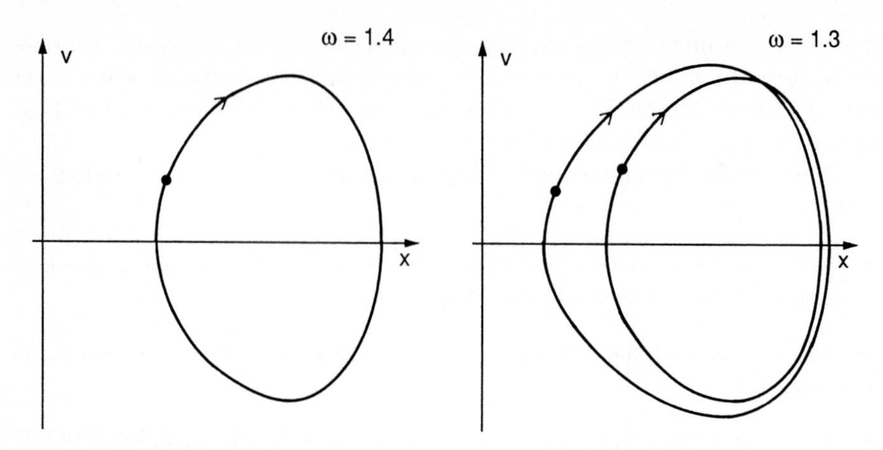

Fig. 8.17. Period-doubling bifurcation in phase space (same orbits as in the preceding figure).

Similar bifurcations can be found by varying the amplitude f of the force for fixed frequency. This variation can be done automatically by the program (compare the menu item *'Compute Feigenbaum diagram'*). Fig. 8.18 shows by way of example the bifurcation diagram (also known as the 'Feigenbaum tree'; for more details see Chap. 9) for $\omega = 1.4$ in the interval $0.34 \leq f \leq 0.40$ (other parameters are the same as above). This bifurcation diagram can be displayed on the screen by activating the menu item *'Show Feigenbaum diagram'*

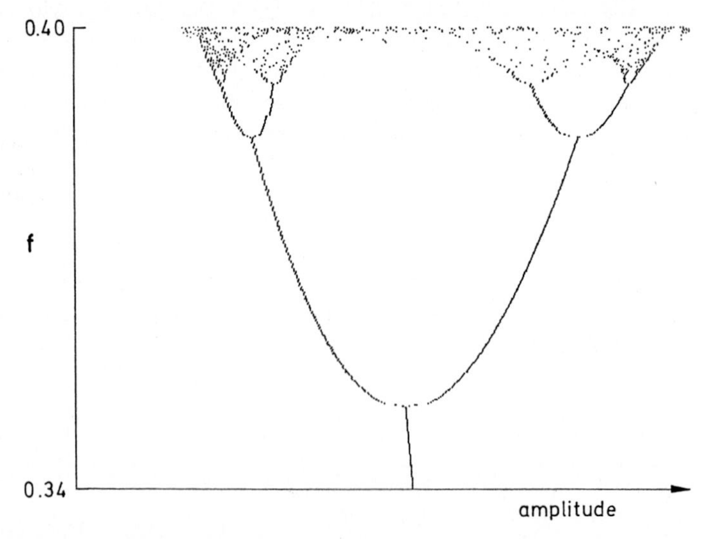

Fig. 8.18. Bifurcation diagram: Amplitude of the oscillation as function of the excitation amplitude f. This figure is stored (file FEIGB.FGB).

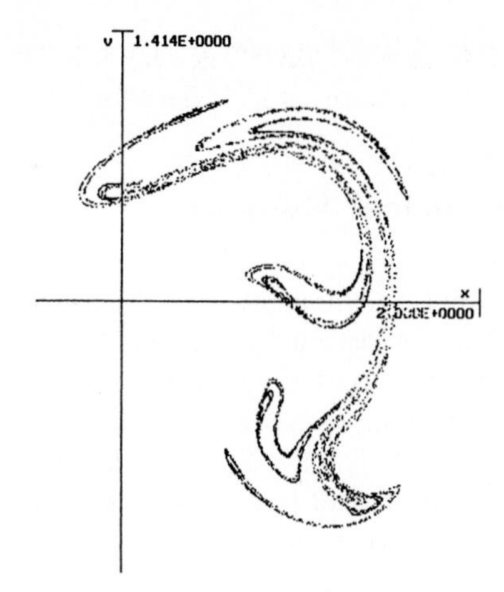

Fig. 8.19. Poincaré section of a strange attractor.

(file FEIGB.FGB). For additional numerical experiments with the double well potential Duffing oscillator see Moon [8.4, Appendix 5.6].

Strange Attractors. Seydel [8.11] has carried out an extensive investigation of the Duffing oscillator for parameter values $r = 1/25$, ${\omega_0}^2 = -1/5$, $\beta = 8/15$, $f = 2/5$, and variable driving frequency ω. Many strange attractors are observed and discussed. These studies can be reproduced and extended. As an example, Fig. 8.19 shows a strange attractor for $\omega = 0.48$, which can be found using a trajectory with the initial condition $(x, v) = (0, 1)$.

For the simplified Duffing oscillator (8.20), additional numerical experiments are suggested by Moon [8.4, Appendix 5.5]).

Periodic recurrences of fine structure in the bifurcation set ('superstructures') can also be observed (see [8.12] for a numerical study of a single well Duffing oscillator).

Numerous numerical examples of the behavior of the Duffing oscillator illustrating 'classical' nonlinear oscillations can be found in the highly recommended book by Hayashi [8.13]. This book contains many diagrams of the stability zones of higher harmonic and subharmonic oscillations, their basins of attraction etc.

8.5 Suggestions for Further Studies

The Quantum Mechanical Duffing Oscillator has recently been studied from several aspects as a typical example of a forced nonlinear oscillator without damping. Rozanov and Smirnov [8.14] discuss the quantum resonance excitation and demonstrate the possibility of hysteresis effects. Tunneling in a periodically driven bistable system is considered by Hänggi and coworkers [8.15, 8.16], and the intricate relation between classical and quantum behavior is investigated in Ref. [8.17, 8.18].

8.6 Real Experiments and Empirical Evidence

Experimental studies of the Duffing oscillator are frequent, since the Duffing oscillator models any one-dimensional oscillator driven by a linear harmonic force in second order expansion. Many different pieces of apparatus are known to model qualitatively and quantitatively a Duffing oscillator, the free and the Ueda oscillator. The examples are chosen from almost all physical field: mechanics, aerodynamics, electronics [8.4, p. 14], magnetism [8.4, p. 100], Josephson elements, plasma waves. Common to many variations is the necessity to build up a double well potential. Typical experimental studies are:

The driven pendulum in a gravitational field for not too large amplitudes, or a rotating dipole in a magnetic field (see, e.g., [8.4, 8.19] and references given there).

Ueda [8.2, 8.3] modeled an electronic circuit with a nonlinear inductor and obtained many beautiful Poincaré maps of strange attractors.

The case of bistable oscillations (${\omega_0}^2 < 0$) discussed above can be experimentally realized by the forced vibration of a cantilever beam in the field of two permanent magnets [8.4, 8.10, 8.20, 8.21].

A simple mechanical system that is very easy to construct has been suggested in Ref. [8.22]. It consists of a glider on an air track which is a attached to a rubber band. The system is described by the Duffing equation. Nonlinear resonance effects can be observed.

Briggs [8.23] describes three different cases in technical detail: (1) a driven inverted pendulum, which makes frequency doubling and chaotic behavior easily visible. (2) A nonlinear electronic oscillator showing typical chaotic waveforms. (3) The spinning magnet, in which a compass needle is free to rotate in a periodically reversing magnetic field perpendicular to the axis. If the frequency of the magnetic field is sufficiently high, the magnet, starting with arbitrary initial conditions, eventually becomes phase locked and rotates with the same frequency as the field. As the frequency is reduced, the time taken to reach phase locking increases until a critical value is reached at which this time becomes infinite and the motion chaotic. In this state, the magnet, typically, makes many revolutions in one direction before reversing unpredictably.

Meissner and Schmidt [8.24] describe a quantitative setup of a spinning magnet in which they vary the magnetic field amplitude and monitor the position by stroboscopic illumination; in addition, they vary the damping of the compass needle systematically. This experiment is excellently suited to teaching elements of chaos such as periodic orbits in phase space, bifurcation diagrams and strange attractors in a student laboratory.

Tufillaro [8.25] investigates nonlinear string vibrations; due to the tricky geometry the restoring force can be expanded in a series and approximated by a cubic restoring component. The numerical solution of this special case, which is nevertheless rich in its motions, allows the discussion of resonance curves and bifurcation diagrams.

Whineray et al. [8.26] realized the Tufillaro model using an air track oscillator in which the linear to cubic coefficient of the restoring force can be continuously adjusted. They measure resonance curves (velocity amplitude versus frequency) and phase diagrams (phase velocity versus frequency).

Khosropour and Millet [8.27] determine the resonance curve for the mechanical case of a string as a function of the driving frequency. A loudspeaker excites the oscillator and a video camera displays the oscillation so that the amplitude can be directly determined on the monitor screen.

Moon [8.4, p. 158] shows the chaos diagram, nonlinear damping versus forcing amplitude, for a nonlinear electronic circuit presenting the richness of periodic, nonperiodic, subharmonic and chaotic motion. He also describes [8.4, p. 197] the measurement of the Lyapunov exponent for chaotic motion of the two-well potential attractor.

References

[8.1] G. Duffing, *Erzwungene Schwingung bei veränderlicher Eigenfrequenz und ihre technische Bedeutung* (Vieweg, Braunschweig 1918)

[8.2] Y. Ueda, *Randomly transitional phenomena in the system governed by Duffing's equation*, J. Stat. Phys. **20** (1979) 181

[8.3] Y. Ueda, *Steady motions exhibited by Duffing's equation: A picture book of regular and chaotic motions*, in: H. Bai-Lin, D. H. Feng, and J.-M. Yuan, editors, *New Approaches to Nonlinear Problems*, SIAM, Philadelphia 1980

[8.4] F. C. Moon, *Chaotic Vibrations* (J. Wiley, New York 1987)

[8.5] N. Minorsky, *Nonlinear Oscillations* (Van Nostrand, New York 1962)

[8.6] A. H. Nayfeh and D. T. Mook, *Nonlinear Oscillations* (Wiley, New York 1979)

[8.7] L. D. Landau and E. M. Lifshitz, *Mechanics* (Pergamon, Oxford 1976)

[8.8] R. L. Kautz, *Chaos in a computer-animated pendulum*, Am. J. Phys. **61** (1993) 407

[8.9] K.-E. Thylwe, *Exact quenching phenomenon of undamped driven Helmholtz and Duffing oscillators*, J. Sound Vib. **161** (1993) 203

[8.10] J. Guckenheimer and P. Holmes, *Nonlinear Oscillations, Dynamical Systems, and Bifurcations of Vector Fields. Springer, New York 1983*,

[8.11] R. Seydel, *Attractors of a Duffing equation — dependence on the exciting frequency*, Physica D **17** (1985) 308

[8.12] U. Parlitz and W. Lauterborn, *Superstructure in the bifurcation set of the Duffing equation* $\ddot{x} + d\dot{x} + x + x^3 = f\ \cos(\omega t)$, Phys. Lett. A **107** (1985) 351

[8.13] C. Hayashi, *Nonlinear Oscillations in Physical Systems* (Princeton University Press, Princeton 1964)

[8.14] N. N. Rozanov and V. A. Smirnov, *Resonant excitation of an anharmonic quantum–mechanical oscillator*, Sov. Phys. JETP **59** (1984) 689

[8.15] F. Großmann, P. Jung, T. Dittrich, and P. Hänggi, *Coherent destruction of tunneling*, Phys. Rev. Lett. **67** (1991) 516

[8.16] F. Großmann, P. Jung, T. Dittrich, and P. Hänggi, *Tunneling in a periodically driven bistable system*, Z. Phys. B **84** (1991) 315

[8.17] N. Ben-Tal, N. Moiseyev, and H. J. Korsch, *Quantum vs. classical dynamics in a periodically driven anharmonic oscillator*, Phys. Rev. A **46** (1992) 1669

[8.18] N. Ben-Tal, N. Moiseyev, S. Fishman, F. Bensch, and H. J. Korsch, *Weak quantum limitation of chaos in a periodically driven anharmonic oscillator*, Phys. Rev. E **47** (1993) 1646

[8.19] D. D'Humiers, M. R. Beasley, B. A. Huberman, and A. Libchaber, *Chaotic states and routes to chaos in the forced pendulum*, Phys. Rev. A **26** (1982) 3483

[8.20] F. C. Moon and P. J. Holmes, *A magnetoelastic strange attractor*, J. Sound Vib. **65** (1979) 285

[8.21] F. C. Moon and P. J. Holmes, *Addendum: a magnetoelastic strange attractor*, J. Sound Vib. **69** (1980) 339

[8.22] T. W. Arnold and W. Case, *Nonlinear effects in a simple mechanical system*, Am. J. Phys. **50** (1981) 220

[8.23] K. Briggs, *Simple experiments in chaotic dynamics*, Am. J. Phys. **55** (1987) 1083

[8.24] H. Meissner and G. Schmidt, *A simple experiment for studying the transition from order to chaos*, Am. J. Phys. **54** (1986) 800

[8.25] N. M. Tufillaro, *Nonlinear and chaotic string vibrations*, Am. J. Phys. **57** (1989) 408

[8.26] S. Whineray, C. Rofe, and A. Ardekani, *The resonant response of a cube-law track oscillator*, Eur. J. Phys. **13** (1992) 201

[8.27] R. Khosropour and P. Millet, *Demonstrating the bent tuning curve*, Am. J. Phys. **60** (1992) 429

9. Feigenbaum Scenario

In this chapter, we will study a very important class of dynamical systems, which is almost ideally suited as an introduction to the basic characteristics of nonlinear dynamics, namely discrete mappings. An example of such a discrete system is provided by a stroboscopic map, where a system is only observed at well–defined time steps t_i. In some situations, the discrete mapping arises directly from the nature of the system, as for instance in population dynamics or kicked or billiard-type systems. In other cases, one uses a discrete approximation to the true continuous evolution.

The dynamics of discrete systems is generated by iterating the mapping equation, which is much easier to study than the continuous time evolution followed by integrating differential equations. Here, we will study the simplest case, namely one-dimensional discrete maps of the unit interval onto itself — more precisely, one-parameter families of one-dimensional maps —, which are by far the most simple dynamical systems which can be imagined.

The dynamics is described by the iteration

$$x_{n+1} = f(x_n, r)\,, \tag{9.1}$$

starting from an initial point x_0. Here, $f(x, r)$ is a user-defined function dependent on a parameter r. In most of the interesting cases, this map is non-invertible and, in many cases, also continuous. For one-dimensional maps, one-parameter families show generic behavior and no new phenomena appear when more parameters are introduced. This is, however, not true for higher dimensional maps, such as, for instance, the two-dimensional case studied in Chap. 11.

Such iterated maps appear quite naturally in nonlinear dynamics. However, despite their simplicity, the iterated maps possess an internal complexity going far beyond the innocent functional dependence given by $f(x, r)$. Numerous scientific articles are devoted to these iterated maps, and many unexpected results have been obtained during the last two decades. One of the most spectacular discoveries is the *universality* of certain important features of these mappings. The characteristic route from regular to chaotic dynamics which appears in these maps — the so-called *'Feigenbaum scenario'* — is discussed in some detail below. In addition, the interested reader is advised to consult the book by Collet and Eckmann [9.1] and references therein, as well as the beautiful and influential early review by May [9.2] and the introductory article by Feigenbaum [9.3]. An extensive discussion can also be found in the textbook by

Schuster [9.4]. For a more mathematical presentation, see the book by Devaney [9.5], which contains many rigorous results on iterated maps.

9.1 One-Dimensional Maps

Iterated maps were first introduced in population dynamics to model biological systems, such as the increase in the number of species from one year to the next. Later on, they appeared in all branches of science. As an example, we will first discuss a simple physical model whose dynamics leads directly to such an iterated map, namely a periodically kicked particle of mass m. The equations of motion read

$$\begin{aligned} \dot{x} &= \frac{p}{m} \\ \dot{p} &= -\beta p + F(x) \sum_{n=-\infty}^{+\infty} \delta\left(\frac{t}{T} - n\right), \end{aligned} \tag{9.2}$$

where β is a frictional coefficient and $F(x)$ the amplitude of the external forcing function. Integrating over a period T from $t = nT - \epsilon$ to $t = (n+1)T - \epsilon$, we obtain in the limit $\epsilon \to 0$

$$\begin{aligned} p_{n+1} &= \mathrm{e}^{-\beta T}\left[p_n + T\,F(x_n)\right] \\ x_{n+1} &= x_n \frac{1}{\beta m}(1 - \mathrm{e}^{-\beta T})\left[p_n + T\,F(x_n)\right], \end{aligned} \tag{9.3}$$

where x_n and p_n represent the coordinate and momentum immediately before the nth kick. The mapping (9.3) is dissipative:

$$\left|\frac{\partial(p_{n+1}, x_{n+1})}{\partial(p_n, x_n)}\right| = \left|\begin{array}{cc} \mathrm{e}^{-\beta T} & \mathrm{e}^{-\beta T} T F'_n \\ \frac{1}{\beta m}(1 - \mathrm{e}^{-\beta T}) & 1 + \frac{1}{\beta m}(1 - \mathrm{e}^{-\beta T}) T F'_n \end{array}\right| = \mathrm{e}^{-\beta T} \tag{9.4}$$

with $F'_n = \partial F/\partial x|_{x_n}$, i.e. the phase space volume is contracted by a constant factor

$$\Delta x_{n+1} \Delta p_{n+1} = e^{-\beta T} \Delta x_n \Delta p_n \,. \tag{9.5}$$

For $\beta = 0$, the mapping is area-preserving, as one would expect. It should be noted that the contraction is much faster in the p-direction. Taking the limit of large damping $\beta \to \infty$ and small mass $m \to 0$ with $\beta m/T = \text{constant} = c$, we obtain $p_n \to 0$ and

$$x_{n+1} = x_n + c\,F(x_n)\,, \tag{9.6}$$

which shows that the dynamics reduces to a one-dimensional map. Interesting dynamical features appear when the mapping is non-invertible, which is most simply found when the forcing function $F(x)$ has an extremum. In particular, for the obvious parabolic choice

$$c\,F(x) = 4rx(1-x) - x\,, \tag{9.7}$$

where the parameter r may be varied, we find the well-known logistic map

$$x_{n+1} = 4rx_n\,(1-x_n) \tag{9.8}$$

discussed below. It is instructive to note that in this case the force $F(x) = -\partial V/\partial x$ is zero for $x_0^* = 0$ and for $x_1^* = 1 - 1/(4r)$. For $r < 0.25$, the point x_0^* is stable (the potential $V(x)$ has a minimum) and x_1^* unstable (maximum of $V(x)$). For $r > 0.25$, the stability properties are exchanged. These minima will reappear as first order fixed points of the discrete mapping (9.8), as discussed in the numerical experiment 9.3.1 below.

Of course, the logistic map (9.8) can also appear in very different situations. An amusing example is the fish-pond: let x_n be the number of fish in our fish-pond in year n, or, more precisely, we measure the ratio of fish to water, so as to estimate the fish population in the following year $n+1$ in a simple way: first, it is evident that there will be no fish next year if there are no fish this year ($x_{n+1} = 0$ for $x_n = 0$). Secondly, there will be no fish next year if there is no water ($x_{n+1} = 0$ for $x_n = 1$). Therefore, being non-negative, $x_{n+1} = f(x_n)$ must have at least one maximum. The simplest choice of such a function is a parabola and we recover the mapping (9.8), where r characterizes all the other parameters of our fish-pond (mortality, food-supply, ...). The naive expectation is that, after some time, the fish population will stabilize. This is, however, not the case, as demonstrated in the numerical experiments below. The dynamics generated by the innocent looking one-dimensional maps is very complicated and has attracted a considerable amount of interest.

In the program described below, mappings of the unit interval $x \in [0,1]$ depending on a parameter r taking the variable x modulo 1 are studied. In some cases discussed in the literature, maps appear in a slightly different form, for instance as

$$y_{n+1} = y_n^2 + c\,, \tag{9.9}$$

which can be transformed by

$$y = 2r(1-2x) \quad \text{with parameter} \quad c = 2r(1-2r) \tag{9.10}$$

to the logistic map in (9.8). It should be noted that (9.9) is the real version of the complex Mandelbrot map discussed in Chap. 11. In addition, several of the two-dimensional maps described in Chap. 11 reduce to the logistic map in some limit, so that an understanding of the dynamics of one-dimensional maps is a prerequisite for a study of the more complicated two-dimensional cases.

9.2 Interacting with the Program

After starting the program FEIGBAUM, the title appears. Pressing a key changes the screen, showing a menu bar in the first line of the screen with the general items for pull down menus:

— **File** — **Parameters** — **Diagram** — **Range** —

The action of the program can be controlled by choosing the general item and the corresponding pull-down menu. The entries in these menus are described below.

- **File**

- *Load:* Enquires the name under which a picture has been stored and reads the data. The screen is replaced by the loaded picture. The results of any previous computation are lost. As a pre-computed example, the files LOGIST.NLD and LOGIST.DAT contain the combined bifurcation- and Lyapunov diagrams shown in Figs. 9.9 and 9.15 below.

- *Save:* This menu item asks for the name under which the current data and picture are to be stored. Suffixes are ignored. After input of a valid name, a file with suffix *.NLD (picture) and a file *.DAT (parameters) are created.

- *Hardcopy:* Upon entering the hardcopy option, one is asked for the printer type (Matrix, Laserjet, Postscript). A hardcopy is then printed. The end of the printing process is indicated by a beep. Selecting a printer can be aborted by ⟨ESC⟩.

- *Quit:* A box offers the choice between leaving or continuing the program. Pressing ⟨ESC⟩ while in the menu offers this box too.

- **Parameters**

- *Function:* The function $f(x, r)$ for the iteration

$$x_{n+1} = f(x_n, r) \tag{9.11}$$

can be specified, where r is a parameter. The quadratic function $f(x, r) = 4rx(1 - x)$ is pre-defined, i.e. the logistic map (9.8).

To accept the function press ⟨ENTER⟩, ⟨ESC⟩ cancels the input. The program checks the expression. If a problem is detected, a message appears and, after pressing a key, the program returns to the input box. The formula can then be corrected.

It should be noted that the program will automatically restrict the resulting values to the unit interval by taking x modulo 1.

- *Parameter* r*:* Input of the value of r.

- *Iteration seed:* The pre-set initial value $x_0 = 0.01$ can be altered.

- *Number of iterations:* The number of iterations can be specified. 100 iterations are pre-set.

- *Degree of iterate:* The kth iterate of the function $f(x,r)$ is plotted. One is asked for the degree k of the iterate. Pre-set is the previous degree increased by one.

- *Delay:* Delay time for slowing down the display in the *'Iteration'* mode. A delay of 50 ms is pre-set.

- *Epsilon:* Input of the accuracy parameter ϵ used in a determination of periods in *'Bifurcation'*, *'Lyapunov exponent'*, or *'Period doubling'*. Pre-set is $\epsilon = 10^{-6}$.

- *Stepsize:* The stepsize of the r-variation used in an automatic detection of the bifurcation points in item *'Period doubling'*. Pre-set is a stepsize of 10^{-4}.

- **Diagram**

- *Function:* The function $f(x,r)$ is displayed (yellow), as well as the bisector $y = x$ (white).

- *Iterates:* The kth iterate of the given function is plotted (blue). The degree k can be modified in the *'Parameters'* menu. Pre-set is the previous degree increased by one.

- *Iteration:* Starts the iteration at an initial condition (*'iteration seed'*) x_0, which can be specified in the *'Parameters'* menu ($x_0 = 0.01$ is pre-set). The current iteration step is displayed in red. The first sequence of iterations can be erased by pressing ⟨DEL⟩.

- *Bifurcation:* Computes a bifurcation diagram. Starting at a parameter $r = r_{min}$ and initial point x_0, the x_n are calculated for $n = 0, \ldots, N$. Searching backwards for the first x_n in an ϵ-neighborhood of x_N:

$$|x_N - x_n| < \epsilon \,. \tag{9.12}$$

When no such x_n is found, the one closest to x_N is used. The $x_{n+1}, \ldots, x_N$ are then considered as the approximate periodic orbit and plotted on the screen. Then, the parameter r is increased and the iteration is started again with $x_0 = |x_N - \epsilon|$.

The values of the number N of iterations and the accuracy parameter ϵ can be specified in the *'Parameters'* menu. The boundaries $r_{\rm min}$ and $r_{\rm max}$, as well as the range shown for the variable x, can be modified in the *'Range'* menu.

At the end of the iteration at $r_{\max}$, or when the iteration is stopped by ⟨ESC⟩ or ⟨ENTER⟩, a new range can be defined by means of the menu item *'Zoom in'*.

It is possible to read numerical values of interest from the screen by activating the menu item *'Determine values'*.

- *Lyapunov exponent:* Computes a diagram of the Lyapunov exponent (denoted as '*Ly*' in the program) as a function of the parameter r. Starting at $r = r_{min}$ with initial point x_0 the x_n are calculated for $n = 0, \ldots, N$. The Lyapunov exponent is calculated from (9.24).

  The values of the number N of iterations and ϵ can be specified in the *'Parameters'* menu. The boundaries $r_{\min}$ and $r_{\max}$, as well as the range shown for *Ly*, can be modified in the *'Range'* menu. If $Ly(\min) = Ly(\max)$, the program automatically scales (and rescales) the *Ly*-axis.

  At the end of the iteration at $r_{\max}$, or when the iteration is stopped by ⟨ESC⟩ or ⟨ENTER⟩, a new range can be defined by means of the menu item *'Zoom in'*.

  It is possible to read numerical values of interest from the screen by activating the menu item *'Determine values'*.

- *Bifurc./Lyapunov:* Simultaneous computation of a bifurcation diagram and the Lyapunov exponent as a function of r.

- *Period doubling:* Following essentially the same strategy as in the computation of a bifurcation diagram, the program computes the numerical values of r at the bifurcation points. Parameters must be assigned as described for the *'Bifurcation diagram'* above. The stepsize for r is only relevant in this computation.

- *Determine values:* Crosshairs appear, which can be moved by the cursor keys to a point of interest and allow a determination of the numerical value of the parameter r by pressing ⟨ENTER⟩.

- **Range**
  The range for a variation of the parameter r, as well as the displayed interval of the variable x in a bifurcation diagram, or *Ly* in a computation of the Lyapunov exponent, can be changed.

- *Zoom in:* A new range can be defined by means of a zoom box, which can be moved by means of the cursor keys. ⟨TAB⟩ switches between coarse and fine adjustment, pressing the ⟨SPACE BAR⟩ toggles the active boundary.

- *r(min) — r(max) — reset r:* Pre-set values are $r_{\min} = 0$, $r_{\max} = 1$. *'Reset r'* resets a modified r-range to the pre-set one.

- *x(min) — x(max) — reset x:* Pre-set values are $x_{\min} = 0$, $x_{\max} = 1$. *'Reset x'* resets a modified x-range to the pre-set one.

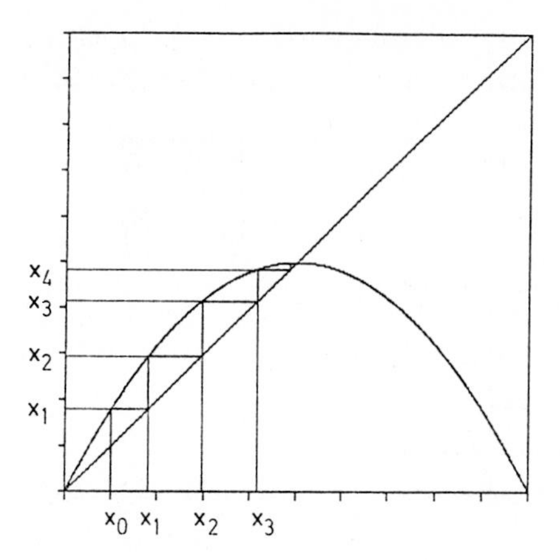

Fig. 9.1. Iteration of the logistic map for $r = 0.5$ and initial point $x_0 = 0.1$.

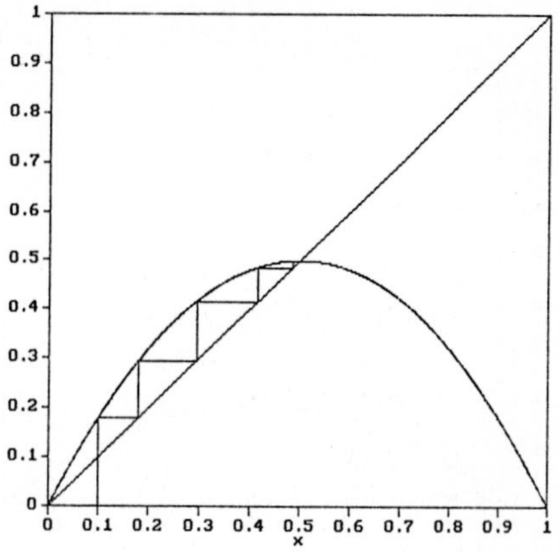

Fig. 9.2. Simplification of the mapping shown in Fig. 9.1.

- *Ly(min) — Ly(max) — reset Ly:* Automatic scaling is pre-set. *'Reset x'* resets a modified *Ly*-range to the pre-set one.
- *Reset all:* Resets all range parameters to the pre-set values.

9.3 Computer Experiments

9.3.1 Period-Doubling Bifurcations

The dynamics of the iterated points will be strongly influenced by the fixed points of the mapping equation $x_{n+1} = f(x_n, r)$, as well as the kth order fixed points, which return to the initial point after k iterations. In such a case the iteration generates a k-periodic orbit $x_0, x_1, x_2, \ldots, x_k = x_0$. It is evident that all the k points of this cycle are also kth order fixed points. Furthermore these points are fixed points of the k-times iterated map

$$f^k(x, r) = \underbrace{f(f(\ldots f}_{k\,-\text{times}}(x, r) \ldots)) \,. \tag{9.13}$$

The stability properties of these fixed points, as well as their bifurcation with varying parameter r, are discussed in Sect. 2.4.4.

We will start our numerical experiments with the so-called *logistic map* (9.8)

$$x_{n+1} = 4\,r\,x_n\,(1 - x_n)\,,$$

which is pre-set in the program. The fascinating dynamical properties of this simple mapping of the unit interval have been analyzed by many authors (for some early examples see [9.2, 9.6, 9.7]). The logistic mapping function $f(x, r) = 4\,r\,x(1 - x)$ is a quadratic polynomial and, hence, the simplest differentiable

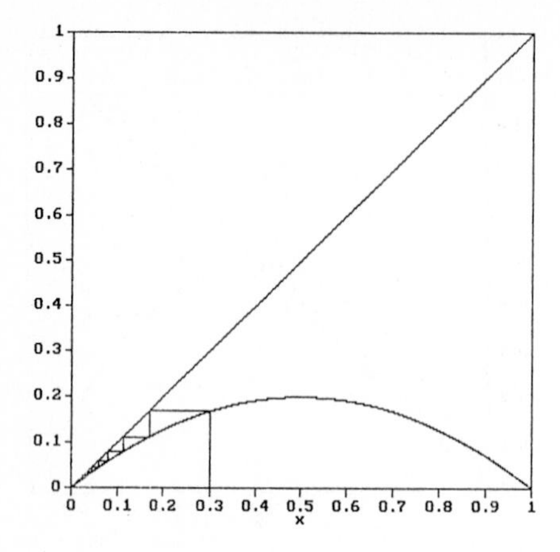

Fig. 9.3. Iteration of the logistic map for $r = 0.2$ and initial point $x_0 = 0.3$.

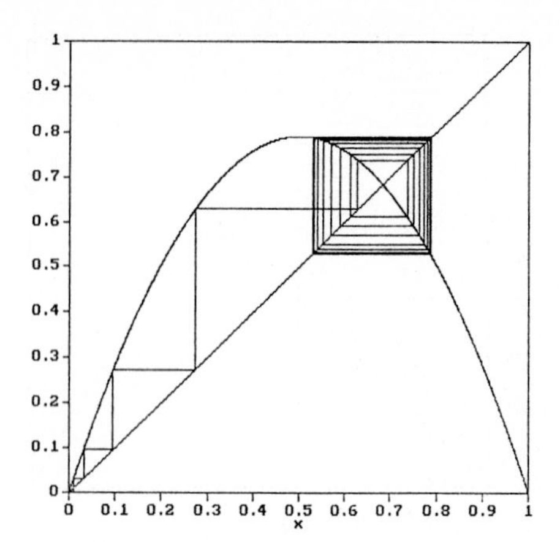

Fig. 9.4. Iteration of the logistic map for $r = 0.79$ and initial point $x_0 = 0.01$.

nonlinear function. Let us first change the pre-set r-parameter to the smaller value of $r = 0.5$. We choose an initial value ('iteration seed') $x_0 = 0.1$ and activate the program item *'Iteration'*. The sequence of iterated points x_n, $n = 0, 1, \ldots$ is 0.100, 0.180, 0.295, 0.416, 0.486, ... converging eventually to $x_1^* = 0.5$, which is a fixed point of the map satisfying

$$x^* = f(x^*, r) \,. \tag{9.14}$$

Graphically, the mapping can be displayed by reading the value of $x_1 = f(x_0)$ from the graph of the function $f(x)$, which is then copied to the horizontal axis by reflection on the bisector of the first quadrant. Choosing this value of x_1 as a new input, the process is repeated, as illustrated in Fig. 9.1. The graph can be simplified by erasing unnecessary detail, as shown in Fig. 9.2, and the mapping is graphically an alternating vertical line to the function $x' = f(x)$ and a horizontal line to the bisector $x' = x$. Such a graphical plot is displayed by the program. By changing the initial seed x_0, one can check that qualitatively the same behavior is found for all initial seeds $0 < x_0 < 1$. The point $x = 0$ is also a fixed point, albeit unstable. Any point in its neighborhood will move away, converging eventually to the fixed point x_1^*.

When r is decreased, this picture remains qualitatively the same with a modified position of the stable fixed point

$$x_1^* = 1 - \frac{1}{4r} \,, \tag{9.15}$$

which appears graphically as an intersection point of $y = f(x)$ with the bisector $y = x$. At a critical value of $r = 0.25$, this intersection disappears from the interval. A numerical test for, e.g., $r = 0.2$ shows that the iterated points converge to the fixed point $x_0^* = 0$, which now appears to be stable (see Fig. 9.3). The analysis of the stability properties is simple: the slope of the mapping

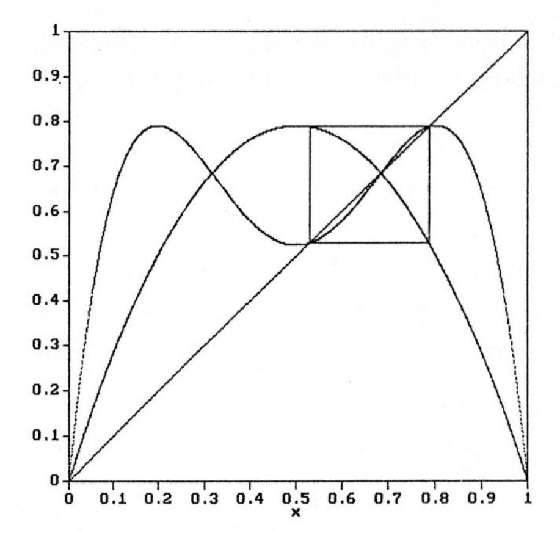

Fig. 9.5. Stable period-two orbit of the logistic map for $r = 0.79$. Also shown is the iterated map $f^2(x)$.

Fig. 9.6. Stable period-four orbit of the logistic map for $r = 0.88$. Also shown is the iterated map $f^4(x)$.

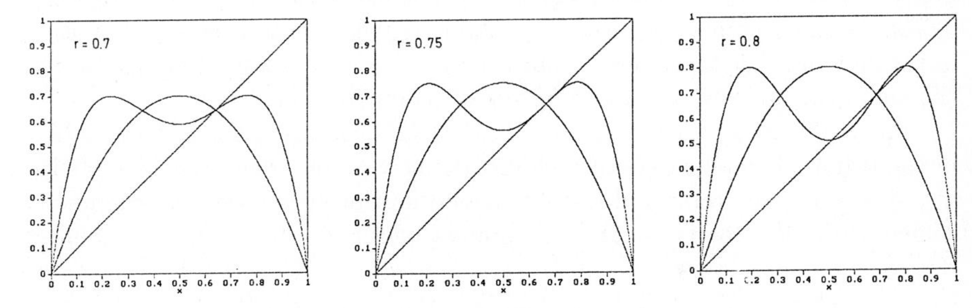

Fig. 9.7. Logistic mapping function $f(x)$ and first iteration $f^2(x) = f(f(x))$ for r-values below (r=0.7), at (r=0.75), and above (r=0.8) the bifurcation point.

function $f(x, r)$ must be less than unity to ensure stability of the fixed point. For the logistic map we have

$$f'(x, r) = 4r(1 - 2x)\,, \tag{9.16}$$

where prime denotes differentiation with respect to x, and, therefore,

$$f'(x_0^*, r) = 4r \quad \text{and} \quad f'(x_1^*, r) = 2 - 4r\,. \tag{9.17}$$

For $r < 0.25$, the fixed point $x_0^* = 0$ is stable ($|f'(0)| = 4r < 1$). At $r = 0.25$ this fixed point becomes unstable and a new stable one, x_1^*, appears. These fixed points are the only ones related to extrema in the underlying potential $V(x)$ discussed in the introduction to this chapter.

When we increase the mapping parameter to $r = 0.79$, the iterated points no longer approach the fixed point x_1^*, which loses stability for $|f'| = 1$, i.e. $r_1 =$

0.75, where a new stable period-two orbit is born. Figure 9.4 shows the iteration for the pre-set value $r = 0.79$. The attracting stable period-two orbit for $r = 0.79$, as well as the iterated map

$$f^2(x) = f(f(x)) = 16r^2x\left[1 - (4r+1)x + 8rx^2 - 4rx^3\right] \tag{9.18}$$

is shown in Fig. 9.5. Again, the fixed points of f^2 can be evaluated in closed form by solving the equation $f^2(x) = x$. The analysis is simplified by observing that the fixed points of f^1 are also fixed points of f^2. Additional ones are found at

$$x^*_{2,3} = \frac{1}{2}\left[1 + \frac{1}{4r} \pm \sqrt{(1+\frac{1}{4r})(1-\frac{3}{4r})}\right], \tag{9.19}$$

which are stable for $r \gtrsim 3/4$. This yields the values $x^*_2 = 0.5291$ and $x^*_3 = 0.7873$, in agreement with the computed orbit. The period-one fixed point $x^*_1 = 1 - 1/(4r) = 0.68354$ is unstable, a fact which can be checked numerically by starting a new iteration at $x_0 = x^*_1$. It should be noted that the stability can be seen in the graphs by looking at the slope of the iterated function at the fixed points, which must be smaller than one in magnitude. At $r_2 = (1+\sqrt{6})/4 \approx 0.862$, the fixed points $x^*_{2,3}$ become unstable, giving rise to a new stable orbit of period four. This transition — illustrated in Fig. 9.7 — leads to a qualitative change in the limiting orbit, a *'bifurcation'* (compare Sect. 2.4.6).

Figure 9.8 shows these so-called *pitchfork bifurcations* of the periodic orbits in the region of small values of r. A further increase of r destroys the stability of the two-periodic orbit and a stable period-four orbit is born. An example is shown in Fig. 9.6 for $r = 0.88$. This process continues and we find an infinite

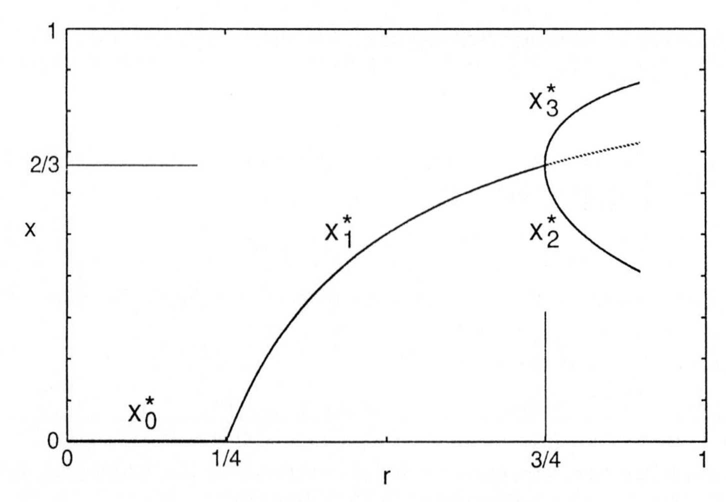

Fig. 9.8. Lowest periodic orbit bifurcations as a function of the parameter r. The unstable orbit is shown as a dashed curve.

Table 9.1. Bifurcation values r_k for bifurcation $2^{k-1} \to 2^k$, differences $r_k - r_{k-1}$, and ratios δ_k for the logistic map.

k	r_k	$r_k - r_{k-1}$	δ_k
1	0.749606		
2	0.862246	0.112640	4.747
3	0.885973	0.023727	4.640
4	0.891087	0.005114	4.662
5	0.892184	0.001097	

sequence of period-doubling bifurcations 1 to 2, 4, 8, ... , 2^k, ... for increasing values of r, ending at $r_\infty = 0.8924864\ldots$.

The critical values r_k, $k = 1, 2, \ldots$ of the parameter r at the bifurcation points for a $2^{k-1} \to 2^k$ period doubling can be computed by means of the program item *'Period Doubling'*. Numerical results are listed in Table 9.1, together with the values of the differences $r_k - r_{k-1}$ and the ratios

$$\delta_k = \frac{r_k - r_{k-1}}{r_{k+1} - r_k} \tag{9.20}$$

for the first five bifurcations. In this computation, we used 5000 iterations, an accuracy parameter $\epsilon = 10^{-5}$ and a stepsize of 10^{-6} for the parameter r.

It has been rigorously shown that these ratios converge to the so-called *Feigenbaum constant*

$$\delta = \lim_{k \to \infty} \delta_k = 4.66920\ldots, \tag{9.21}$$

which is in reasonable agreement with the values obtained numerically. The deviations are due to the finite number of digits used in the computation. The convergence of the bifurcation parameters is therefore exponential [9.7]

$$r_k = r_\infty - \mathit{const}\, \delta^{-k}. \tag{9.22}$$

Most remarkably, this convergence property and, in particular, the precise value of δ, are *universal* [9.8, 9.9, 9.10], i.e. independent of the choice of the mapping function $f(x, r)$, provided it is differentiable having a single quadratic maximum. The numerical values of r_∞ and *const* in (9.22) are *not* universal and depend on the mapping function. It should be stressed, however, that this universality is *not* due to a restriction of the whole dynamics to a small neighborhood of the maximum. The origin is much deeper, because the dynamics explores most of the whole interval. Other universal properties are addressed below.

9.3.2 The Chaotic Regime

In the chaotic regime $r_\infty = 0.8924864\ldots < r < 1$, the iterated points behave erratically. Strange attractors are observed. The chaotic intervals move together

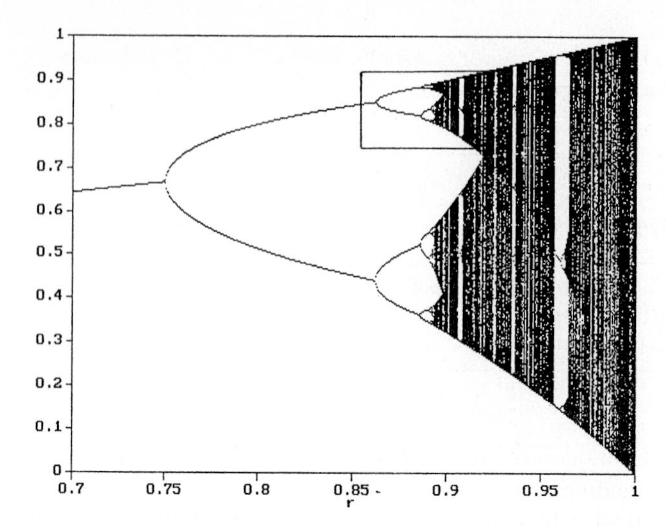

Fig. 9.9. Bifurcation diagram for the logistic map. This figure is stored on the disk.

by inverse bifurcations, showing characteristic windows with periodic p-cycles ($p = 3, 5, 6, \dots$) which bifurcate again.

Let us first look at a diagram displaying the dynamics of the x_n for large values of n as a function of the parameter r. Such a bifurcation diagram can be computed by means of the program item *'Bifurcation'*. The range $0.7 < r < 1$ is shown in Fig. 9.9, where 2000 iterations are used. This plot is pre-computed

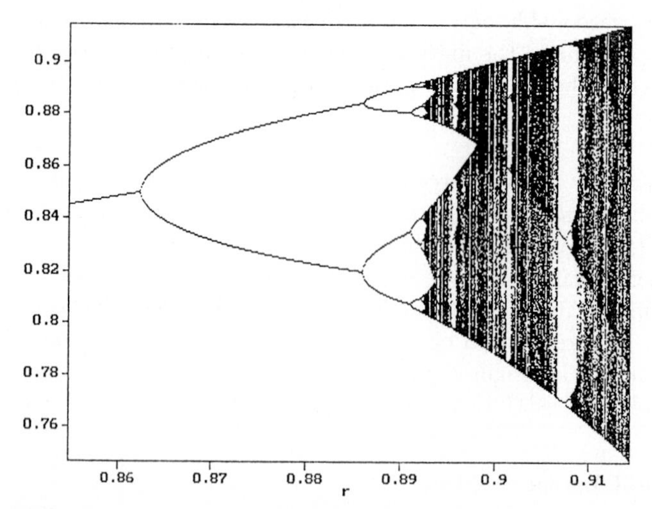

Fig. 9.10. Bifurcation diagram for the logistic map. Magnification of the rectangle in Fig. 9.9. This figure is stored on the disk.

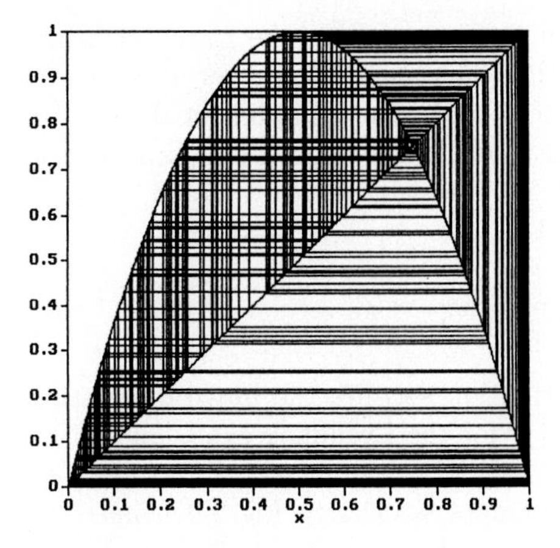

Fig. 9.11. Chaotic sequence of iterated points of the logistic map for $r = 1$.

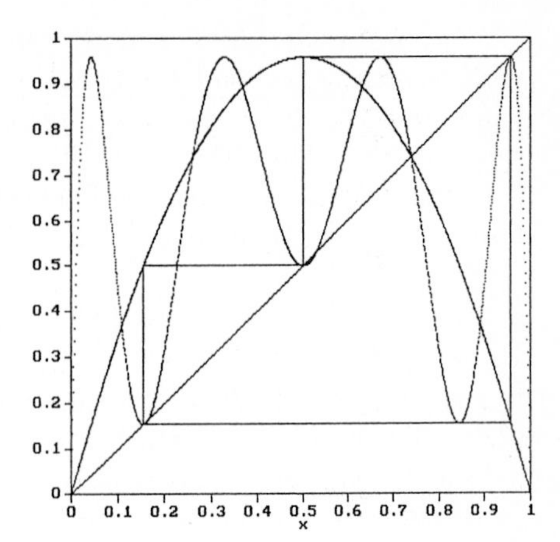

Fig. 9.12. Stable orbit in the period-three window embedded in the chaotic region ($r = 0.957968$).

and can be displayed by loading the file LOGIST. The qualitative similarity of the bifurcation tree to the diagram for the Duffing oscillator shown in Fig. 8.18 should be noted. In the region $r < r_\infty \approx 0.8925$, we see the period-doubling bifurcation sequence as described above. For larger values of r exceeding r_∞, chaotic bands are observed where the iterated points appear chaotic. In fact, it can be shown that in this case the attractor is a fractal set, a so-called *strange attractor*.

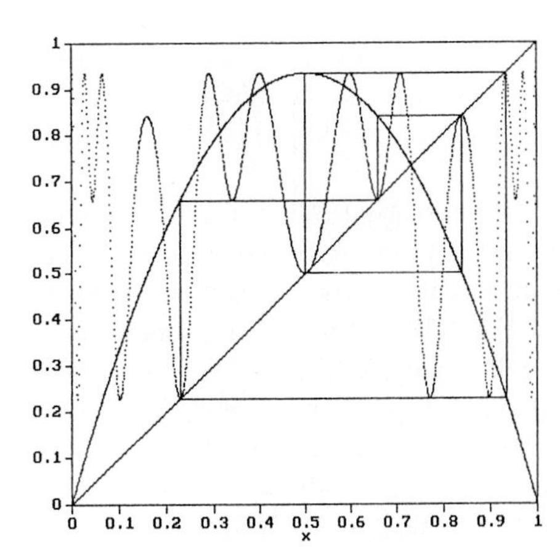

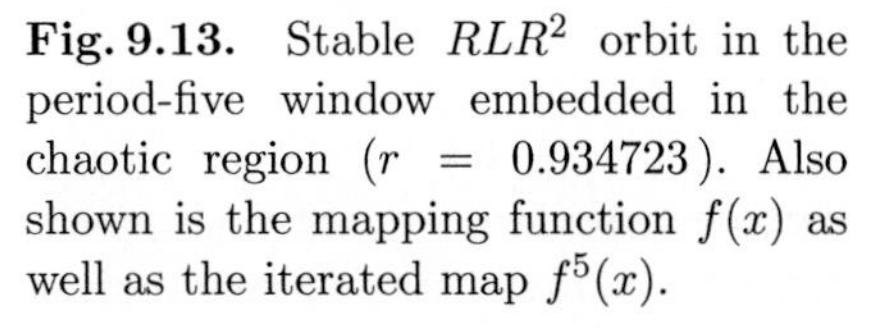

Fig. 9.13. Stable RLR^2 orbit in the period-five window embedded in the chaotic region ($r = 0.934723$). Also shown is the mapping function $f(x)$ as well as the iterated map $f^5(x)$.

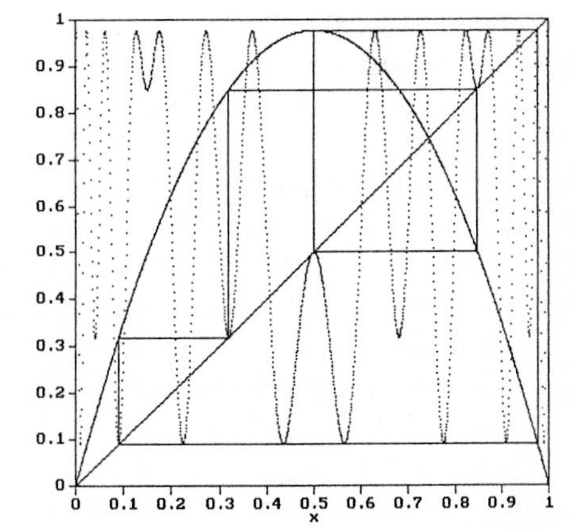

Fig. 9.14. Stable RL^2R orbit in the period-five window embedded in the chaotic region ($r = 0.976427$). Also shown is the mapping function $f(x)$ as well as the iterated map $f^5(x)$.

As can be seen from Fig. 9.9, the interval containing the attractor grows with increasing parameter r, covering the whole unit interval for $r = 1$. The mapping is ergodic in this case. Figure 9.11 shows the iterations of the map in this case. The accumulated distribution of the iterated points approaches the invariant density $\varrho(x)$ discussed in Sect. 2.4.3, which can be evaluated in closed form for the logistic mapping with $r = 1$:

$$\varrho(x) = \frac{1}{\pi\sqrt{x(1-x)}} . \tag{9.23}$$

It can easily be checked that this distribution satisfies the integral equation (2.68). Furthermore, the result $\lambda_L = \ln 2$ for the Lyapunov exponent can be rederived from the integral (2.69).

Windows, in addition to chaotic dynamics, appear in Fig. 9.9, where the attractor consists of only a few points and the dynamics appears to be regular. The most obvious windows are a period-three and period-five window. Fig. 9.12 shows a stable period-three orbit and Figs. 9.13 and 9.14 show two stable period-five orbits. Closer inspection shows the appearance of many more periodic orbits. The position and organization of these periodic windows has been investigated by Metropolis et al. [9.6]. The r-parameters, r_s, for periodic orbits passing through the maximum of the mapping function at $x_0 = 0.5$ up to period seven is given in Table 9.2 (adapted from [9.4]). All these orbits are *superstable* cycles, as discussed in Sect. 9.3.5 below. The period-two and -four orbits for the logistic mapping are inside the bifurcation region $r < r_\infty$, while the others are inside the windows. The k-periodic orbits $x_0, x_1, \ldots, x_j, \ldots, x_k = x_0$ can be classified in terms of a binary symbolic code $RL\ldots$, where R or L stands for $x_j > x_0$ or $x_j < x_0$, respectively.

Also shown in the table are the results for the sin-map $x_{n+1} = r\,\sin(\pi x_n)$. The ordering of the orbits is the same for both maps, another universal feature of one-dimensional mappings. This *structural universality* is independent of the order of the maximum, in contrast to the *metric universality* of the Feigenbaum constant.

The bifurcation tree in Fig. 9.9 shows an interesting self-similarity. A magnification obtained using the zoom feature of the program is shown in Fig. 9.10. One observes a striking structural and (approximate) metric agreement between both diagrams. A further magnification of Fig. 9.10 will again be (almost) self-similar, and perfect self-similar mapping will appear in the limit of an infinite series of such magnifications.

As an example of the structural similarity, we observe the 'image' of the period-three window in the magnification in Fig. 9.10, which is actually only one half of the period-six window close to $r = 0.906889$.

9.3.3 Lyapunov Exponents

As discussed in Sect. 2.4.3, the Lyapunov exponent provides a quantitative measure of the chaoticity of the dynamics in terms of the divergence of neighboring

Table 9.2. Periodic orbits for the logistic map (log) and the sin-map (sin) .

Period	r_s (log)	r_s (sin)	Symbolic Code
2	0.809017	0.777734	R
4	0.874640	0.846382	RLR
6	0.906889	0.881141	RLR^3
7	0.925442	0.900491	RLR^4
5	0.934723	0.910923	RLR^2
7	0.943554	0.921335	RLR^2LR
3	0.957968	0.939043	RL
6	0.961142	0.943588	RL^2RL
7	0.971511	0.956844	RL^2RLR
5	0.976427	0.963366	RL^2R
7	0.980548	0.968783	RL^2R^3
6	0.984384	0.973566	RL^2R^2
7	0.987758	0.978251	RL^2R^2L
4	0.990068	0.982035	RL^2
7	0.992244	0.985781	RL^3RL
6	0.994442	0.989202	RL^3R
7	0.996187	0.991915	RL^3R^2
5	0.997567	0.994472	RL^3
7	0.998634	0.996661	RL^4R
6	0.999396	0.998265	RL^4
7	0.999849	0.999451	RL^5

orbits. For the case of one-dimensional maps $x_{n+1} = f(x_n, r)$, the Lyapunov exponents can easily be computed by means of (2.64)

$$\lambda_L(r) = \lim_{n\to\infty} \frac{1}{n} \sum_{j=0}^{n-1} \ln |f'(x_j, r)| . \tag{9.24}$$

The program provides an automatic computation of the Lyapunov exponent (denoted by '*Ly*' in the program) as a function of the parameter r of the map. As an example, Fig. 9.15 shows $\lambda_L(r)$ for the logistic map (9.8). This plot is pre-computed and can be displayed by loading the file LOGIST. In the stability regions λ_L is negative, in the chaotic regions positive, and zero at the bifurcation points. We furthermore observe a characteristic singularity inside the period-k intervals, where the Lyapunov exponent goes to $-\infty$. These singularities correspond, of course, to the superstable orbits, where one of the derivatives $f'(x_j, r)$ in (9.24) is zero.

In the chaotic regime $r > r_\infty$, the overall increase following approximately the law

$$\lambda_L \approx const\, (r - r_\infty)^\tau \tag{9.25}$$

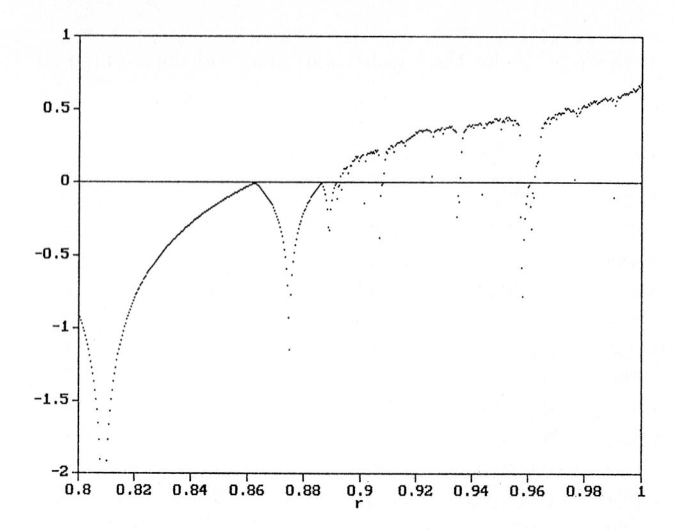

Fig. 9.15. Lyapunov exponent for the logistic map. This figure is stored on the disk.

with a universal constant $\tau = 0.4498\ldots$ [9.20] is interrupted by the stability windows with negative λ_L.

9.3.4 The Tent Map

The so-called *tent map*

$$x_{n+1} = r\,(1 - |2x - 1|)\;, \tag{9.26}$$

which is piecewise linear with slope $\pm 2r$. This mapping function is not differentiable at the maximum.

For the tent map, many results can be obtained in closed form (see, e.g., [9.4, Chap. 2.2]). For $r < 0.5$, we find a stable fixed point at $x = 0$. For r-values exceeding the critical value of $r = 0.5$, the fixed point $x = 0$ becomes unstable and an additional one appears at $x = 2r/(1+2r)$. One easily convinces oneself that the iterated tent map $f^k(x)$ is also piecewise linear, consisting of intervals with slope $|\mathrm{d}^k f/\mathrm{d}x^k| = 2kr$. Hence, the slope at all fixed points exceeds unity for $r > 0.5$ and all fixed points – also those of higher order – are unstable. Examples are shown in Figs. 9.16 and 9.17.

Figures 9.18 and 9.19 show as an example iterations for $r = 0.4$ and $r = 0.8$. In the latter case, the iterations are attracted by an interval on the x-axis. The bifurcation diagram in Fig. 9.20 displays this behavior in more detail. The boundary of the attracting region for $r > 0.5$ is bounded by the critical motion passing through the maximum value $f(0.5) = r$ of the tent map. This yields $2r(1-r) \le x \le r$. As an exercise, the reader can investigate the dynamical origin of the empty region embedded in this zone.

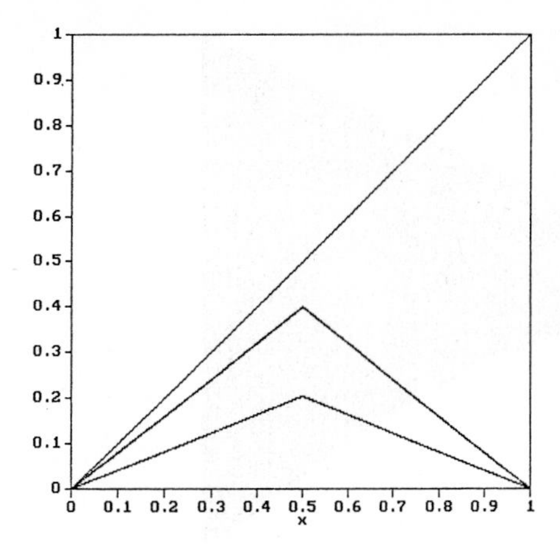

Fig. 9.16. Tent map for $r = 0.4$ and iterated map $f^4(x)$.

Fig. 9.17. Tent map for $r = 0.8$ and iterated map $f^4(x)$.

Finally, one can calculate the Lyapunov exponent λ_L for the tent map. From (9.24), we directly obtain

$$\lambda_L = \lim_{n\to\infty} \frac{1}{n} \sum_{j=0}^{n-1} \ln |f'(x_j, r)| = \ln (2r) , \tag{9.27}$$

since the magnitude of the slopes has the constant value $|f'(x)| = 2r$. It should be noted that λ_L is negative below the critical value of $r = 0.5$, zero at the bifurcation point, and positive above in the chaotic regime. The theoretical prediction (9.27) can be checked numerically.

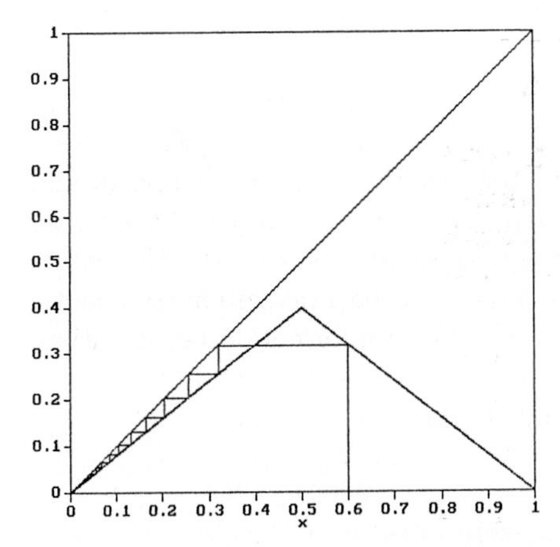

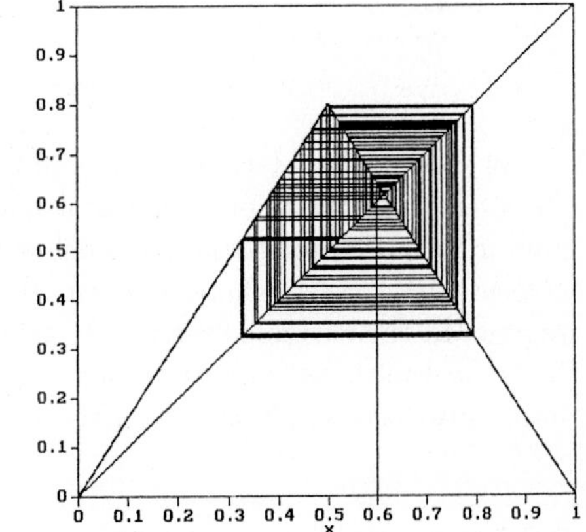

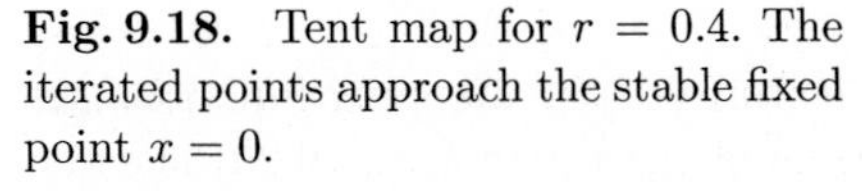

Fig. 9.18. Tent map for $r = 0.4$. The iterated points approach the stable fixed point $x = 0$.

Fig. 9.19. Tent map for $r = 0.8$. No stable fixed point exists.

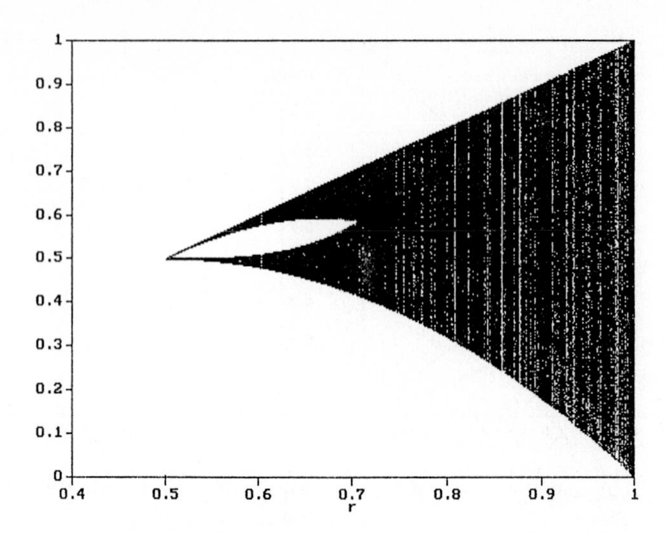

Fig. 9.20. Period-doubling bifurcation tree for the tent map. This figure is stored on the disk.

Table 9.3. Bifurcation values r_k for bifurcation $2^{k-1} \to 2^k$ and ratios δ_k for the quartic map.

k	r_k	$r_k - r_{k-1}$	δ_k
1	0.831992		
2	0.951669	0.119677	8.185
3	0.966291	0.014622	7.278
4	0.968300	0.002009	7.305
5	0.968575	0.000275	

In view of the relation $\lambda_L = -\ln(2\overline{I})$ between the Lyapunov exponent and the average loss of information as discussed in Sect. 2.4.3 (compare (2.66)), we gain information for parameters below the critical value of $r = 0.5$ (the points converge to $x = 0$), whereas information is lost at a constant rate in the chaotic regime. In the extreme case $r = 1$, we lose one bit of information per iteration. If, for instance, we only know that x_n is in the lower half of the interval, this information is completely lost after a single iteration.

9.3.5 Suggestions for Additional Experiments

Different Mapping Functions. Some features of the mapping functions are universal, as, for instance, the limiting value of the ratios δ_k in (9.20) is equal to the Feigenbaum constant $\delta = 4.6692016091\ldots$ for differentiable mappings having a quadratic maximum. This can be tested numerically, e.g., for the sin-map [9.3]

$$x_{n+1} = r\sin(\pi x_n)\,, \tag{9.28}$$

the mapping

$$x_{n+1} = r x_n(1 - x_n^2)\,, \tag{9.29}$$

or the mapping

$$x_{n+1} = r\,(1 - |2x - 1|^m)\;. \tag{9.30}$$

The first two mappings show a quadratic maximum and are expected to show the same limiting bifurcation scenario as the logistic map. In particular, the limiting value of the ratios δ_k should be identical. Numerical results for the sin-map (9.28) listed in Table 9.3 confirm this metric universality.

For the sin-map, the periodic windows embedded in the chaotic region show the same structural organization as the logistic map. The r-values for such periodic orbits are listed in Table 9.2 above and show the same ordering of the symbolic sequences, i.e. structural universality.

The mapping function (9.30) reduces for $m = 1$ to the tent map (9.26) and for $m = 2$ to the logistic map (9.8), both of which are studied above. The behavior is quantitatively, and even qualitatively, different for other values of m. The bifurcation values for the case $m = 4$, the quartic map

$$x_{n+1} = r\left(1 - (2x - 1)^4\right)\,, \tag{9.31}$$

are listed in Table 9.3 for the first five bifurcations. In this computation, we used the same conditions as for the logistic map in Table 9.1 (5000 iterations,

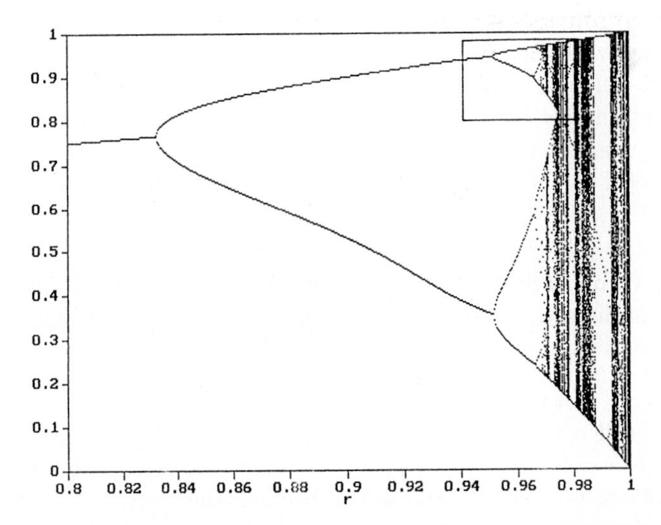

Fig. 9.21. Period-doubling bifurcation tree for the quartic map. This figure is stored on the disk.

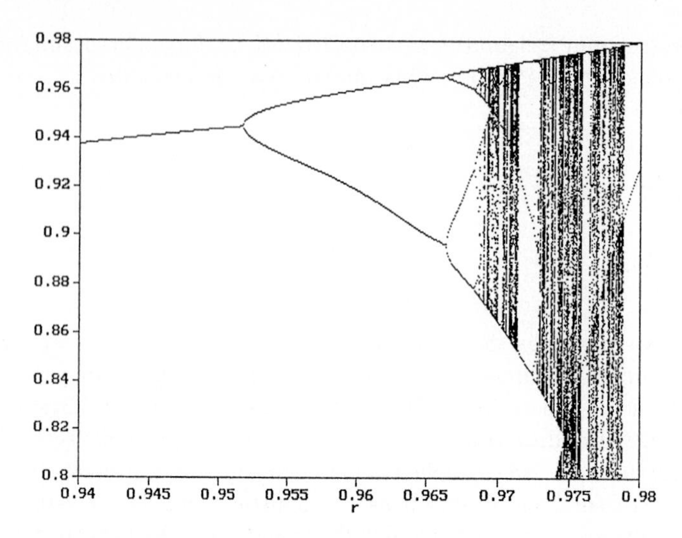

Fig. 9.22. Period-doubling bifurcation tree for the quartic map. Magnification of the rectangle in Fig. 9.21. This figure is stored on the disk.

accuracy parameter $\epsilon = 10^{-5}$ and stepsize 10^{-6} for parameter r). Again, the δ_k sequence converges, albeit to a value $\tilde{\delta} \approx 7.3$, which differs from the Feigenbaum constant $\delta \approx 4.67$.

Finally, Fig. 9.21 shows a bifurcation diagram for the quartic map and Fig. 9.22 a magnification of this plot. We again observe self-similarity. In comparison with the logistic map in Figs. 9.9 and 9.10, we observe the same window organization as discussed above. As a numerical experiment, the reader can localize numerically the periodic orbits for superstable cycles for the quartic map and compare with Table 9.2.

The reader may also compute the Lyapunov exponent in analogy to the logistic map in Sect. 9.3.3 for different mapping functions. As an example, the Lyapunov exponent for the sin-map (9.28) is shown in Fig. 9.23 as a function of the parameter r, which appears to be *very* similar to the logistic case shown in Fig. 9.15.

Periodic Orbit Theory. Periodic orbits play a very important role in the dynamics of discrete one-dimensional maps of the interval. A large number of rigorous results has been obtained. Here, we list only a few important features, which may guide and inspire further numerical experiments.

– The so-called *superstable* orbits or *supercycles* are k-periodic orbits passing through a point x_c with $f'(x_c) = 0$ (a *critical point*). In this case, the stability of the orbit is guaranteed because the derivative of f^k vanishes. Such supercycles are used e.g. in renormalization methods and in proofs of universality. For mapping functions with a single quadratic maximum, it has been shown, e.g., that the distances d_n of the fixed point closest to x_c

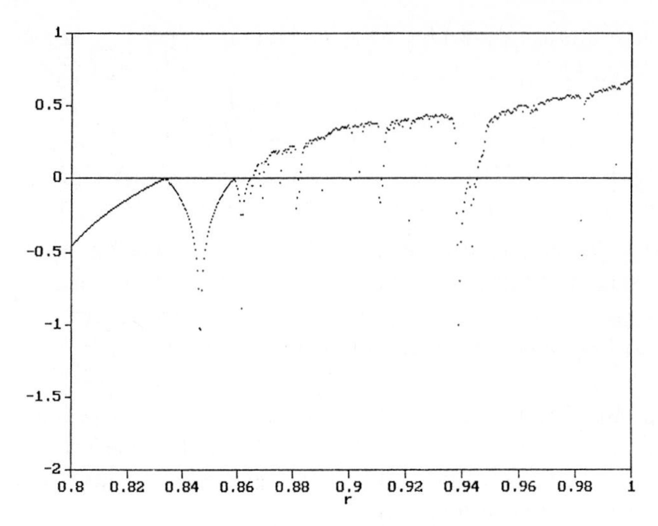

Fig. 9.23. Lyapunov exponent for the sin-map.

for 2^n-periodic supercycles scale as

$$\frac{d_n}{d_{n+1}} = -\alpha \tag{9.32}$$

for large values of n, with a universal value of $\alpha = 2.5029078750\ldots$. For more details see, e.g., Ref. [9.4].

– The implications of the existence of an orbit with period k on the existence of others with different period l is thoroughly understood from *Sarkovskii's Theorem* [9.11, 9.1, 9.4, 9.5], under the *single* (!) condition that $f(x)$ be continuous: a period-k orbit implies the existence of all orbits with period l, provided that $k \rhd l$, where the ordering $\rhd$ is given by

$$\begin{aligned} &3 \rhd 5 \rhd 7 \cdots \rhd 2 \cdot 3 \rhd 2 \cdot 5 \rhd \cdots \rhd 2^2 \cdot 3 \rhd 2^2 \cdot 5 \rhd \cdots \rhd 2^3 \cdot 3 \rhd 2^3 \cdot 5 \rhd \cdots \\ &\cdots \rhd 2^m \cdot 3 \rhd 2^m \cdot 5 \rhd \cdots\cdots \rhd 2^n \rhd \cdots\cdots \rhd 2^3 \rhd 2^2 \rhd 2 \rhd 1 \;. \end{aligned}$$

First, all odd numbers (except 1) appear in increasing order, followed by 2 times all odd numbers, and then all powers of two times an odd number. The powers of two missing in this list are added at the end in decreasing order. The list contains all natural numbers.

It should be noted that the theorem says nothing about stability. A trivial consequence is that the appearance of an orbit of period 2^k implies the existence of all orbits with period 1, 2, 4, ..., 2^k, and, if a period different from a power of 2 is observed, then there exist infinitely many periods.

Period three appears at the end of the ordering chain and, therefore, its existence implies the existence of *all* periods, which is sometimes stated as 'period three implies chaos' [9.12].

– Various results are known for mapping functions $f(x)$ with negative *Schwarzian derivative*

$$\mathrm{S}f(x) = \frac{f'''(x)}{f'(x)} - \frac{3}{2}\left(\frac{f''(x)}{f'(x)}\right)^2 . \tag{9.33}$$

The negativity of $\mathrm{S}f$ is conserved under the composition of mappings. An example is the logistic map with $\mathrm{S}f(x) = -6(1-2x)^{-2} < 0$.

For $\mathrm{S}f < 0$, one has at most $N_\mathrm{c} + 2$ attracting periodic orbits, where N_c is the number of critical points of $f(x)$, i.e. $N_\mathrm{c} = 1$ for a single maximum. For the special case of the logistic map, it can furthermore be shown, that there exists at most one attracting periodic orbit [9.5].

Exploring the Circle Map. In the above experiments, the properties of maps of the unit interval were explored. Maps of the circle, where the variable x is taken modulo 1, show some interesting new features. An example of these maps is the so-called one-dimensional *baker map*

$$x_{n+1} = 2\,x_n \mod 1 , \tag{9.34}$$

which is chaotic [9.5].

Another prominent example is the *circle map*,

$$x_{n+1} = f(x_n, r, K) = x_n + r - \frac{K}{2\pi}\sin(2\pi x_n) \mod 1 , \tag{9.35}$$

which appears as a simplification of the equations of motion for a periodically kicked rotator with orientation angle $2\pi x$: here, K determines the degree of nonlinearity and r gives the rotation rate. The dynamics is sensitively dependent on the two parameters r and K, as is for instance evident from the plot of the Lyapunov exponent on the parameter plane shown in [9.4, plate XVI].

For $K > 1$, the map is noninvertible and period-doubling bifurcations and chaotic behavior can be observed, densely interwoven, however, with regular regions in the parameter plane.

For $0 < K < 1$, *mode locking* occurs i.e. the iterated orbit approaches a periodic limit cycle. In the parameter plane $(K,\, r)$, this mode locking occurs inside the so-called *Arnold tongues*, which are areas with a rational value, p/q, of the winding number

$$w = \lim_{n\to\infty} \frac{f^n(x_0) - x_0}{n} . \tag{9.36}$$

It should be noted that in (9.36), the function f is *not* taken modulo 1. In these tongues, the attractor is non-chaotic and has period q. The order of the appearance of the different modes follow a Farey tree [9.4] organization. The simplest orbits are period-one attractors for $r < K/2\pi$. Period-two orbits ($w = 1/2$) appear in a tongue close to the line $r = 1/2$. Numerical examples for mode-locking into period-three ($w = 1/3$) and period-seven ($w = 3/7$) are shown in Figs. 9.24 and 9.25. Further details regarding the circle map can be found in [9.4] or [9.13].

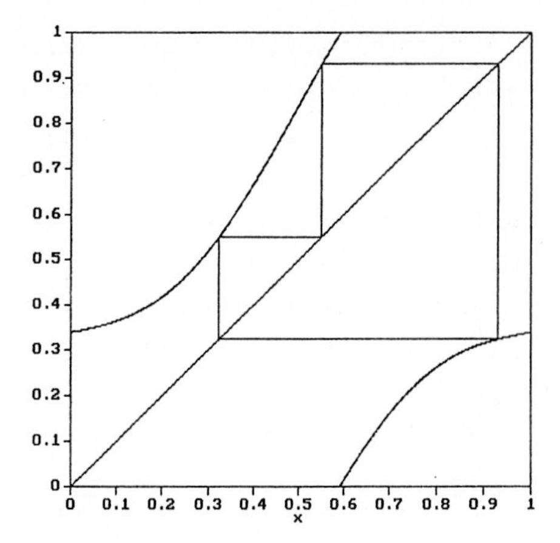

Fig. 9.24. Mode-locking for the circle map ($K = 0.8$ and $r = 0.34$) to a period-3 orbit with winding number $w = 1/3$.

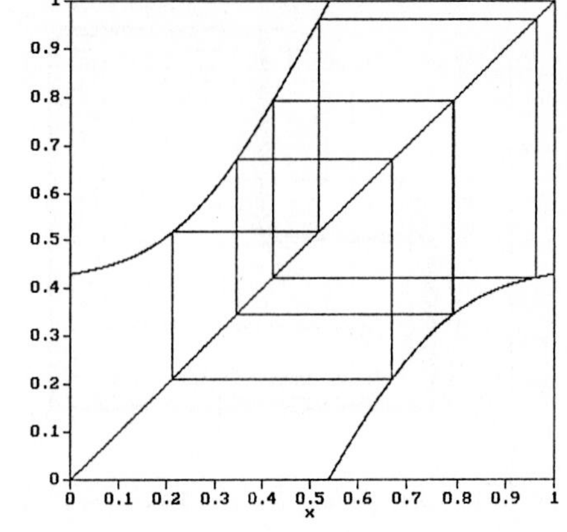

Fig. 9.25. Mode-locking for the circle map ($K = 0.8$ and $r = 0.43$) to a period-7 orbit with winding number $w = 3/7$.

9.4 Suggestions for Further Studies

The Power Spectrum — i.e. the Fourier transform of the iterates — contains valuable information on the dynamics, especially because such a spectrum is often obtained in experiments. Power spectra and their universal features for iterated maps are discussed in [9.1] and [9.4].

Influence of Noise on the dynamics or on the power spectra is of interest, because it will always be present experimentally. This can be modeled by adding a noise term to the mapping equation, e.g. a random Gaussian variable. Despite the quenching of finer details (depending on the noise level), the sharp transition to chaotic motion in the logistic map is still observable [9.14, 9.4].

Intermittency (compare Sect. 2.4.4) can also be observed in one-dimensional maps. In such a case no stable fixed points exist, but the orbit is 'attracted' by a region, where the mapping function is closely parallel to the bisector $y = x$. The trajectory spends quite a long time in this region until it is finally ejected and (almost) randomly re-injected into the intermittency zone after a longer excursion. Examples of intermittent behavior are found for the logistic map for r-values slightly below the period-three window ($r \lesssim (1+\sqrt{8})/4$) and $f^3(x)$ is almost tangent to the bisector (see Fig. 9.26). Another example is the circle map (9.35) for $K = 0.8$ and $r = 0.15$, as shown in Fig. 9.27. See [9.4] for further discussions.

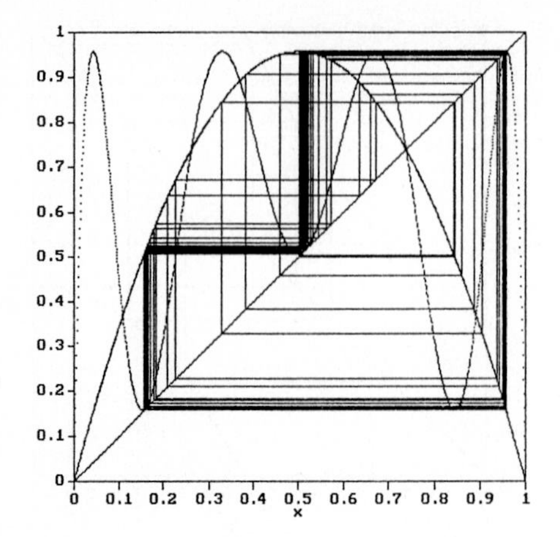

Fig. 9.26. Intermittency for the logistic map ($r = 0.957$), where the iterated points are close to a period-three orbit.

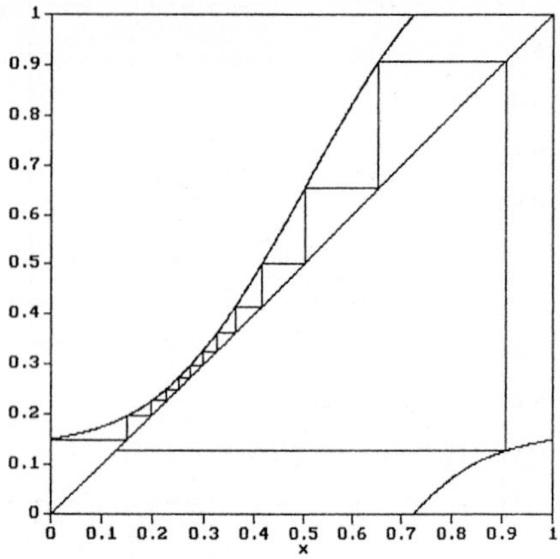

Fig. 9.27. Intermittency for the circle map ($K = 0.8$ and $r = 0.15$).

9.5 Real Experiments and Empirical Evidence

The evolution of a dynamical system can be described either in continuous time (a flow), or in discrete time (a mapping). The pendulum is a system that is dynamically determined by a flow in phase space, although its behavior can be conveniently analyzed in terms of a discrete mapping (compare the Poincaré map used, e.g., for the double pendulum in Chap. 5). On the other hand, the sequence of drip intervals of a leaky faucet is naturally described by a discrete map.

Therefore, mappings are mainly used as a convenient means of data analysis in conjunction with various experiments (see list below) in order to achieve a better insight into chaotic behavior. In addition, through this method many relevant terms such as 'stable' and 'unstable' fixed points, 'bifurcation', 'period doublings', 'attractors', 'basins of attraction',... can be taught. In some experiments, the maps are directly — on line — measured or indirectly generated — off line —- from raw data. Some experiments:

- The current $I(t)$ versus $I(t + \Delta t)$ in solid state lasers is measured to show deterministic or chaotic fluctuations in the output intensity (see Ref. [9.15, Fig. 9]).

- The displacement x_{n+1} versus x_n of a magnetoelastic ribbon is studied as a function of experimental parameters (see Ref. [9.16, Fig. 1]).

- The voltage $U(t_n + \tau)$ versus $U(t_n)$ in electrical conduction measurements in barium sodium niobate (see Fig. 1 in Ref. [9.17]) shows oscillatory and chaotic behavior.

- Each impact between ball and membrane (see Ref. [9.18, Fig. 4]) is displayed in an impact map showing orbits of period two or four or strange attractors.
- The voltage $U(t_{n+1})$ versus $U(t_n)$ in the chaotic oscilloscope monitoring a nonlinear electronic circuit (see Ref. [9.19, Fig. 1]).
- Two slightly different signals from a nonlinear electronic circuit, which could be described by a discrete noninvertible mapping, are compared in order to determine the Lyapunov exponent experimentally [9.20].

Briggs [9.21] describes the Feigenbaum machine, which produces an iteration modeled by the logistic map. This is essentially an analog computer to perform the iteration by producing and storing electronic signals.

Núñez Yépez et al. [9.22] describe an advanced undergraduate experiment on the chaotic behavior of a dripping faucet. It can be used for demonstration and measurements. The experimental set-up is described in detail: the dripping is registered by a light barrier and the electronic signal read by a computer. Plots of the time intervals between drips, T_{n+1} versus T_n, reveal certain patterns which can be assigned to periodic or complex chaotic behavior; strange attractors and routes to chaos can be discussed.

Cahalan et al. [9.23] repeat the same experiment, but make general remarks and provide a technical description of some experimental problems such as water reservoirs, flow valves, volume flow rates, drop rates, water purity, which critically influence the experimental observation. T_n versus T_{n+1}–diagrams or three-dimensional plots (T_n, T_{n+1}, T_{n+2}) are presented, discussed, and modeled by the logistic equation.

References

[9.1] P. Collet and J.-P. Eckmann, *Iterated Maps on the Interval as Dynamical Systems* (Birkhäuser, Basel 1980)

[9.2] R. May, *Simple mathematical models with very complicated dynamics*, Nature **261** (1976) 459 (reprinted in: B.-L. Hao , *Chaos* (World Scientific, Singapore 1984) and P. Cvitanović, *Universality in Chaos* (Adam Hilger, Bristol 1984))

[9.3] M. J. Feigenbaum, *Universal behaviour in nonlinear systems*, Los Alamos Science **1** (1980) 4 (reprinted in: B.-L. Hao, *Chaos* (World Scientific, Singapore 1984))

[9.4] H. G. Schuster, *Deterministic Chaos* (VCH, Weinheim 1988)

[9.5] R. L. Devaney, *An Introduction to Chaotic Dynamical Systems* (Addison–Wesley, New York 1987)

[9.6] N. Metropolis, M. L. Stein, and P. R. Stein, *On finite limit sets for transformations on the unit interval*, Jour. of Combinatorial Theory **15** (1973) 25 (reprinted in: B.-L. Hao, *Chaos* (World Scientific, Singapore 1984) and P. Cvitanović, *Universality in Chaos* (Adam Hilger, Bristol 1984))

[9.7] S. Grossmann and S. Thomae, *Invariant distributions and stationary correlation functions of one-dimensional discrete processes*, Z. Naturf. A **32** (1977)

1353 (reprinted in: P. Cvitanović, *Universality in Chaos* (Adam Hilger, Bristol 1984))

[9.8] M. J. Feigenbaum, *Quantitative universality for a class of nonlinear transformations*, J. Stat. Phys. **19** (1978) 158 (reprinted in: B.-L. Hao, *Chaos* (World Scientific, Singapore 1984))

[9.9] M. J. Feigenbaum, *The universal metric properties of nonlinear transformations*, J. Stat. Phys. **21** (1979) 669 (reprinted in: B.-L. Hao, *Chaos* (World Scientific, Singapore 1984) and P. Cvitanović, *Universality in Chaos* (Adam Hilger, Bristol 1984))

[9.10] O. E. Lanford III, *A computer-assisted proof of the Feigenbaum conjectures*, Bull. Am. Math. Soc. **6** (1982) 427 (reprinted in: P. Cvitanović, *Universality in Chaos* (Adam Hilger, Bristol 1984))

[9.11] A. N. Sarkovskii, *Coexistence of cycles of a continuous map of a line into itself*, Ukr. Mat. Z. **16** (1964) 61

[9.12] T.-Y. Li and J. A. Yorke, *Period three implies chaos*, Ann. Math. Monthly **82** (1975) 985 (reprinted in: B.-L. Hao, *Chaos* (World Scientific, Singapore 1984))

[9.13] J. Frøyland, *Introduction to Chaos and Coherence* (IOP Publishing, Bristol 1992)

[9.14] J. P. Crutchfield, J. D. Farmer, and B. A. Huberman, *Fluctuations and simple chaotic dynamics*, Phys. Rep. **92** (1982) 45

[9.15] C. Bracikowski and R. Roy, *Chaos in a multimode solid-state laser system*, Chaos **1** (1991) 49

[9.16] T. Shinbrot, C. Grebogi, J. Wisdom, and J. A. Yorke, *Chaos in a double pendulum*, Am. J. Phys. **60** (1992) 491

[9.17] S. Martin, H. Leber, and W. Martienssen, *Oscillatory and chaotic states of the electrical conduction in barium sodium niobate crystals*, Phys. Rev. Lett. **53** (1984) 303

[9.18] T. M. Mello and N. M. Tufillaro, *Strange attractors of a bouncing ball*, Am. J. Phys. **55** (1987) 316

[9.19] M. T. Levinsen, *The chaotic oscilloscope*, Am. J. Phys. **61** (1993) 155

[9.20] J. C. Earnshaw and D. Haughey, *Lyapunov exponents for pedestrians*, Am. J. Phys. **61** (1993) 401

[9.21] K. Briggs, *Simple experiments in chaotic dynamics*, Am. J. Phys. **55** (1987) 1083

[9.22] H. N. Núñez Yépez, A. L. Salas Brito, C. A. Vargas, and L. A. Vincente, *Chaos in a dripping faucet*, Eur. J. Phys. **10** (1989) 99

[9.23] R. F. Cahalan, H. Leidecker, and G. D. Cahalan, *Chaotic rhythms of a dripping faucet*, Comput. in Phys. Jul./Aug. (1990) 368

10. Nonlinear Electronic Circuits

Nonlinear electronic networks can be used as a laboratory set-up of nonlinear systems. The dynamics directly generates an electric signal, which can be easily handled for further analysis. Such an electronic circuit is a physical system of the real world. It is, however, on account of its electronic nature, also similar to a computing device and, therefore, electronic circuits are also used as analog computers to model more elaborate experimental set-ups in different areas of physics. The circuit studied in this computer program has been used by Mitschke and Flüggen to model chaotic behavior in hybrid optical systems [10.1].

10.1 A Chaos Generator

This computer program simulates a *'Chaos Generator'*. The nonlinear electronic circuit is a simple RLC circuit in series, where a signal is amplified, squared, and fed back into the circuit without delay, as shown in Fig. 10.1. Without any feedback and $R_{\mathrm{m}} = 0$, the circuit oscillates with a frequency

$$\omega = \sqrt{\frac{1}{L}\left(\frac{1}{C_{\mathrm{m}}} + \frac{1}{C}\right)}\,. \tag{10.1}$$

For $R_{\mathrm{m}} > 0$, any oscillation is damped out and decays to zero.

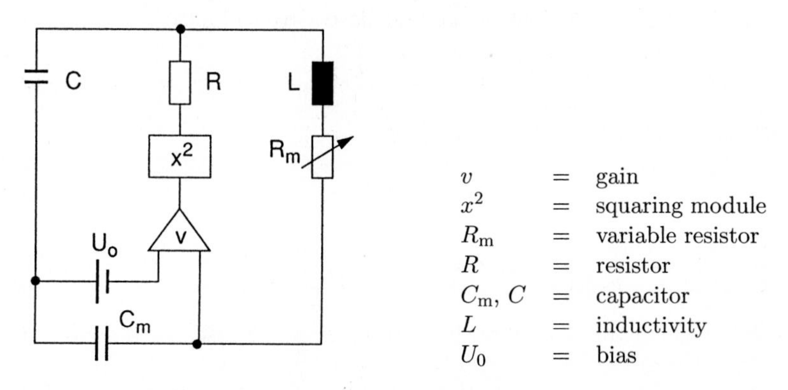

Fig. 10.1. The *'Chaos Generator'*.

The feedback branch of the circuit consists of an amplifier (characterized by a bias voltage U_0 and gain v) and a squaring module. The resistance R includes the output impedance of the amplifier. The voltage $U_{C_\mathrm{m}} =: U$ and the current $I_\mathrm{m} =: I \sim U'$ (a prime denotes time differentiation) at the capacitor C_m can be monitored in the program.

The network contains three complex impedances and will thus lead to a differential equation of third order, which can easily be derived: applying Kirchhoff's law to the loop connecting C_m, C, R gives

$$\frac{Q}{C} + R\,(Q' + Q'_\mathrm{m}) - v^2(U - U_0)^2 = 0\,, \tag{10.2}$$

where the prime denotes time differentiation. Q and Q_m are the charges at C and C_m, respectively, and U is the voltage at capacitor C_m. For the outer loop connecting C, L, R_m, C_m, one obtains

$$\frac{Q}{C} - LQ''_\mathrm{m} - R_\mathrm{m}Q'_\mathrm{m} - \frac{Q_\mathrm{m}}{C_\mathrm{m}} = 0\,. \tag{10.3}$$

Subtracting the time derivative of (10.3) multiplied by RC from the difference of (10.2) and (10.3) leads to a third order differential equation for $U = Q_\mathrm{m}/C_\mathrm{m}$:

$$U''' + a\,U'' + b\,U' + c\,U = cv^2\,(U - U_0)^2 \tag{10.4}$$

with coefficients

$$\begin{aligned} a &= \frac{1}{RC} + \frac{R_\mathrm{m}}{L} \\ b &= \frac{1}{LC}\left(1 + \frac{R_\mathrm{m}}{R} + \frac{C}{C_\mathrm{m}}\right) \\ c &= \frac{1}{LCRC_\mathrm{m}}\,. \end{aligned} \tag{10.5}$$

For a theoretical analysis, it is convenient to rescale the time by $t = b^{-1/2}\,\tilde{t}$, which normalizes the coefficient of the first derivative to unity:

$$\dddot{U} + \beta\ddot{U} + \dot{U} = F(U) \tag{10.6}$$

with a positive dimensionless friction coefficient

$$\beta = ab^{-1/2} \tag{10.7}$$

and

$$F(U) = cb^{-3/2}\left(-U + v^2(U - U_0)^2\right)\,. \tag{10.8}$$

Here, a dot denotes the derivative with respect to the rescaled dimensionless time $\tilde{t}$.

The dynamical behavior of the system is determined by the function $F(U)$, which is a quadratic polynomial dependent on three parameters in the present case. The introduction of a new dimensionless variable, x, through the linear transformation

$$U = A - Bx \tag{10.9}$$

brings the quadratic function $F(U)$ into a standard form. Choosing

$$A = \frac{1}{2v^2}\left(1 + 2v^2U_0 + \sqrt{1 + 4v^2U_0}\right) \tag{10.10}$$

$$B = v^{-2}\sqrt{1 + 4v^2U_0}\,, \tag{10.11}$$

the differential equation appears as

$$\dddot{x} + \beta\ddot{x} + \dot{x} = f(x)\,, \tag{10.12}$$

where the function

$$f(x) = \mu x(1 - x) \tag{10.13}$$

dependent on a single parameter

$$\mu = cb^{-3/2}\sqrt{1 + 4v^2\,U_0} \tag{10.14}$$

agrees with that which appears in the logistic map (compare Chap. 9). The dynamics of (10.12) or (10.15), having a 'logistic function' (10.13), has been studied in Ref. [10.2]. The reader should be aware of the fact that the parameter μ depends implicitly on the parameters C, C_{m}, R, R_{m}, L, v, U_0 characterizing the electronic circuit.

It is instructive to rewrite the third order differential equation (10.12) as a system of three first order equations

$$\begin{aligned}\dot{x} &= y\\ \dot{y} &= z\\ \dot{z} &= -\beta z - y + f(x)\,,\end{aligned} \tag{10.15}$$

which is of the form

$$\frac{\mathrm{d}\mathbf{r}}{\mathrm{d}\tilde{t}} = \mathbf{v}(\mathbf{r}) \tag{10.16}$$

discussed in Chap. 2.3.

Some characteristic features of the differential equation (10.15) — or the equivalent forms (10.4), (10.6), and (10.12) — are:

- The volume τ in phase space is contracted with constant rate

$$\frac{\dot{\tau}}{\tau} = \operatorname{div}\mathbf{v} = -\beta\,. \tag{10.17}$$

- The stationary solutions of the system are given by $f(x) = 0$, i.e. $x_- = 0$ or $x_+ = 1$, where the fixed points $x_\mp$ are mapped onto the stationary solutions

$$U^*_\pm = \frac{1}{2v^2}\left(1 + 2v^2U_0 \pm \sqrt{1 + 4v^2U_0}\right) \tag{10.18}$$

in terms of the electronic circuit. The upper of these stationary values, U^*_+, is unstable. The lower, U^*_-, can be stable, depending on the parameters. This will be studied in more detail in the numerical experiment in Sect. 10.4.1.

- The variable resistance R_m serves as a control parameter. For large values of R_m, the current I is approximately zero and U remains at its fixed point U^*_-. Varying R_m changes the behavior of the system. When R_m decreases, a closed orbit appears in phase space.
- Decreasing R_m further, one observes a sequence of period-doubling bifurcations. For even smaller R_m values, the system becomes chaotic.

For parameters not very different from the pre-set values modeling a laboratory set-up (see Sect. 10.4.4), the nonlinear dynamics of this *'Chaos Generator'* is less complex than the Duffing oscillator studied in Chap. 8. In most cases, one can clearly observe a sequence of period-doubling bifurcations with transition into chaotic behavior. The system is therefore well suited to demonstrate this route into chaos, in particular because it allows a parallel control by experimental measurements.

10.2 Numerical Techniques

Numerical solutions of the differential equation (10.4) can be obtained in the program by three different numerical algorithms [10.3, 10.4]:

- The *Euler* method, which is a simple first order method (the error term is $O(\Delta t^2)$), as described in Sect. 5.2, is fast, but inaccurate.
- A method proposed by *Heun* to solve a differential equation $\mathrm{d}y/\mathrm{d}t = f(y,t)$ is a second order method (the numerical error is $O(\Delta t^3)$). It requires an evaluation of the function $f(y,t)$ twice per time step Δt. Its propagation scheme

$$y_{i+1} = y_i + \frac{\Delta t}{2}\{\, f(y_i,t_i) + f(y_i + \Delta t\, f(y_i,t_i),\, t_i + \Delta t)\,\} \tag{10.19}$$

  can easily be extended to the three-dimensional case.
- The fourth order *Runge-Kutta* method (the error term is $O(\Delta t^5)$) also discussed in Sect. 5.2, is relatively slow, but accurate.

The constant time step for the numerical integration (the 'step size') can be chosen by the user. One should carefully check the accuracy of the numerical integration by running the program with reduced values of the step size.

It is of interest to compare the actual performance of the electronic circuit with the numerical simulation. The computer is, typically, slower by several orders of magnitude. For practical purposes, such an electronic circuit can be used as an analog computer to model more complicated nonlinear devices [10.1]. The high speed of the analog model makes the study of the dynamics by means of an oscilloscope very convenient. On the other hand, a detailed numerical control of the influence of parameters, the determination of high accuracy data or the suppression of external noise can be more precisely achieved using a digital computer.

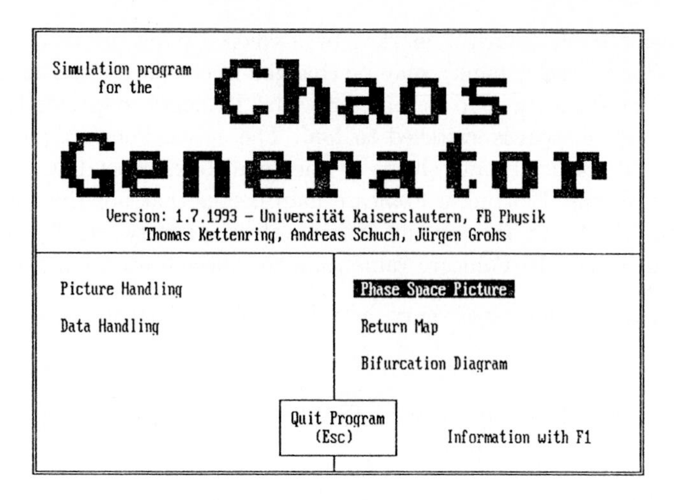

Fig. 10.2. Main menu of program CHAOSGEN.

10.3 Interacting with the Program

The program starts with the *Main Menu* shown in Fig. 10.2. On the left, picture and data handling are controlled. On the right, different iteration modes can be selected. The program calculates phase space pictures, return maps, and bifurcation diagrams, as discussed below. Pressing ⟨F1⟩ provides information.

Main Menu

- **Picture Handling** — Pictures can be saved, loaded, displayed or erased. One can print pictures, enter a comment or – for bifurcation diagrams – select a section. ⟨ESC⟩ cancels the action and switches to the main menu. To toggle the selected printer, move the cursor onto the corresponding item and press any key (except ⟨ESC⟩ and ⟨F1⟩). The picture is printed via LPT1. Pictures are stored as files ∗.PIC, the parameters as files ∗.DAT. It is not necessary to enter the file name extensions. Pictures can only be stored or shown if there is a buffered picture. Buffering can be done while calculating. To quit, press any key.

- **Data Handling** — Suitable system configurations may be saved, loaded or deleted. The appropriate pictures can not be manipulated.

- **Phase Space Picture** — provides phase space visualization in many variations. For the horizontal axis it is possible to choose t, U, U', U'', and for the vertical axis one may choose U, U', U''. Move cursor to the desired item and press ⟨ENTER⟩. A grey text background indicates the active setting. When U (U' or U'') is plotted versus the time t, it is efficient to use the trigger, i.e. the phase is fixed at the left edge of the screen. The value of

U (trigger level), which is marked by a colored point, as well as the sign of the slope (trigger flank), may be changed freely. The center of the menu shows the Poincaré map parameters. The Poincaré map is active if the corresponding item is switched to *'on'*. The active Poincaré parameter is chosen automatically and shown in the small box, depending on the axis representations. Plotting a Poincaré map means that not the whole curve is plotted, but only those points where a third quantity (small box) crosses a certain value, the Poincaré value (in a specified direction, if desired).

Looking at the phase space curve as a three-dimensional object in (U, U', U'') space,

- *'Poincaré Map Off'* plots a two-dimensional projection of the curve, e.g. on the (U, U')-plane,
- *'Poincaré Map On'* plots the intersection points of the curve with a plane parallel to such a two-dimensional plane, e.g. the (U, U')-plane.

By using the picture buffer, one may plot various representations into the same picture in different colors, combining the whole phase space curve with Poincaré maps having different Poincaré values.

- **Initial Values** — can be chosen as the adjusted initial values (mostly in transient state) or the previously computed values. During the computation, one may take over the current values as new initial values. To start computations, move cursor to *'Compute and Display'* and press ⟨ENTER⟩. *'Main Menu'* or ⟨ESC⟩ switch back to the main menu.

- **Change Parameters** — permits one to select the algorithm for integrating the differential equations and to change v, U_0, R_{m}, R, C, C_{m}, L, as well as the initial values, current values, and the scales for t, U, U', and U''. Parameters to change are:

  1. Scale of representation for t, U, U', U'' (screen range for the menu items *'Phase Space'*, *'Bifurcation Diagram'*, and *'Return Map'*) : Lower limit and upper limit.
  2. Values of U, U', U'' : Initial value (i.e. fixed starting values) and current value.
  3. Poincaré values of U, U', U'' : A fixed value which is compared with the current values of U, U', U'' in program modes *'Phase Space'* (only if *'Poincaré Map'* is active), *'Return Map'*, and *'Bifurcation Diagram'*.
  4. Parameters of the electronic circuit : Amplification, voltage, ...; preset values are chosen to model a real electronic circuit assembled using standard components, as discussed in more detail below:

v	gain	$1.2/\sqrt{\mathrm{V}}$
R	resistance	3300Ω
R_{m}	variable resistance	50Ω
C	capacitor	$47 \cdot 10^{-9}$ F
C_{m}	capacitor	$47 \cdot 10^{-9}$ F
L	inductivity	0.1 H
U_0	bias	4 V

5. Step size for integration. A convenient time scale is provided by $\omega^{-1} = \sqrt{LC}$, which is equal to $6.8556 \cdot 10^{-5}$ s for the pre-set parameter values. The pre-set time step is chosen as $0.05\sqrt{LC}$.
6. Method of integration: *Euler* (fast, relatively inaccurate), *Heun* (pre-set) , or *Runge-Kutta* (slow, accurate).

- **Computing and Display** — starts computation and displays the computed results in a desired representation. Pressing ⟨F1⟩ displays information about the hot keys:

Back to menu:	⟨Esc⟩
Clear screen:	⟨Del⟩
Change color:	⟨Enter⟩
Halt calculation:	⟨H⟩
Save picture:	⟨S⟩ (→ buffering the picture)
Show buffered picture:	⟨P⟩
Increase/decrease R_{m}:	
Fine adjustment	⟨←⟩ , ⟨→⟩
Coarse adjustment	⟨Ctrl ←⟩ , ⟨Ctrl →⟩
Trigger on/off:	⟨T⟩
Toggle trigger flank:	⟨-⟩
Adjust trigger level:	⟨↑⟩ , ⟨↓⟩
Restart with initial values:	⟨Home⟩
Set initial to current values:	⟨End⟩
De-/Increase display delay:	⟨F7⟩/ ⟨F8⟩
Display time	⟨F10⟩
Help	⟨F1⟩

Attention: *'Save Picture'* and *'Show Buffered Picture'* use RAM memory (buffer). To preserve the buffered picture, it must be saved using the item *'Picture Handling'* in the main menu.

- **Return Map** — maps the system parameter of one cycle versus the previous cycle when a selected parameter passes the Poincaré value. All combinations are possible. One can call *'Change Parameters'* to alter the configuration. *'Make Diagram'* starts computation. While showing the diagram the following hot keys can be used:

1st ⟨Esc⟩:	enables one to choose a section, a frame is shown
⟨Cursor Keys⟩:	move a corner of the frame
⟨Space Bar⟩:	switches over to the opposite corner
⟨Tab⟩:	switches between coarse and fine adjustment
⟨Enter⟩:	confirms the chosen section
⟨S⟩:	buffers current picture; a previously buffered picture will be destroyed.
2nd ⟨Esc⟩:	section remains unchanged, exit to return map menu

- **Bifurcation Diagram** — computes the behavior of the system as a function of the resistance R_m. The range of R_m ($R_1 \geq R_m \geq R_0$), as well as the direction of the variation, can be adjusted. The values of the selected variable, e.g. the voltage U, are plotted if the prescribed conditions — the Poincaré values, e.g. $U' = 0$ — are met. In order to magnify finer scale features, R_m is varied and displayed on a logarithmic scale.

  The calculated picture can be copied into the internal buffer by pressing ⟨S⟩. A previously buffered picture will be destroyed. At the end of the calculation, a key has to be pressed. The calculation can be aborted by pressing ⟨Esc⟩. In both cases crosshairs appear, which can be moved by the cursor keys and allow one to determine the numerical value of R_m by pressing ⟨Enter⟩. If ⟨Esc⟩ is pressed again, a box appears, which allows one to change the boundaries of the picture. The following hot keys can be used:

1st ⟨Esc⟩:	stops calculation and shows crosshairs
⟨Cursor Keys⟩:	move the crosshairs
⟨Tab⟩:	switches between coarse and fine adjustment
⟨Enter⟩:	displays value of R_m at current position
2nd ⟨Esc⟩:	shows a frame
⟨Cursor Keys⟩:	move a corner of the box
⟨Space Bar⟩:	switches to the opposite corner
⟨Tab⟩:	switches between coarse and fine adjustment
⟨Enter⟩:	confirms the chosen section
3rd ⟨Esc⟩:	section remains unchanged, exit to bifurcation diagram menu

  During the first 'transient' oscillations, the influence of initial conditions is damped and the system approaches the attractor. Therefore the first N points are not displayed. The value of N can be modified in the parameter menu. If the chosen value is too small, hysteresis effects may appear. The pre-set value is $N = 5$. For a fast overview, $N = 0$ can be chosen. An input of $N = -1$ activates an automatic adaption of the number of transient periods.

  On the left *'No automatic saving'* or *'Save automatically to file'* can be chosen. Since the computing time may be long (some hours), the automatic saving mode is convenient. *'No automatic saving'* is preset. If the automatic save mode is activated and no file is defined, the file select box appears and a file name can be specified. The file name is displayed on the screen in a box below. To change the autosave file, move the cursor onto this box and press ⟨ENTER⟩.

10.4 Computer Experiments

10.4.1 Hopf Bifurcation

In the limit of large values of the resistance R_m, the damping term dominates and the oscillation $U(t)$ approaches the fixed point U_-^* given in (10.18). For smaller values of R_m, the circuit starts to oscillate. This is demonstrated in the (U, U')–phase space plots shown in Figs. 10.3 and 10.4 using the pre-set parameter values. Results are shown for two values of the resistance $R_\mathrm{m} = 1000\,\Omega$ and $500\,\Omega$. In the first case, the trajectory (started with the pre-set initial conditions $(U, U', U'') = (0, 0, 1)$) spirals inward, approaching the stable fixed point at $U_-^* = 2.6477$, which is in numerical agreement with (10.18). It should be noted that this value does only depend on the system parameters v and U_0. U_-^*, in particular, is independent of the variable resistance R_m. For $R_\mathrm{m} = 500\,\Omega$, the behavior is qualitatively different: the orbits are attracted by a limit cycle, as shown for an initial condition of $(U, U', U'') = (U_-^*, 5000, 1)$.

The situation resembles somewhat the limit cycle found in the well-known damped and forced harmonic oscillator, as briefly discussed in the introduction of Chap. 8 where, in the long time limit, the system oscillates with the frequency of the external force. However, it should be stressed that there is *no* such external force in the present case. The frequency of the limit cycle is an intrinsic property of the system, depending, of course, on the values of its parameters.

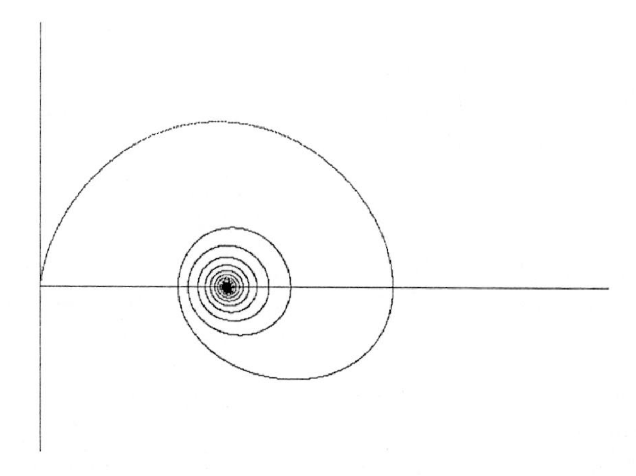

Fig. 10.3. (U, U') – phase portrait for large damping ($R_\mathrm{m} = 1000\,\Omega$). The trajectory is attracted by a fixed point.

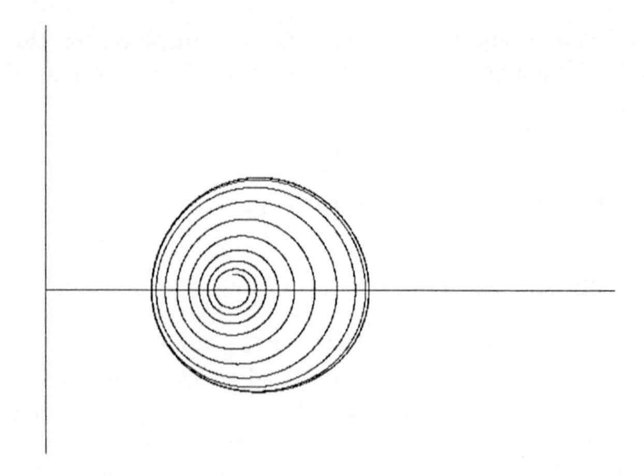

Fig. 10.4. (U, U') – phase portrait for reduced damping ($R_{\mathrm{m}} = 500\,\Omega$). The trajectory is attracted by a limit cycle.

The transition from a point attractor to a limit cycle (by decreasing R_{m}) is an example of a *Hopf bifurcation* as discussed in Sect. 2.4.6. To explore the transition in more detail, the dynamics of the system can be studied as a function of the resistance R_{m}. One observes a (constant) stable fixed point up to a value of $R_{\mathrm{m}} \approx 771\,\Omega$, where a limit cycle appears, increasing in amplitude when the dissipation is further reduced.

The bifurcation can be analyzed theoretically using the transformed differential equations (10.15) – (10.16). Linearizing the force term at the fixed point $(x^*, 0, 0)$, we obtain the linear system

$$\frac{\mathrm{d}}{\mathrm{d}\tilde{t}} \begin{pmatrix} x - x^* \\ y \\ z \end{pmatrix} = \begin{pmatrix} 0 & 1 & 0 \\ 0 & 0 & 1 \\ f'(x^*) & -1 & -\beta \end{pmatrix} \begin{pmatrix} x - x^* \\ y \\ z \end{pmatrix} \tag{10.20}$$

(here prime denotes differentiation with respect to x). The eigenvalues λ of the matrix — given by the solutions of

$$\lambda^3 + \beta\lambda^2 + \lambda = f'(x^*) \tag{10.21}$$

with $f'(x) = \mu(1-2x)$ — determine the stability properties of the fixed point. These stability properties change at the bifurcation points, where the eigenvalues λ crosses the axis $\mathrm{Re}(\lambda) = 0$. There are two possibilities:

- A single eigenvalue crosses $\mathrm{Re}(\lambda) = 0$. This happens at $f'(x^*) = 0$ and a stable solution appears for $f'(x^*) < 0$. Since $f'(x^*_-) = f'(0) = \mu < 0$, the fixed point x^*_- is always unstable for the system considered here.

- A pair of complex conjugate eigenvalues crosses $\mathrm{Re}(\lambda) = 0$. This happens at $f'(x^*) = -\beta$, where we find eigenvalues $\lambda = \pm\mathrm{i}$ and $-\beta$. The condition

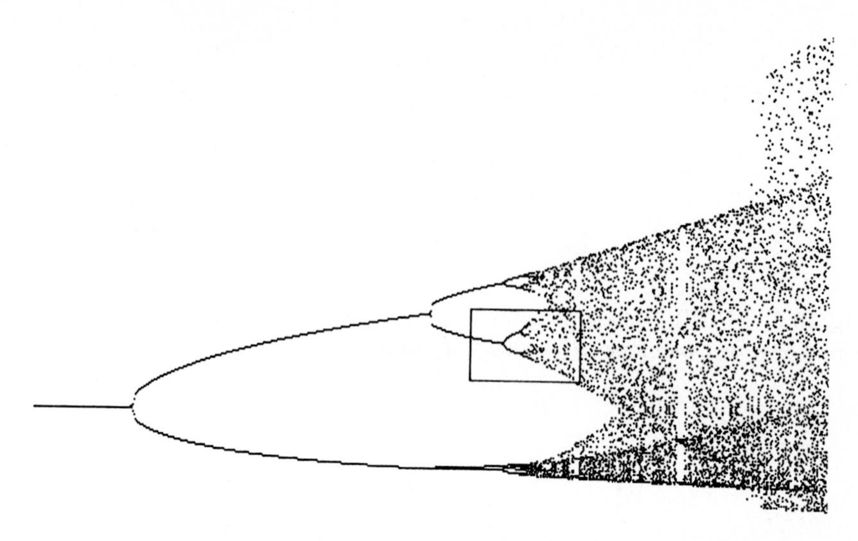

Fig. 10.5. Bifurcation diagram: amplitude of voltage U as a function of $\log R_m$ varying from 250 (left) to 80 Ω (right). This figure is stored (files BIFURC.PIC, –.DAT).

is satisfied for the fixed point $x_+^* = 1$ in the present case, which yields the bifurcation condition $\mu = \beta$. The fixed point x_+^* becomes unstable and a stable limit cycle appears, i.e. we have a supercritical Hopf bifurcation.

Because μ and β are both functions of R_m given in (10.14) and (10.7), such a Hopf bifurcation is to be expected for a certain value of R_m, which can be calculated as 770.6 Ω for the pre-set parameters values, which is in accord with observations obtained from the above computer experiment.

10.4.2 Period Doubling

A bifurcation diagram shows the resistance (the control parameter R_m) on the horizontal axis (logarithmic scale) and a system parameter on the vertical axis which characterizes the behavior of the circuit. The system parameter may, for example, be the time between two zero passages of U', or the value of U or U'' at this zero passage. One should allow the resonant circuit to oscillate for several periods without taking data in order to avoid transient effects in the diagram.

Figure 10.5 shows, as an example, the voltage U as a function of R_m in the interval $250\,\Omega \geq R_m \geq 80\,\Omega$ for the pre-set parameter values.

For a resistance $R_m > 220\,\Omega$, we find the limit cycle that has emerged from the fixed point at large resistance through a Hopf bifurcation, as described above. The trajectory approaches a closed loop in phase space, as shown in Fig. 10.6. In the bifurcation diagram 10.5, this limit cycle appears as a single point, because data are only recorded for positive slope of U'. With decreasing dissipation, we observe a succession of further bifurcations, where the period

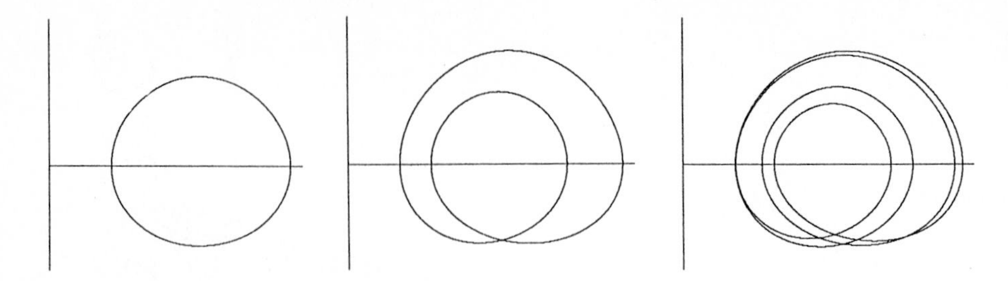

Fig. 10.6. (U, U') – phase portraits of period-doubling bifurcations: oscillations of period one, two, and four ($R_{\mathrm{m}} = 250$, 150, 130 Ω).

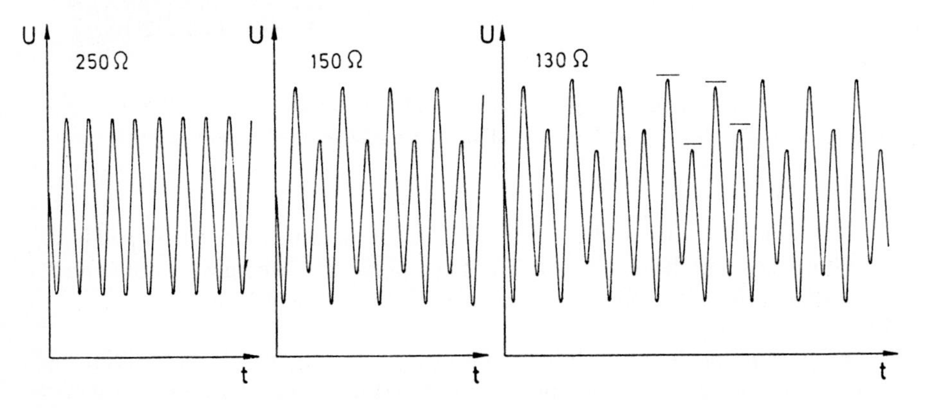

Fig. 10.7. Time dependence of the voltage $U(t)$ showing period-doubling bifurcations from period one, two, and four.

doubles. At $R_{\mathrm{m}} \approx 220\,\Omega$, the period-one limit circle becomes unstable and a period-two oscillation is observed, followed by a bifurcation to period-four. These orbits are also shown in Fig. 10.6, or, more precisely, the projections onto the (U, U')-plane for $R_{\mathrm{m}} = 250$, 150, and 130 Ω. The period-doubling bifurcations are also visible in the oscillations of the voltage $U(t)$ as a function of time in Fig. 10.7. This bifurcation process continues up to a a critical value of $R_{\mathrm{m}} \approx 120\,\Omega$, where the dynamics becomes chaotic.

In order to establish a *quantitative* agreement with the period doubling sequence found in the Feigenbaum scenario for one-dimensional maps in Chap. 9, the reader might be tempted to try a 'measurement' of the critical values $R_{\mathrm{m,k}}$ of the resistance at the $2^{k-1} \to 2^k$ bifurcation points in a computer experiment. Then, the limit of the ratio

$$\Delta = \lim_{k\to\infty} \frac{R_{\mathrm{m,k}} - R_{\mathrm{m,k-1}}}{R_{\mathrm{m,k+1}} - R_{\mathrm{m,k}}} \tag{10.22}$$

could be compared with the Feigenbaum constant $\delta \approx 4.669$ discussed in Chap. 9 (compare (9.20) and (9.21)). For several reasons, this is, however, *not* straightforward.

— First, the brute force method to determine the bifurcation values $R_{\mathrm{m,k}}$ is not as simple as it seems to be. The reader may try to pin down the value for the first bifurcation from period-one to period-two for the pre-set parameters. A few test runs of phase space orbits will show a bifurcation near $R_{\mathrm{m}} \approx 220\,\Omega$. To ensure a converged result, one then increases the accuracy by reducing the timestep to, say, $0.01\sqrt{LC}$ and selecting the more accurate Runge-Kutta integration. It then comes as a surprise that a kind of hysteresis is observed. At large R_{m}, we find a period-one orbit. Decreasing slowly the values of R_{m} (by mean of the cursor keys), it will bifurcate to a period-two cycle at about $215\,\Omega$. Increasing R_{m} again will yield a seemingly stable period-two cycle up to about $221\,\Omega$. The same can be found in an automatically computed bifurcation diagram. The solution follows the unstable branch for some time until it realizes its instability.

— Secondly, it is possible to reduce these numerical 'hysteresis' effects by waiting long enough to allow the system to settle down to the stable branch. This may lead to extremely long simulation times, because the unstable orbit is only weakly repelling close to a bifurcation.

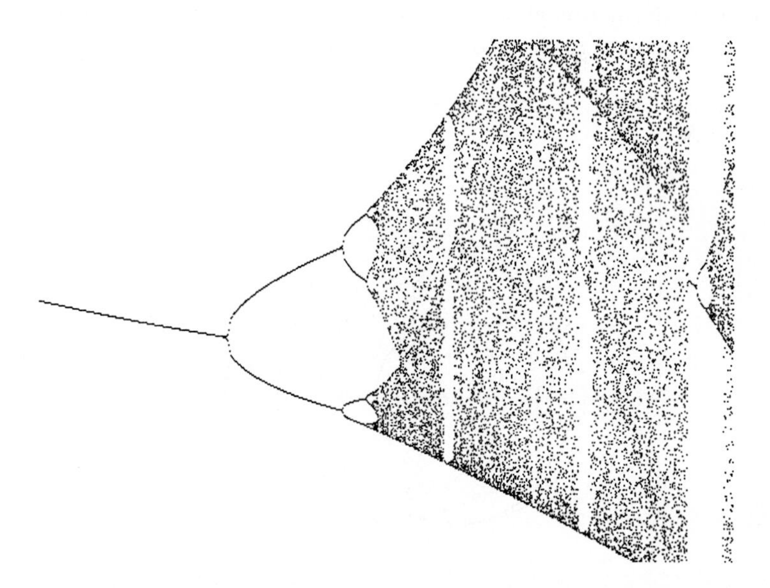

Fig. 10.8. Bifurcation diagram: amplitude of voltage U as a function of $\log R_{\mathrm{m}}$ varying from 133 (left) to 114 Ω (right). Magnification of the rectangle in Fig. 10.5.

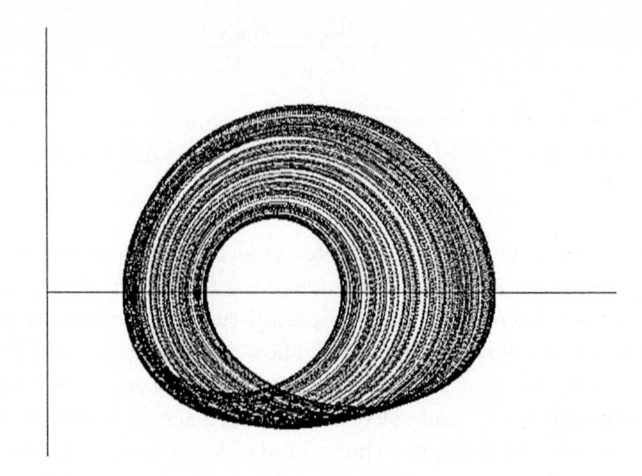

Fig. 10.9. (U, U') – phase portrait of a strange attractor in the chaotic regime ($R_{\mathrm{m}} = 100\,\Omega$).

— In order to determine the limit ratios of the differences between the bifurcation points in (10.22), a number of these values must be known with sufficiently high precision. This is definitely a *very* time consuming task on a PC — at least when brute force techniques are used.

Further information about such numerical problems and more refined techniques can be found in the book by Parker and Chua [10.5].

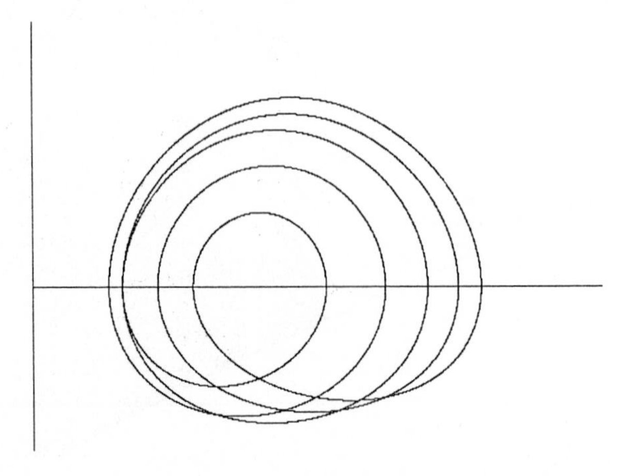

Fig. 10.10. (U, U') – phase portrait of a stable limit cycle in the period-five window embedded in the chaotic region ($R_{\mathrm{m}} = 99.5\,\Omega$).

Another phenomenon which appears in the bifurcation diagram is the approximate self-similarity visible in Fig.10.5. A magnification of this figure is shown in Fig.10.8, which closely resembles the original diagram. This can be analyzed quantitatively by a renormalization mapping. The period-doubling bifurcation sequence closely resembles the logistic map discussed in Chap. 9. This is analyzed in more detail in terms of a return map in the following numerical experiment.

In the chaotic regime, we find chaotic bands with strange attractors interrupted by periodic windows, as discussed in Chap. 9 for the logistic map. The most prominent cycles in these windows have periods six and five. Apart from a few exceptions discussed below, this window organization is again consistent with the structural universality found for one-dimensional maps. As an example, a phase portrait of a strange attractor is shown in Fig. 10.9, which is very similar to the Rössler attractor explored in Sect. 12.3.5. Fig. 10.10 finally shows a stable limit cycle in the period-five window embedded in the chaotic region.

10.4.3 Return Map

The dynamics of the system can be approximately described by a one-dimensional map, the so-called return map, which maps a system parameter in a Poincaré map versus its values in the preceding cycle. Such a map reduces the dynamics by projection to a lower-dimensional space. The program allows a numerical construction of a return map by activating the item *'Return Map'*. Fig. 10.11 shows a return map of U_{n+1} (vertical) versus U_n for a resistance $R_\mathrm{m} = 105\,\Omega$ (pre-set parameter values).

Data are recorded when U' passes through zero with negative slope. For the case $R_\mathrm{m} = 105\,\Omega$, the map is close to a one-dimensional line and can be described to a good approximation as a one-dimensional iteration $U_{n+1} = U_n$ having a

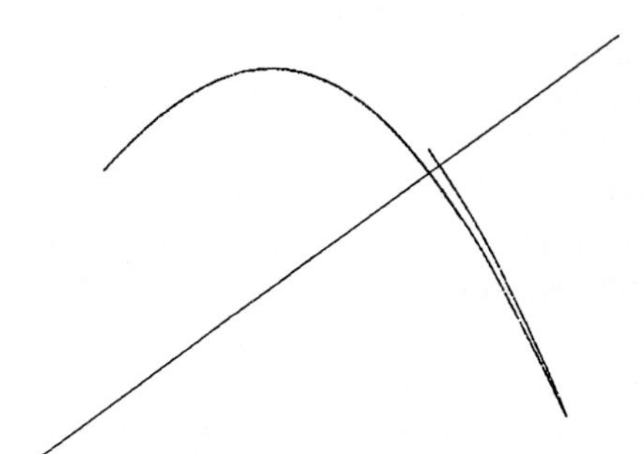

Fig. 10.11. Return map of U_{n+1} (vertical) versus U_n for resistance $R_\mathrm{m} = 105\,\Omega$ (pre-set parameter values). Shown is the range $4\,\mathrm{V} < U < 6.5\,\mathrm{V}$.

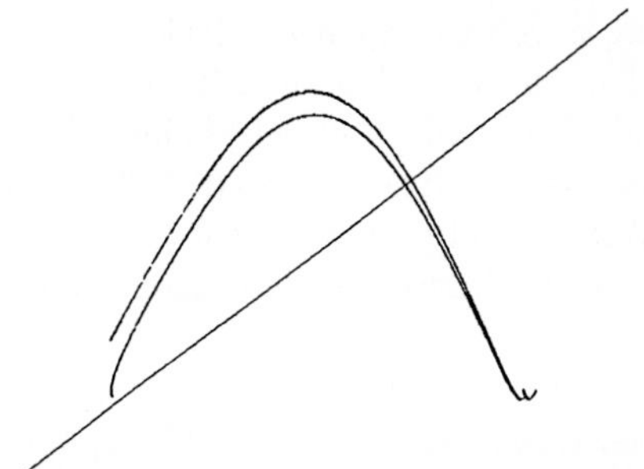

Fig. 10.12. Return map of U_{n+1} (vertical) versus U_n for resistance $R_\mathrm{m} = 62\,\Omega$ (pre-set parameter values). Shown is the range $2.5\,\mathrm{V} < U < 7.5\,\mathrm{V}$.

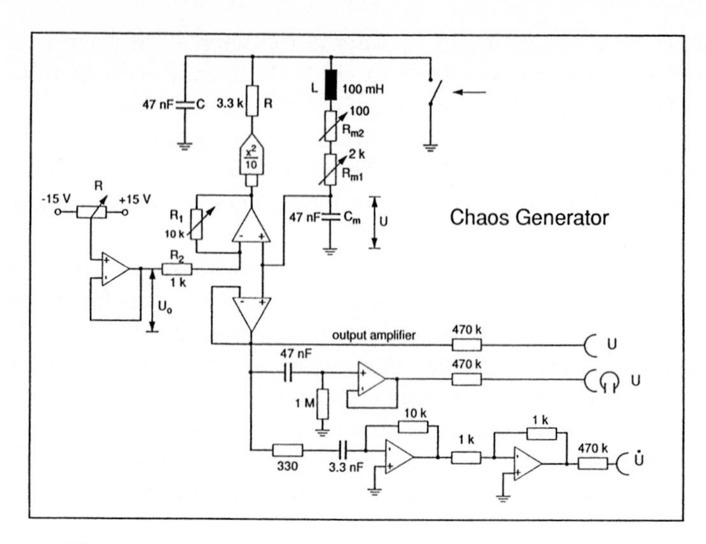

Fig. 10.13. Electronic circuit used in a laboratory set-up of the chaos generator.

quadratic maximum. Hence, the dynamics can be approximately described by a logistic map, which is discussed in Chap. 9. For decreasing values of R_{m}, the seemingly one-dimensional line splits, showing that it is multiply folded in phase space. In addition, a minimum appears at larger values of U_n. This is illustrated in Fig. 10.12 for $R_{\mathrm{m}} = 62\,\Omega$. Deviations from the behavior of the logistic map are thus expected in this parameter region (see below).

10.4.4 Suggestions for Additional Experiments

Comparison with an Electronic Circuit. An electronic oscillator [10.1] was set up for students in an advanced laboratory. A RLC circuit in series, containing a voltage squaring device as a nonlinear element, shows a broad variety of nonlinear behavior. Figure 10.13 shows the circuit used in the experimental set-up, which closely follows Fig. 10.1 (LF 356 H operational amplifiers and analog multipliers ICL 8013 CC or AD 534 used as a $\mathrm{x}^2/10$ squaring device). Voltage and current at the capacitor C_{m} can be monitored on an oscilloscope. It is also impressive to demonstrate the different oscillations acoustically via a loudspeaker.

The control parameter for the system is represented by a variable damping resistance $R_{\mathrm{m}} = R_{\mathrm{m1}} + R_{\mathrm{m2}}$, where R_{m1} is used for coarse and R_{m2} for fine adjustment. The bias U_0 can be varied and the resistance R_1 allows one to control the amplification v. As a function of R_{m}, the electronic circuit shows normal oscillations, frequency doubling periods, bifurcation as a route to chaos and chaotic motion, as well as irregular amplitude steps. Within the chaotic regime, windows of regular dynamics are observed.

In addition to this experimental set-up, one can use the simulation program CHAOSGEN which models exactly the same RLC circuit. The predictions of the numerical simulation can be compared with qualitative observations and quantitative measurements. Of interest are:

- The typical features of a period-doubling bifurcation, both for phase space trajectories and the corresponding time-dependent amplitudes.
- The localization of the bifurcation points when the resistance R_{m} is varied.
- The transition to chaotic motion
- The experimentally derived data are used to determine the Feigenbaum constant.
- The appearance of the characteristic succession of regular windows within the chaotic region.
- The influence of a variation of other system parameters such as, for instance, the amplification v, which can be controlled by the resistance R_1.
- A Fourier transformation of the experimental data easily shows frequency doubling. As an example, Fig. 10.14 displays the frequency (power) spectra of a period-one, period-two, and a chaotic oscillation of this electronic circuit. Also shown are the corresponding phase portraits. For periodic motion the frequency spectra are sharply peaked and the frequency halving (period doubling) is clearly visible. A broadening of the frequency spectrum indicates the onset of chaotic motion and the power spectra are a superposition of broad band noise and peaks due to periodic components of the oscillation.

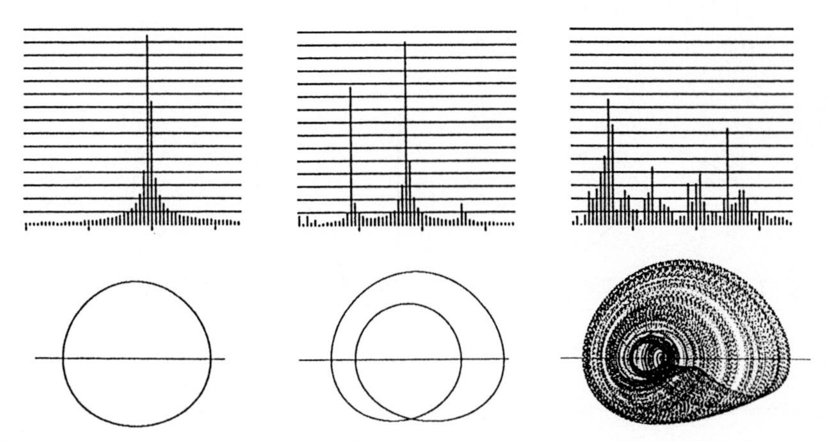

Fig. 10.14. Frequency spectra and phase portraits of period-one, period-two, and chaotic oscillation. The broadening of the characteristic frequencies is due to the limited time window during data sampling.

Deviations from the Logistic Mapping. As demonstrated above, the dynamical behavior of a chaos generator is very similar to that of the logistic equation, which was discussed in detail in Chap. 9. However, a three-dimensional flow should behave like a two-dimensional mapping rather than a one-dimensional one. Therefore, some deviations from the logistic equation can be expected. Some such deviations are discussed by Mitschke and Flüggen [10.1]:

- An amplitude step accompanied by a hysteresis appears at the onset of the period-three window.

- Windows not consistent with the characteristic window organization for quadratic one-dimensional maps appear. Beyond the first period-three window, one observes a second period-three window with a bifurcation to period-six.

- For low values of R_m additional structures appear in the return map.

As pointed out in Ref. [10.1], some of these deviations can also be observed for the two-dimensional Hénon map (see Chap. 11), which is also volume contracting with constant rate as the phase space flow in the present case (compare (10.17).

Boundary Crisis. In the investigation of the return map, deviations from the one-dimensional quadratic map are observed when the resistance R_m is decreased. It is interesting to study the oscillation in this region in more detail. Figs. 10.15 and 10.16 show phase portraits for three values of R_m at 49 Ω, 32 Ω, and 27 Ω (pre-set parameter values). One observes an extra 'hook' in the phase portrait for small values of R_m, which approaches $(U, U') = (U_+^*, 0)$ at a critical value of R_m. This is an example of a *boundary crisis*, as discussed in Ref. [10.6].

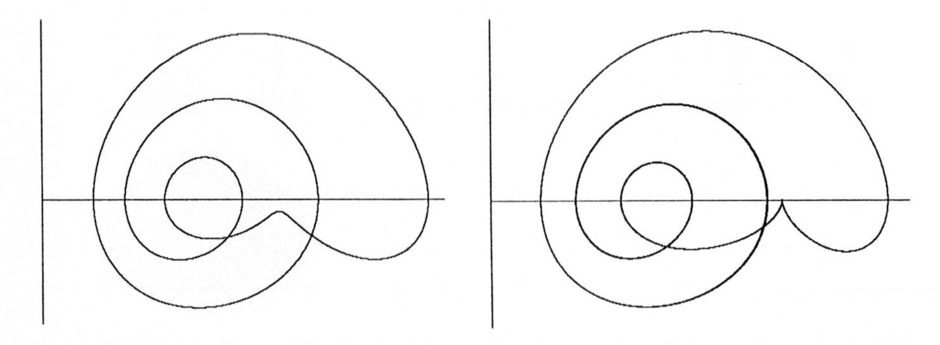

Fig. 10.15. (U, U') – phase portraits of the stable period-three orbit for small values of the resistance ($R_\mathrm{m} = 49$ and 32 Ω).

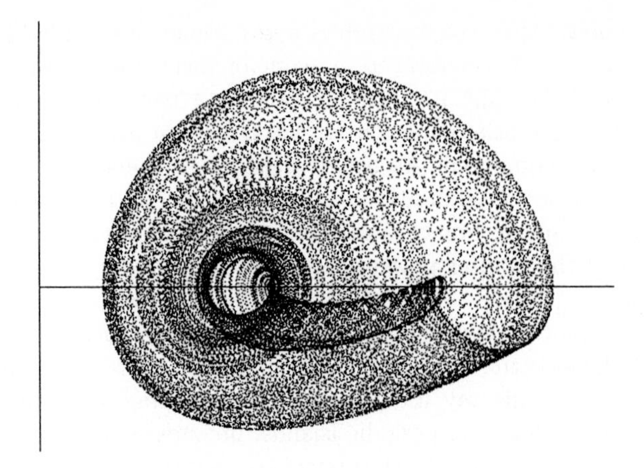

Fig. 10.16. (U, U') – phase portraits of a strange attractor appearing at small values of $R_{\rm m} = 27\,\Omega$.

10.5 Real Experiments and Empirical Evidence

The known experiments use electronic circuits and nonlinear elements or tricky feedback mechanisms.

Briggs's [10.7] chaos generator consists of a resistor, an inductor, and a silicon diode in series, driven by sinusoidal voltage. A diode acts approximately as a small capacitance when reverse biased and as a constant voltage source when forward biased; therefore, the reverse capacitance of the diode is a nonlinear function of the reverse voltage. The waveform of the diode voltage for the circuit (one channel) and that of the driving frequency (second channel) are shown on an oscilloscope. If the applied frequency is close to the resonant frequency of the inductor and reverse capacitance of the diode, then, as the applied voltage is increased, the diode voltage is seen to undergo period doublings, eventually becoming chaotic.

In the original publication by Testa et al. [10.8], who also used a series of RLC circuits, where the voltage across a Si-diode is the nonlinear element, all technical details and parameters of the electronic components are mentioned, and the differential equation for the circuit is described and solved numerically; the first measured bifurcation diagram shows the subharmonic sequences, the bifurcation threshold, the onset of chaos, band merging, noise-free windows, etc.

As already described above, we used an electronic circuit in an advanced student laboratory, where the students utilized the features of the numerical simulation program CHAOSGEN in parallel with measurements of the real electronic circuit.

Levinsen [10.9] used a high frequency oscilloscope operated in the xy-mode and a photodiode, which detects the light emitted from the trace on the oscilloscope and feeds the original signal back to the oscilloscope. In spite of its apparent simplicity, the system exhibits extraordinarily complex behavior. In addition, he modeled this chaotic oscilloscope by means of an equivalent circuit consisting of a diode, capacitor, and resistor in a parallel circuit. Numerical solutions of the differential equation for the circuit (a driven van der Pol–like equation) are also compared. As usual, the measurements are made using a two channel oscilloscope (excitation signal, response) and displayed as period doubling in phase space, Poincaré mapping, Feigenbaum tree. Of interest is a detailed study of the Fourier power spectra in the periodic, quasiperiodic, and chaotic cases.

For a driven RLC circuit containing a diode as a nonlinear capacitor, Moon [10.10, p. 164] displays the driving voltage versus driving frequency plane. When data are presented in this way, it is possible — by choosing electronic parameters — to look for subharmonics, periodic islands, precursors to chaos, and chaotic regimes. Moon [10.10, p. 285] further refers to a series of experiments in which a circuit consists of coupled two neon bulb circuits. A single circuit can manifest relaxation oscillations. When coupled together, the two circuits can exhibit stationary, periodic, or chaotic dynamics made visible to the observer by the flashing neon bulb. Sometimes tunnel diodes were used instead of neon bulbs monitored by an oscilloscope.

References

[10.1] F. Mitschke and N. Flüggen, *Chaotic behavior of a hybrid optical bistable system without time delay*, Appl. Phys. **B35** (1984) 59

[10.2] P. Coullet, C. Tresser, and A. Arnéodo, *Transition to stochasticity for a class of forced oscillators*, Phys. Lett. A **72** (1979) 268

[10.3] J. Stoer and R. Burlisch, *Introduction to Numerical Analysis* (Springer, New York 1983)

[10.4] W. H. Press, B. P. Flannery, S. A. Teukolsky, and W. T. Vetterling, *Numerical Recipes* (Cambridge University Press, Cambridge 1986)

[10.5] T. S. Parker and L. O. Chua, *Practical Numerical Algorithms for Chaotic Systems* (Springer, New York 1989)

[10.6] C. Grebogi, E. Ott, and J. A. Yorke, *Crisis, sudden changes in chaotic attractors and transients to chaos*, Physica D **7** (1983) 181

[10.7] K. Briggs, *Simple experiments in chaotic dynamics*, Am. J. Phys. **55** (1987) 1083

[10.8] J. Testa, J. Perez, and C. Jeffries, *Evidence for universal chaotic behaviour of a driven nonlinear oscillator*, Phys. Rev. Lett. **48** (1982) 714 (reprinted in: B.-L. Hao, *Chaos* (World Scientific, Singapore 1984) and P. Cvitanović, *Universality in Chaos* (Adam Hilger, Bristol 1984))

[10.9] M. T. Levinsen, *The chaotic oscilloscope*, Am. J. Phys. **61** (1993) 155

[10.10] F. C. Moon, *Chaotic Vibrations* (J. Wiley, New York 1987)

11. Mandelbrot and Julia Sets

11.1 Two-Dimensional Iterated Maps

As already pointed out in Chap. 9, discrete iterated maps appear almost routinely in studies of nonlinear dynamical systems, e.g. as Poincaré maps. Because they are discrete, such maps are much simpler to study (both numerically and analytically) than continuous differential equations. In general, the maps can be written as

$$\mathbf{r}_{n+1} = \mathbf{F}\,(\mathbf{r}_n, \mathbf{c})\,, \tag{11.1}$$

where $\mathbf{r} = (r_1, \ldots, r_N)$ is the state vector of the system — for example, a vector in N-dimensional phase space — and $\mathbf{c} = (r_1, \ldots, r_M)$ denotes a number of M parameters.

Of central interest is the iterative behavior of $\mathbf{r}_n$, starting from an initial point $\mathbf{r}_0$. This behavior can be extremely complicated, showing an intricate dependence on the parameters $\mathbf{c}$. There are 'k-periodic cycles' $\mathbf{r}_{n+k} = \mathbf{r}_n$, 'attractors', which attract all iterated sequences starting from their 'basin of attraction', 'strange attractors' showing a complicated Cantor set structure, etc. These features show characteristic structural changes ('bifurcations') when the parameters $\mathbf{c}$ are varied.

The most popular example of iterated maps is the logistic map $x_{n+1} = 4\lambda x_n(1 - x_n)$ discussed in Chap. 9, which is a one-dimensional quadratic map depending on a single parameter, λ, which leads to the typical scenario of sequential period-doubling bifurcations followed by a chaotic regime. The behavior of higher dimensional maps is much richer. Here, we study two-dimensional quadratic maps

$$\begin{aligned} x_{n+1} &= a_0 + a_1 x_n + a_2 y_n + a_3 x_n y_n + a_4 x_n^2 + a_5 y_n^2 \\ y_{n+1} &= b_0 + b_1 x_n + b_2 y_n + b_3 x_n y_n + b_4 x_n^2 + b_5 y_n^2 \;. \end{aligned} \tag{11.2}$$

It should be noted that the logistic map appears as a special case of (11.2). Another important subclass is the class of quadratic complex valued maps, i.e. quadratic maps (11.2) satisfying the Cauchy-Riemann differential equations

$$\frac{\partial x_{n+1}}{\partial x_n} = \frac{\partial y_{n+1}}{\partial y_n} \quad , \quad \frac{\partial x_{n+1}}{\partial y_n} = -\frac{\partial y_{n+1}}{\partial x_n} \tag{11.3}$$

which impose the analyticity constraints

$$a_1 = b_2\,,\ a_2 = -b_1\ ,\ a_4 = -a_5 = b_3/2\ ,\ b_4 = -b_5 = -a_3/3 \tag{11.4}$$

in the present case.

A typical representative of quadratic complex maps is the so-called Mandelbrot map

$$z_{n+1} = z_n^2 + c\,, \tag{11.5}$$

where $z = x+iy = (x,y)$ and the parameter $c = a_0+ib_0$ are complex numbers. For the real and imaginary parts written explicitly in the form (11.2), we obtain $a_4 = 1$, $a_5 = -1$, and $b_3 = 2$ for the non-zero parameters.

The peculiarities of the Mandelbrot map (11.5) are explored and discussed in many original articles and textbooks (see, e.g., [11.1]–[11.10]). Here, we confine ourselves to a discussion of some basic features.

Let us first look at the set of all starting points $z_0 = (x_0, y_0)$, for which the sequence z_n goes to infinity. Its complement contains, in particular, all periodic cycles of the mapping (11.5). The boundary of this set is denoted as the *Julia set* $\mathcal{J}$ [11.11]–[11.14]. It has been shown that the Julia set is contained inside the circle $|z| < 2$, i.e., in order to decide whether the sequence z_n approaches infinity, one has to iterate until $|z_n|$ exceeds the value of two. If this is the case, the sequence goes to infinity.

The structure of the Julia set depends intricately on the parameter c, and one observes a surprising variety of structurally different shapes. It has been shown that the Julia set is either connected or a Cantor set, i.e. a dust of points which is nowhere dense. More details will be discussed in the numerical experiment *'Mandelbrot and Julia Sets'* below. Fig. 11.12 shows several examples of Julia sets.

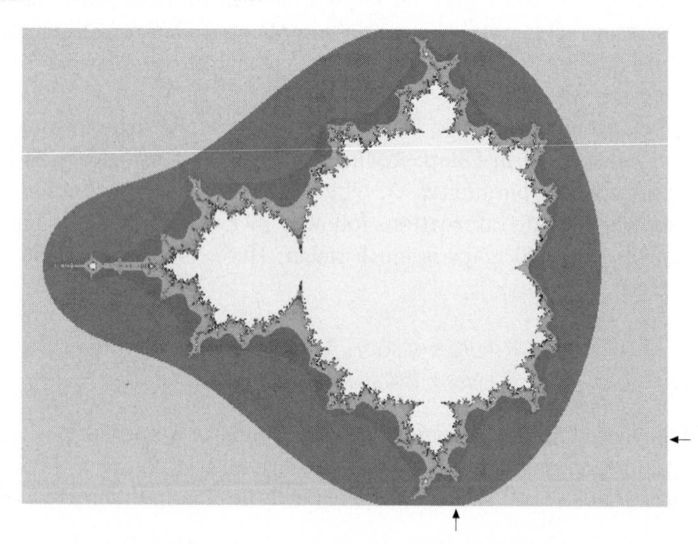

Fig. 11.1. The Mandelbrot set $\mathcal{M}$. Shown is the region $-2.1 < \mathrm{Re}\,c < 1.0$, $-1.15 < \mathrm{Im}\,c < 1.15$. This figure is stored (files MDLBRVGA.MDB, –.DAT).

In 1979 B. B. Mandelbrot discovered that all varieties of structurally different Julia sets can be cataloged by an extremely complicated two-dimensional object, the *Mandelbrot set* $\mathcal{M}$. This is the set of all points $c = (a_0, b_0)$ in parameter space, for which the iteration of (11.5) — starting at $z_0 = (0, 0)$ — does *not* grow without limit. A first impression of the Mandelbrot set can be obtained from Fig. 11.1 (see also the following Sect. 11.4.2). The Mandelbrot set — shown as a white region in Fig. 11.1 — possesses an extremely intricate fractal structure. It can be roughly described as a cactus carrying side lobes of different size arranged in a specific way. These side lobes carry smaller side lobes, and so on, down to infinitely small scales.

Outside the Mandelbrot set, we have $z_0 \to \infty$, and we can use the simple criterion $|z_n|^2 = x_n^2 + y_n^2 > 4$ for any iteration n to decide that a given value of the parameter c does not belong to the Mandelbrot set. In addition, the number n, where this criterion is first satisfied, can be used to measure the speed of divergence, the 'lifetime'. In the program, this value is coded in color and generates a fascinating esthetic picture, which can be only crudely reproduced in black and white in Fig. 11.1. It should be noted that the contour lines of such a plot can be interpreted as equipotential lines when the Mandelbrot set is taken to be a charged conductor [11.5, 11.15].

A sequence of magnifications of certain parts of the Mandelbrot set reveals an approximate self-similarity. Quite often, one discovers almost identical microscopic copies of the whole Mandelbrot set as shown in Fig. 11.2. Some of

Fig. 11.2. A magnification of the Mandelbrot set $\mathcal{M}$ shows a microscopic replica of the whole set. Shown is a tiny region $-0.0081 < \mathrm{Re}\, c < 0.0066$, $-0.7945 < \mathrm{Im}\, c < 0.7934$ at a position marked by arrows in Fig. 11.1. This figure is stored (files MANDLSON.MDB, –.DAT).

these copies are located free in space. This is, however, an illusion, since it has been proved that the Mandelbrot set is connected.

The Mandelbrot set provides a one page catalogue of the infinite number of structurally different Julia sets. One finds a different type of Julia set in each side lobe. In addition, it has been shown that a Julia set is connected if, and only if, the parameter c is an element of the Mandelbrot set.

The Mandelbrot map (11.5) shows the behavior of analytic complex quadratic iterated maps. The general quadratic map (11.2) is, however, much richer and shows many other features. The program also allows a study of the dynamics of such iterated real valued maps. The coloring is related to the boundary and structure of the basin of attraction of the point ∞. The user should, however, be aware of the fact that the program provides a coloring of the plane by counting the number of iterations necessary to reach the distance

$$d_n = \sqrt{x_n^2 + y_n^2} \geq \Gamma\,, \tag{11.6}$$

where the value of Γ can be prescribed by the user. For the Mandelbrot map (11.5), $d_n \geq \Gamma = 2$ yields $(x_n, y_n) \to \infty$, which is in general not true for other maps. Here, one has to chose an appropriate value of Γ, as well as a reasonable scaling of the color steps. In addition, it should be realized that the variation of the parameters a_0 and b_0 done by the program (menu entry *'Mandelbrot'*) may be of limited value for other maps.

Also of particular interest is the behavior of the iterated points (x_n, y_n), the appearance of strange attractors, Smale's horseshoes, ... In addition to the period-doubling bifurcations found for one-dimensional quadratic maps, sequences of period tripling, quadrupling or n-tupling are also found, as well as the coexistence of various limit cycles. An important difference between the analytic Mandelbrot map and general two-dimensional quadratic maps should be emphasized: the quadratic complex maps possess two fixed points in addition to ∞, which may coincide in special cases. For the Mandelbrot map (11.5) these are the points

$$z_\pm = \tfrac{1}{2}\left(1 \pm \sqrt{1-4c}\,\right). \tag{11.7}$$

For the case of a general quadratic map (11.2), we find more than two fixed points (typically up to four). The number, type, and configuration of these fixed points determine the basic features of the dynamics.

More details on general two-dimensional quadratic mappings and a discussion of some numerical experiments can be found in Sect. 11.4.3 below.

The program MANDELBR allows numerical experiments with the iterated quadratic maps (11.2):

- 'Lifetime charts' of the iterated points can be computed and presented graphically in a color code (*'Julia-Set'*).
- The iteration of a single point can be displayed (*'Follow-Iteration'*).

- The dependence of the dynamics on the parameters $(a_0, b_0) = c$ can be visualized (*'Mandelbrot-Set'*).
- By opening windows in the displayed pictures, in a suggestive way one can explore the connection between the behavior during iteration — the Julia set — and the parameter dependence — the Mandelbrot set — (*'Julia Windows'*).
- The intricate richness of the dynamics on smaller and smaller scales can be explored by sequential magnification of regions of interest (*'Zoom In/Out'*).
- Various algorithms for generating the colored maps can be selected. The pictures can be stored, loaded, and printed. It is also possible to generate sequences of pictures automatically.

Details are given in the following sections.

11.2 Numerical and Coloring Algorithms

Numerically, the program iterates equations (11.2) for (x_n, y_n), starting from an initial value (x_0, y_0) up to a maximum number, n_{max}, of iterations. The iteration is stopped when a point is outside the circle $x_n^2 + y_n^2 \leq \Gamma^2$ and the corresponding value of n is called the 'lifetime'. The program generates a lifetime chart in parameter space (i.e. lifetime as a function of the parameters (a_0, b_0)) or in (x_0, y_0) space. This lifetime chart is represented in color code: lifetimes in specified intervals are graphically represented by colors. The coloring code can be controlled by the menu entries *'Coloring'* and *'Palette'*, where the user can select linear, exponential or logarithmic coloration. In addition, one can choose between a coloring of lifetime intervals, or a cyclic change of colors at each alteration of the lifetime. Points which did not reach the iteration boundary Γ are colored black (note that these points appear in white in the figures).

The program allows a selection of different coloring algorithms, which have both advantages and shortcomings. First of all, one can distinguish between 'exact' and 'approximate' methods:

- An 'exact' algorithm finally generates a correctly colored chart, where each pixel on the screen shows the color corresponding to its lifetime. Such a method ultimately carries out the numerical iteration for each point on the screen. The different algorithms (*'Straight Forward'*, *'Refinement'*) differ in the chosen sequence of computed points. It should be noted, however, that additional error sources exist; for instance, an inappropriate choice of the iteration boundary Γ or the limited precision of the numerical iteration, which may be important in cases of extreme magnification (the numerical precision can be altered in the menu item *'Accuracy'*).
- An 'approximate' algorithm may lead to wrongly colored points. Some methods are based on interpolation, where certain points are never iterated

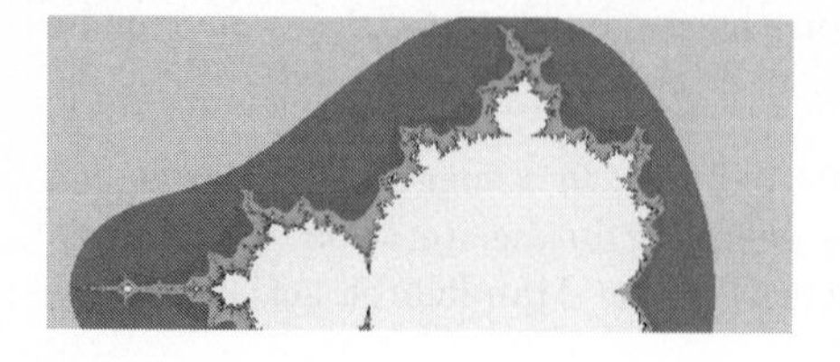

Fig. 11.3. Coloring by *'Straightforward'*.

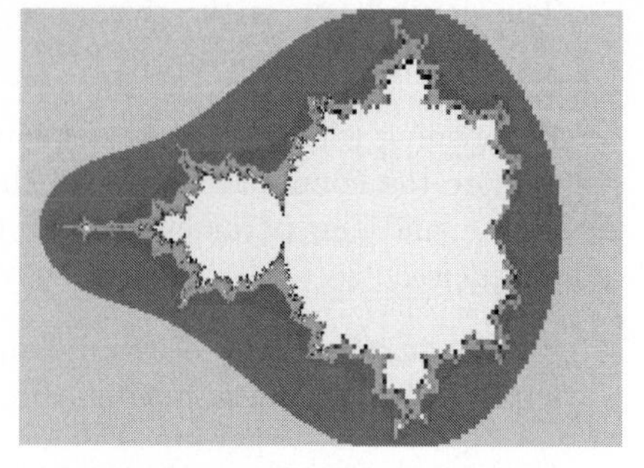

Fig. 11.4. Coloring by *'Refinement'*.

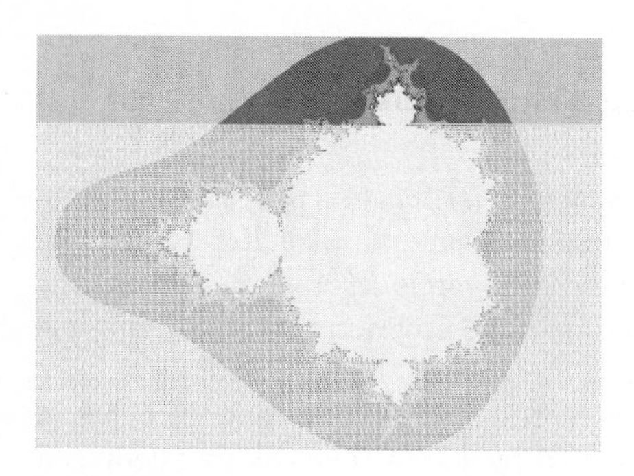

Fig. 11.5. Coloring by *'Interpolation'*.

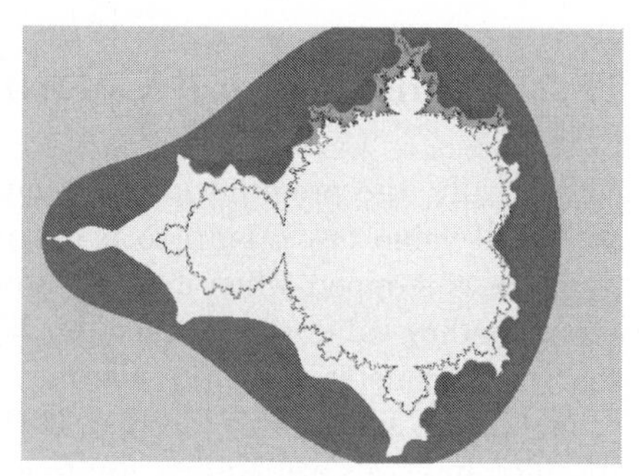

Fig. 11.6. Coloring by *'Turtle Fill'*.

(*'Interpolation'*). Other methods follow certain search strategies to detect the boundaries of the colored regions (*'Turtle Fill'* or *'Square Fill'*). Typically, these algorithms do not discover island structures embedded in a monochrome sea.

The visual impression of the different coloring algorithms is illustrated in Figs. 11.3–11.7. More details are given below. Generally — though calculating basically the same colored picture — the different algorithms generate completely different visual impressions during the time of computation. It should be noted that these coloring algorithms (additional ones can be found in Ref. [11.8]) can also be used for other computational purposes.

Finally, it should be pointed out that the program does *not* make use of methods which are only valid for speeding up the computation for the special case of the Mandelbrot map (11.5), such as the so-called cartioid test, where the numerical iteration is bypassed in the interior of a curve inside the Mandelbrot set.

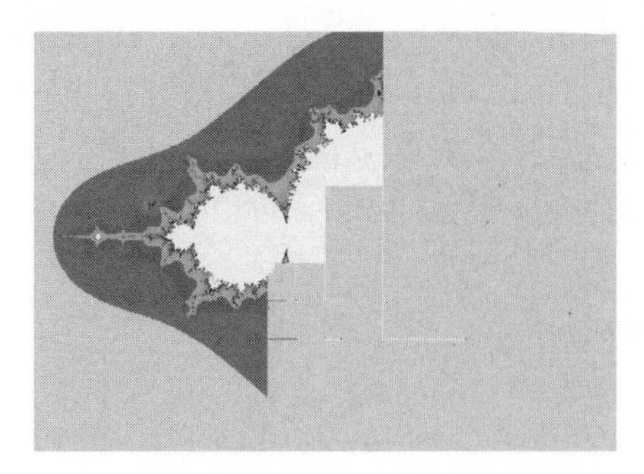

Fig. 11.7. Coloring by *'Square Fill'*.

11.3 Interacting with the Program

The main menu of MANDELBR shown in Fig. 11.8 allows one to choose an item using cursor keys and ⟨ENTER⟩, or by pressing the appropriate highlighted character (hot key). In the following, these hot keys are marked by ⟨ ⟩. The program responds to the hot keys even when the main menu is not displayed.

- **E⟨X⟩ecute** — starts the calculation with current parameters. The calculation can be interrupted by pressing a key. If this is a hot key, the corresponding function is executed.
- **⟨M⟩andelbrot** — allows one to change the parameters (shown area, maximum iteration depth) for calculating the Mandelbrot set. The iteration depth can be chosen by the user or estimated automatically by the program.

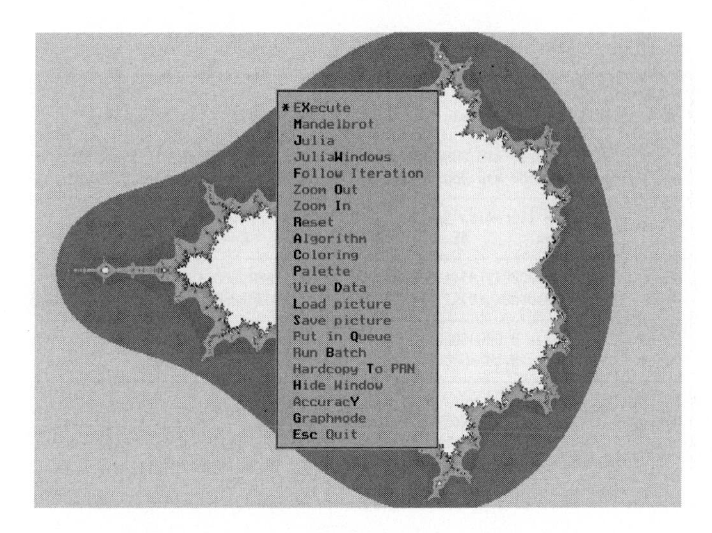

Fig. 11.8. Main menu of program MANDELBR.

The computation is started after selecting the corresponding key, whereas ⟨Esc⟩ leads back to the main menu.

- **⟨J⟩ulia** — allows one to enter an arbitrary second-order polynomial to be used as an iteration formula (not only Julia sets). In the Julia menu shown in Fig. 11.9, the user can choose:
  - the coefficients of the polynomial,
  - the maximum iteration depth,
  - the iteration border Γ,
  - a display of the c- or z-plane ($c = (a_0, b_0)$, $z = (x, y)$) (the lower part of the mask has a section for plotting z_0 with $c =$ const and one for plotting c with $z_0 =$ const),
  - shown area of the c- or z-plane, respectively,
  - the fixed value (c or z_0, according to selected mode).

  The computation is started with *'Plot z (0)'* or *'Plot c'*. ⟨Esc⟩ calls the main menu. With $\Gamma = 2$, $x_0 = 0$, $y_0 = 0$ and $x' = \mathrm{Re}(c) + x^2 - y^2$, $y' = \mathrm{Im}(c) + 2xy$, one obtains the Mandelbrot (*'Plot c'*) or the Julia set (*'Plot z (0)'*).

- **Julia⟨W⟩indows** — selects the fixed value (c or z_0 according to mode) in the graphics using crosshairs. After ⟨Enter⟩ has been pressed, the program displays a small window at the crosshair position and computes the iteration of the complementary mode. ⟨Esc⟩ leaves the *JuliaWindow* mode. The last picture computed with *JuliaWindows* can be displayed on the whole screen by entering the menu item *Julia* and starting the corresponding calculation.

```
Julia Set

x:= z_real(n)     y:= z_imag(n)     x':= z_real(n+1)     y':= z_imag(n+1)

x'= c_real+0.0000000·x+0.0000000·y +0.0000000·xy +1.0000000·x²+-1.000000·y²
y'= c_imag+0.0000000·x+0.0000000·y +2.0000000·xy +0.0000000·x²+0.0000000·y²

        Maximum Iteration Depth                 Iteration Border
           nmax =      45                         Γ = 2.0000000

Shown  | -1.550000000<x(0)<1.5500000000 | -2.100000000<c_real<1.0000000000
Area   | -1.150000000<y(0)<1.1500000000 | -1.150000000<c_imag<1.1500000000

Fixed  |   c_real= 0.000000000          |   x(0)= 0.000000000
Value  |   c_imag= 0.000000000          |   y(0)= 0.000000000

Method |   Plot z(0)  = x(0) + i·y(0)   |   Plot c  = c_real + i·c_imag
       |                                |
       | z(0)-→ min{ n≤nmax| |z(n)|> Γ } | c -→ min{ n≤nmax | |z(n)|> Γ }
```

Fig. 11.9. Julia menu of program Mandelbr.

- **⟨F⟩ollow Iteration** — stepwise observation of iteration. The user has to select a point in the diagram using the crosshair. If the screen shows the z-plane, a little cross appears at this point ($= z_0$). After every key depression, the program makes one iteration step and the cross jumps to the next point. ⟨ESC⟩ leads back to the menu. If the c-plane is displayed, a Julia window, in which the iteration is traced, will be calculated first. ⟨ESC⟩ makes the window disappear.

- **Zoom⟨I⟩n** — magnifies a section. A section of the diagram can be selected using the cursor keys. The ⟨SPACE BAR⟩ switches between the upper left and lower right corner and ⟨TAB⟩ between coarse and fine adjustment. With ⟨ENTER⟩ the section is accepted, and the iteration starts immediately.

- **Zoom⟨O⟩ut** — The current picture is reduced by a factor two and the user can move it to another site on the screen. A new section of the whole screen can then be defined, as described in *'Zoom In'*.

- **⟨R⟩eset** — The displayed section is reset to values given previously in the corresponding mask (Julia or Mandelbrot). The picture is cleared and the color palette uses the default settings.

- **⟨A⟩lgorithm** — A new method of scanning the picture can be activated. ⟨ENTER⟩ starts the iteration. One of the following methods can be selected:

  - *Straight ⟨F⟩orward:* All points of the screen are calculated from the upper left to the lower right (precise but slow).
  - *⟨R⟩efinement:* All points of the screen from low to high resolution are calculated, thus providing a quick overview of the picture (precise but slow).
  - *⟨I⟩nterpolation:* In the first step, every second point of every second line is calculated. The remaining points are then interpolated, i.e. if two adjacent points have the same color, the point in-between has the same color, otherwise this point is calculated. This results in many wrongly-colored points, but the general impression is correct.
  - *⟨T⟩urtle Fill:* The program searches for an as yet uncolored point and calculates its color. It then follows the monochrome border containing this point. The inner area of this border is filled with the corresponding color. The turtle algorithm is not able to discover differently colored 'islands' in monochrome 'seas', and it often misses filigree structures: fast and elegant, but not very precise.
  - *⟨S⟩quare Fill:* The program calculates all points on the border of a large square. If all points have the same color, the square is filled with this color. Otherwise, the square is divided into four sub-squares and the same process restarts. If the size of the square is 2×2 points, all points are calculated exactly. This algorithm has characteristics similar to *'Turtle Fill'*, but is more precise.

- **⟨C⟩oloring** — switches between four methods of transferring the iterated values into colors:
  - *⟨D⟩istribute* the colors between 1 and $n_{\max}$ ($n_{\max}$ is set in the parameter mask, the distribution is linear). An interval of subsequent values of lifetime obtains the same color. This results in a loss of information, but for complex areas the structure is clearer, whereas mod-coloring would only appear as a cluster of colored points.
  - *⟨M⟩od n* uses a pixel color modulo n (n may be changed here). The color is changed cyclically at each alteration of lifetime.
  - *⟨E⟩xponential* coloration similar to the first method, but exponentially scaled.
  - *⟨L⟩ogarithmic* coloration similar to the first method, but logarithmically scaled.
- **⟨P⟩alette** — It is possible to change the appearance of the colors on the screen: ⟨←⟩ and ⟨→⟩ select the columns (color number), R (red part of color), G (green), B (blue), I (intensity), and EGA (EGA number of color). The value of the current column can be changed using ⟨↑⟩ and ⟨↓⟩. The color number can also be selected using ⟨TAB⟩ and ⟨SHIFT TAB⟩. ⟨ENTER⟩ accepts and ⟨ESC⟩ aborts the setting of colors, ⟨F1⟩ gives brief information about all keys.
- **⟨S⟩ave picture** — enables storing of pictures (file *.MDB) and all appropriate parameters (text file *.DAT) by using the file-select box (Appendix B.2). This item is invoked automatically after a picture has been iterated entirely. The parameters are: type of picture (Mandelbrot, Julia), displayed section, constant parameter, maximum iteration depth, algorithm, mode of coloring, graphics mode (CGA, EGA, VGA, etc.), palette, date, and time of computation. If the picture is stored in an incomplete state, some other parameters are additionally stored so that it is possible to complete the diagram later (not possible with algorithm *'Square Fill'*).
- **⟨L⟩oad picture** — loads a diagram and its parameters. If the picture is incomplete, the iteration is continued automatically. Pictures MANDELBR.MDB (or MDLBRVGA.MDB for VGA graphics), MANDSON.MDB, JULIA.MDB, and VARIANT.MDB for the examples shown in Figs. 11.1, 11.2, 11.12, and 11.21 are already computed.
- **View ⟨D⟩ata** — displays a parameter file.
- **Run ⟨B⟩atch** — enables one to calculate a series of pictures automatically. First, the pictures to be computed have to be indicated using *'Put in Queue'*. Then, after *'Run Batch'* has been invoked, all pictures are calculated and stored automatically.
- **Hardcopy ⟨T⟩o PRN** — prints picture.

- **⟨H⟩ide Window** — hides the menu window. Hot keys will be interpreted immediately, all other keys display the menu again.

- **⟨G⟩raphmode** — switches between low (640×200 pixels), medium (640×350), and high (640×480) resolution. The 640×480 mode is only possible with VGA graphics. In both cases, EGA and VGA, the 640×350 mode is the default setting. For VGA, the program may be forced to use the 640×480 mode at the beginning with the command line parameter *VGA*: *mandelbr VGA*.

- **Accurac⟨Y⟩** — the following accuracy routines are only valid when computing pictures of the Mandelbrot set:

  - ⟨1⟩6-bit routine
  - ⟨3⟩2-bit routine
  - ⟨6⟩4-bit routine
  - ⟨S⟩tandard

  The standard iteration routine uses the mathematical coprocessor. If there is none, the same precision is obtained by software floating point procedures. The 16-, 32-, and 64-bit routines are integer routines. This is profitable for systems without a coprocessor in computing the Mandelbrot set, since the software emulation of coprocessor precision is, of course, rather slow.

- **Quit** — quits the program (same as ⟨Esc⟩).

11.4 Computer Experiments

11.4.1 Mandelbrot and Julia sets

The Mandelbrot map (11.5) $z_{n+1} = z_n^2+c$ with initial value $z_0 = (0,0)$ increases without bounds for $c \notin \mathcal{M}$. Inside the Mandelbrot set $\mathcal{M}$, the iterated sequence z_n remains bounded. Its behavior differs, however, depending on the location of the parameter c inside $\mathcal{M}$. It is an interesting experiment to follow the iterated sequence (menu entry *'Follow Iteration'*) and explore the different attractors which appear e.g. in the different side lobes of $\mathcal{M}$. It should be noted that for real values of c, the Mandelbrot map is equivalent to the logistic map studied in Chap. 10 and all phenomena explored there will reappear.

As pointed out at the beginning of this chapter, the Mandelbrot set $\mathcal{M}$ can be considered as a one-page catalogue of the infinite number of structurally different types of Julia sets $\mathcal{J}$. Let us recall that the Julia set for a given value of c, $\mathcal{J}_c$, is the boundary of the set of all starting points $z_0 = (x_0, y_0)$, for which the sequence z_n goes to infinity. The program allows a study of Julia sets by opening windows at selected points inside or outside the Mandelbrot set. The

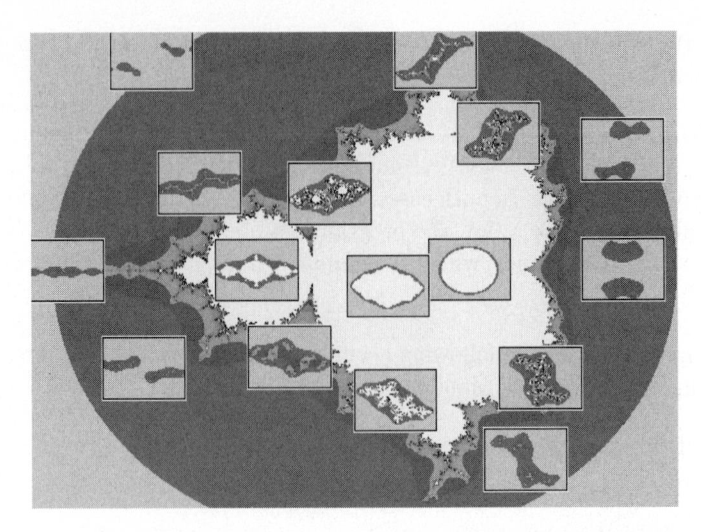

Fig. 11.10. Julia windows in the Mandelbrot set.

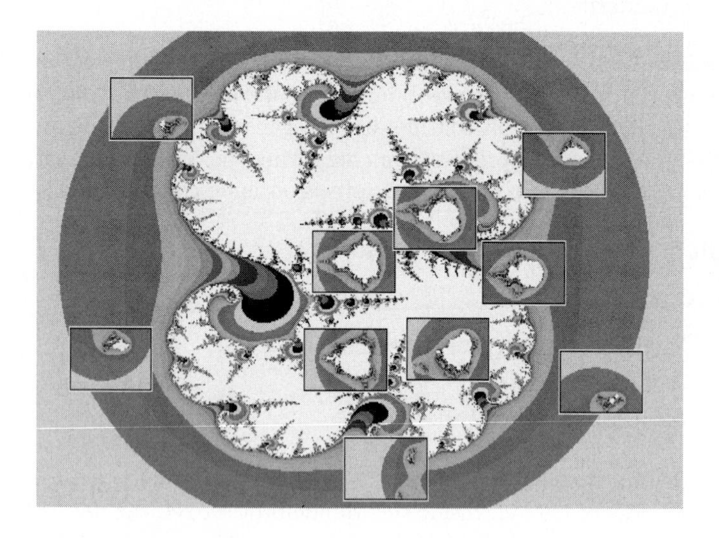

Fig. 11.11. Mandelbrot windows in the Julia set. This figure is stored on the disk.

program computes the Julia set $\mathcal{J}_c$ and displays it in the window. The lifetime coloring scheme as discussed above is again used. Fig. 11.10 shows an example.

The following basic features of the Julia sets $\mathcal{J}$ are of interest and can be explored numerically (see Ref. [11.4, 11.8], where additional results can also be found):

- $\mathcal{J}$ is nonempty and closed.

- $\mathcal{J}$ is contained inside the circle $|z| < \Gamma = 2$.

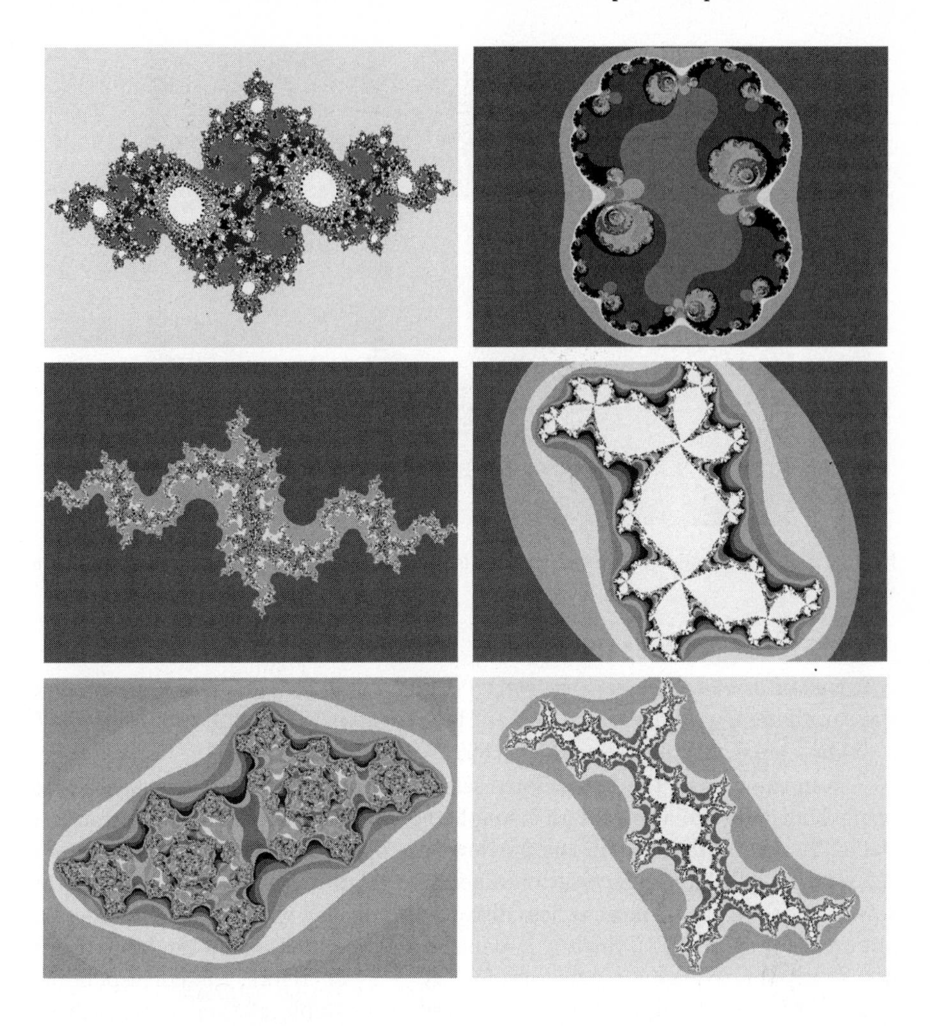

Fig. 11.12. Julia sets for various values of the parameter c (see text). Figure (a) can be loaded (files JULIA.MDB, –.DAT); figures (b) and (f) are stored on the disk.

- $\mathcal{J}$ is either connected (for $c \in \mathcal{M}$) or a Cantor set (for $c \notin \mathcal{M}$).
- The periodic attractors do *not* belong to $\mathcal{J}$.
- The image and the pre-image of $\mathcal{J}$ are equal to $\mathcal{J}$.
- The boundary of the basin of attraction of any attractive fixed point is equal to $\mathcal{J}$.

Fig. 11.13. Generalized Mandelbrot set for $z_0 = (1,0)$. This figure is stored on the disk.

A few examples of Julia sets are shown in Fig. 11.12 for the following values of the parameter $c = (a_0, b_0) = (0.745, -0.1125)$, $(-0.9211, 0.2603)$, $(0.2723, 0.5305)$, $(-0.5827, -0.5205)$, and $(-0.1158, 0.8600)$ in sequential order.

As in the case of the Mandelbrot set discussed above, windows can also be opened within a Julia set. As an example, Fig. 11.11 shows the Julia set for $c = (0.28972, -0.016702)$, where the windows are opened at selected initial points z_0 and display a Mandelbrot set obtained for this initial value. The Mandelbrot sets $\mathcal{M}_{z_0}$ appear to be more or less distorted compared with the original set $\mathcal{M}$, obtained for $z_0 = (0,0)$. Such a z_0-dependence becomes understandable if one notes that the sequence z_n diverges for $|z_0| > 2$. Therefore, the corresponding Mandelbrot set is empty. Fig. 11.13 shows an example of such a generalized Mandelbrot set for $z_0 = (1,0)$.

11.4.2 Zooming into the Mandelbrot Set

The Mandelbrot set $\mathcal{M}$ shown in Fig. 11.1 possesses a fascinating structure, and a magnification of interesting regions — in the vicinity of its boundary in particular — yields pictures of breathtaking beauty. A few examples are shown here, no attempt at a systematic approach being made.

Fig. 11.14 (parameter region $-1.96865397 \leq a_0 \leq -1.96865333$ and $-2.74 \cdot 10^{-7} \leq b_0 \leq 2.22 \cdot 10^{-7}$) is a magnification of a region close to the real c-axis. We observe a tiny copy of the Mandelbrot set itself, which shows its internal self-reference. Closer comparison with the original Mandelbrot set in Fig. 11.1 proves that the self-similarity is only approximate (note the differences in organization and scaling of the side lobes). Such approximate self-replicas are not

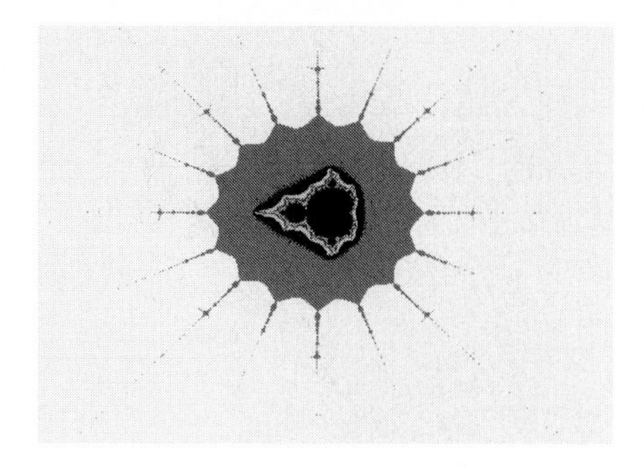

Fig. 11.14. Magnification of the Mandelbrot set. For the parameters, see text.

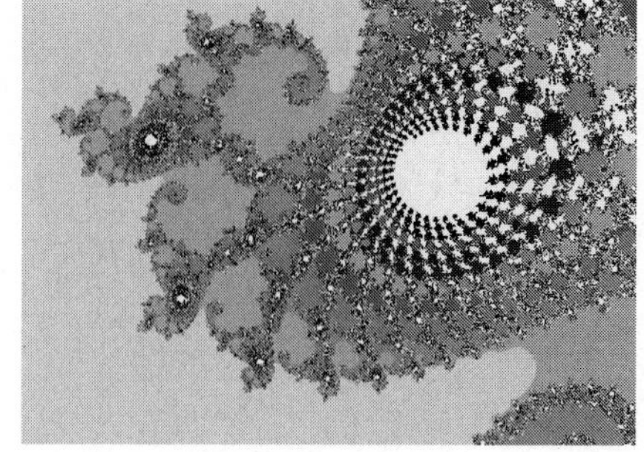

Fig. 11.15. Magnification of the Mandelbrot set. For the parameters, see text.

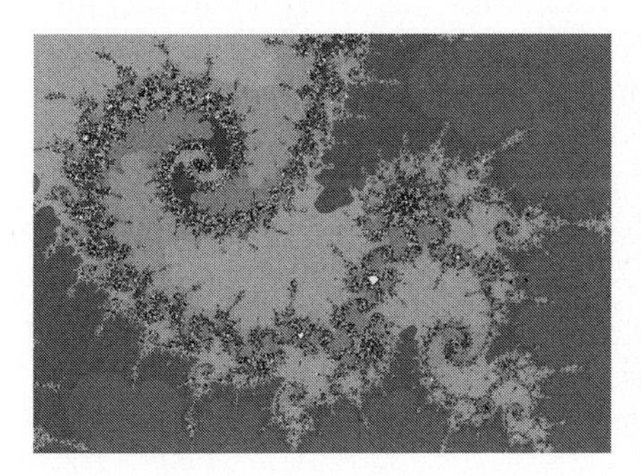

Fig. 11.16. Magnification of the Mandelbrot set. For the parameters, see text. This figure is stored on the disk.

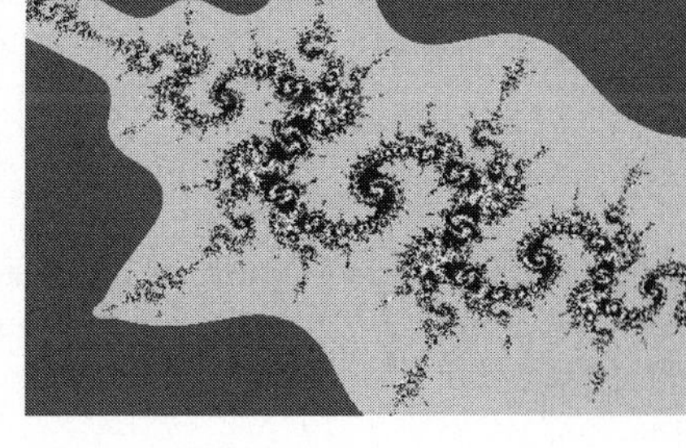

Fig. 11.17. Magnification of the Mandelbrot set. For the parameters, see text.

at all rare. In fact, there is an abundance of them, which may be more or less distorted. Some are seemingly isolated and others appear in highly organized configurations, as shown in the spiral structure of Fig. 11.15 (parameter region $-0.750 \leq a_0 \leq -0.746$, $0.098 \leq b_0 \leq 0.101$).

Sea-horse structures appear in Fig. 11.16 (shown is the parameter region $-1.2623 \leq a_0 \leq -1.2593$, $-4.13 \cdot 10^{-2} \leq b_0 \leq -3.91 \cdot 10^{-2}$) as well as in the valley shown in Fig. 11.17 (parameter region $-1.264480 \leq a_0 \leq -1.264443$ $4.3953 \cdot 10^{-2} \leq b_0 \leq 4.3984 \cdot 10^{-2}$), where they form a dragon-like superstructure. Closer inspection by means of further magnification again reveals microscopic copies of the Mandelbrot set inside the sea-horses, which is also demonstrated by Fig. 11.18 (parameter region $-0.7466 \leq a_0 \leq -0.7444$,$-0.1133 \leq b_0 \leq -0.1120$). Superstructures of spirals are finally shown in Fig. 11.19 (parameter region $-1.232 \leq a_0 \leq -1.226$, $0.1008 \leq b_0 \leq 0.1059$).

Fig. 11.18. Magnification of the Mandelbrot set. For the parameters, see text.

Fig. 11.19. Magnification of the Mandelbrot set. For the parameters, see text.

The grey-toned black and white pictures in Figs. 11.14–11.19 can only provide a vague idea of the beauty of the colored pictures (some of them are stored and can be loaded by the program (menu item *'Load Picture'*)) and the reader is invited to explore the Mandelbrot landscape personally.

11.4.3 General Two-Dimensional Quadratic Mappings

The Mandelbrot map studied above is typical of analytic complex maps. As already pointed out, the analysis of general two-dimensional mappings is much more difficult. Special cases of quadratic maps which appear in the literature are:

1. Perturbations of analytic complex quadratic maps are investigated by Rössler and coworkers [11.7, 11.9], who studied the mapping

$$\begin{aligned} x_{n+1} &= a_0 + a_1 x_n + x_n^2 - y_n^2 \\ y_{n+1} &= b_0 + 2 x_n y_n \end{aligned} \tag{11.8}$$

   For $a_1 = 0$ we recover the Mandelbrot map (11.5) and small values of a_1 lead to a distorted Mandelbrot set [11.7, 11.9].

2. Henon's map [11.16]

$$\begin{aligned} x_{n+1} &= 1 + y_n - a x_n^2 \\ y_{n+1} &= b x_n \end{aligned} \tag{11.9}$$

   (typical parameter values found in the literature are $a = 1.4$, $b = 0.3$) is one of the prototypes for systems showing complex dynamics. The Jacobi determinant of this mapping is equal to b, i.e. the mapping contracts for

$|b| < 1$. The mapping (11.9) possesses a strange attractor and for certain parameter values [11.18], a Smale horseshoe as well. For more details see Ref. [11.1] and [11.16]–[11.2].

3. The Burgers map

$$\begin{aligned} x_{n+1} &= (1-\nu)x_n - y_n^2 \\ y_{n+1} &= (1+\mu)y_n + x_n y_n \end{aligned} \tag{11.10}$$

can be obtained from a discretization of the differential equations which appear in hydrodynamics. The basic features of this map are:

- For $\nu < 2$, the point $(0,0)$ is a hyperbolic point with the x-axis as a stable manifold.
- There are two fixed points at $(-\mu,\ \pm\sqrt{\nu\mu}\,)$, which are stable for $\mu < 0.5$.

For more details see [11.20] and references given there.

4. The mapping

$$\begin{aligned} x_{n+1} &= y_n \\ y_{n+1} &= b + a y_n - x_n^2 \end{aligned} \tag{11.11}$$

shows a Hopf bifurcation, which is not present, e.g., in the Henon map (parameter values used in the literature are, e.g., $a = -1.6$, $b = 1.8$). The mapping (11.11) allows a numerical study of all phenomena associated with a Hopf bifurcation (see Ref. [11.17, pp. 160, 165] and [11.19, 11.21]).

5. The 'delayed logistic map'

$$x_{n+1} = \mu x_n (1 - x_{n-1}) \tag{11.12}$$

arises in certain problems in population dynamics. The term $1 - x_n$ in the logistic map (see Chap. 9) is replaced by the value of preceding generation $1 - x_{n-1}$. This map can be rewritten as a two-dimensional mapping

$$\begin{aligned} x_{n+1} &= y_n \\ y_{n+1} &= \mu y_n (1 - x_n) \end{aligned} \tag{11.13}$$

and is an example for a Neimark bifurcation. The fixed points at

$$x = y = (\mu - 1)/\mu \tag{11.14}$$

lose their stability at $\mu = 2$. At this point the eigenvalues are the sixth roots of unity, i.e. we have a supercritical Neimark bifurcation (compare Sect. 2.4.5). This bifurcation can be studied numerically by iterating the map for $\mu = 1.2, 1.7, 1.9, 2.1,$ and 2.2 (compare, e.g., Ref. [11.22, Sect. 8.5] and [11.17]).

6. The mapping

$$\begin{aligned} x_{n+1} &= a x_n (1 - x_n) - x_n y_n \\ y_{n+1} &= b x_n y_n \end{aligned} \qquad (11.15)$$

(interesting parameter values used in the literature are $a = 3.6545$, $b = 3.226$) describes a prey-predator model in population dynamics (see, e.g., [11.21, 11.23]). Here, x_n and y_n denote the population of the prey and the predator, respectively. For vanishing y_n we recover the logistic equation, discussed in detail in Sect. 9. The growth rate of the predator is taken to be proportional to the number of preys. This mapping shows a rich structure of bifurcation sequences and a variety of strange attractors.

7. The Cremona mapping

$$\begin{aligned} x_{n+1} &= x_n \cos\alpha - (y_n - x_n^2) \sin\alpha \\ y_{n+1} &= x_n \sin\alpha + (y_n - x_n^2) \cos\alpha \end{aligned} \qquad (11.16)$$

is area-preserving (unit Jacobi determinant) and invertible. It models the Poincaré mapping of the four-dimensional phase space flow of a Hamiltonian system describing, e.g., the motion of a particle in a two-dimensional potential (compare Chap. 2). The Cremona map has an elliptic and a hyperbolic fixed point. The map is especially suited for exploring the finer details by magnification, e.g. the infinite hierarchy of nested stability islands with similar structure. It may be useful to note that the inverse map is given by

$$\begin{aligned} x_{n+1} &= x_n \cos\alpha + y_n \sin\alpha \\ y_{n+1} &= -x_n \sin\alpha + y_n \cos\alpha + (x_n \cos\alpha + y_n \sin\alpha)^2 , \end{aligned} \qquad (11.17)$$

since, for a discrete map, one cannot simply take a 'negative stepsize' for backward propagation. For more results on this mapping see Ref. [11.21], [11.24]–[11.26].

8. The Ushiki map

$$\begin{aligned} x_{n+1} &= (a - x_n - b y_n) x_n \\ y_{n+1} &= (a - c x_n - y_n) y_n \end{aligned} \qquad (11.18)$$

is useful for demonstrating the importance of the dimension of an attractor and of the embedding space. Attractors for this map appear as a multiply folded two-dimensional surface in three-dimensional space, which is then projected onto a plane (see e.g. [11.27]).

Table 11.1. c-values for the centers of the components $c^{(k)}$ with k-periodic attractors.

k	c			
1	0			
2	0	-1.000		
3	0	-1.755	$-0.123 \pm 0.745\,i$	
4	0	-1.000	-1.311	$+0.282 \pm 0.530\,i$
			-1.941	$-0.157 \pm 1.032\,i$

Additional interesting two-dimensional mappings can be found in the book by Koçak [11.21], where some of the maps described above are also discussed.

11.5 Suggestions for Additional Experiments

Components of the Mandelbrot Set: A particularly interesting subset of the Mandelbrot set is the open set

$$\mathcal{M}' = \left\{ c \in \mathbb{C} \,\middle|\, p_c(z) = z^2 + c \text{ has a finite attractive cycle} \right\}, \tag{11.19}$$

(see Ref. [11.5], Sect. 4), which has infinitely many connected components characterized by the period, k, of the corresponding cycle. There can be only one such cycle for each value of c. The main cartioid contains all c for which the mapping p_c has a stable fixed point ($k = 1$). Each component has a center $c^{(k)}$, whose corresponding cycle is superstable, i.e. $p_c^k(0) = 0$, which is a polynomial in c of degree 2^{k-1}. The centers up to $k = 4$ are listed in Table 11.1 (compare Ref. [11.5], p. 48) and can be related to the different lobes of the Mandelbrot set. The program allows a numerical study of the various components and of the behavior of the iterated points. The menu item *'Follow Iteration'* will be useful.

Distorted Mandelbrot Maps: Rössler and coworkers [11.7, 11.9] studied the stability properties of the Mandelbrot map with respect to small perturbation. An additional parameter a is introduced (see (11.8)), where $a = 0$ gives the undisturbed Mandelbrot map. The perturbed map is non-analytic (it violates the conditions (11.4)).

Two examples of distorted maps are shown in Figs. 11.20 and 11.21 for parameter values $a = 0.01$ and $a = 0.3$, respectively. For small a, the Mandelbrot set is very close to the original one, and is stable with respect to small perturbations. Increasing values of a lead to stronger deformations. This a-dependence can be explored in more detail. It is also interesting to study the behavior of the detailed structures seen in magnification, as well as the Julia sets under

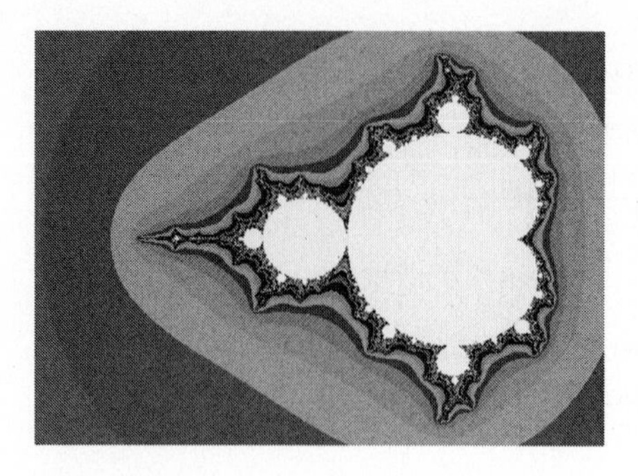

Fig. 11.20. Mandelbrot set for a distorted Mandelbrot map ($a = 0.01$).

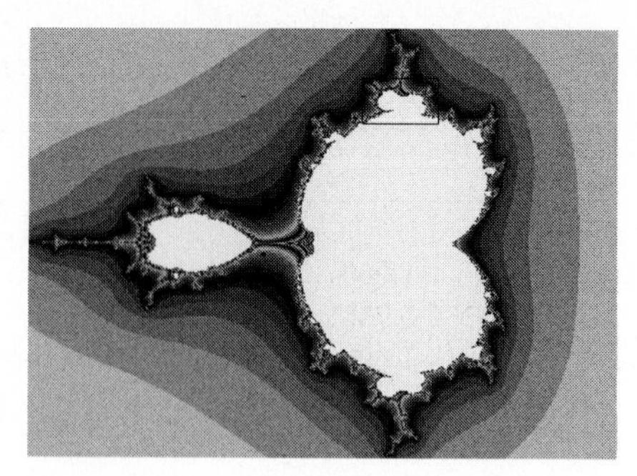

Fig. 11.21. Mandelbrot set for a distorted Mandelbrot map ($a = 0.3$); the figure can be loaded from files VARIANT.MDB, –.DAT.

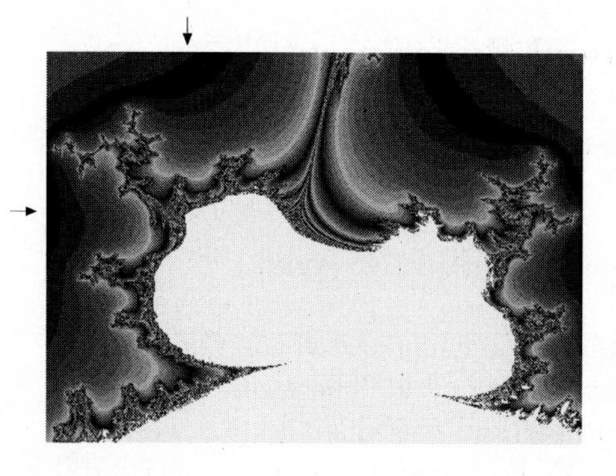

Fig. 11.22. Magnification of a distorted Mandelbrot set. For the parameters, see text.

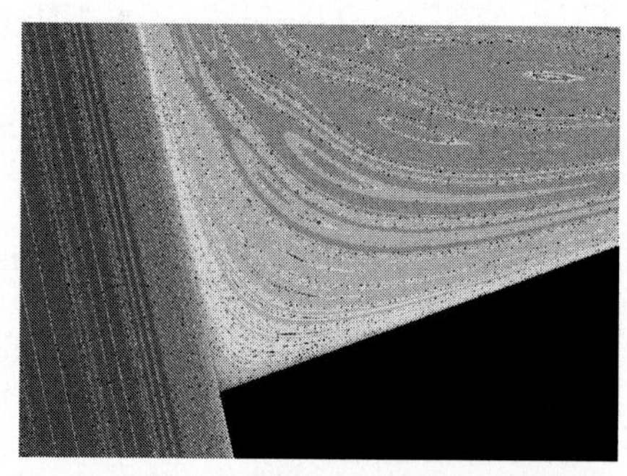

Fig. 11.23. Further magnification of Fig. 11.22. For the parameters, see text.

perturbation. As an example, Figs. 11.22 shows a magnification of the distorted Mandelbrot map shown in Fig. 11.21 (parameter region $-0.3464 \leq a_0 \leq 0.0235$, $0.6677 \leq b_0 \leq 0.9124$). A further magnification in Fig. 11.23 (parameter region $-0.2511 \leq a_0 \leq -0.2509$, $0.8103 \leq b_0 \leq 0.8107$) reveals a pattern which is very different from the plots found for the undistorted map.

Two Julia sets for the same distorted map are finally shown in Figs. 11.24 and 11.25 for a c-parameter $c = (a_0, b_0) = (-0.8361, 5.816 \cdot 10^{-6})$ and $(-1.377, 5.816 \cdot 10^{-6})$, respectively.

Further Experiments: The book by Becker and Dörfler [11.10] is almost entirely devoted to numerical experiments with chaotic mappings, in particular the Mandelbrot map. The reader will find there many suggestions for additional investigations.

Fig. 11.24. Julia set for a distorted Mandelbrot set. For the parameters, see text.

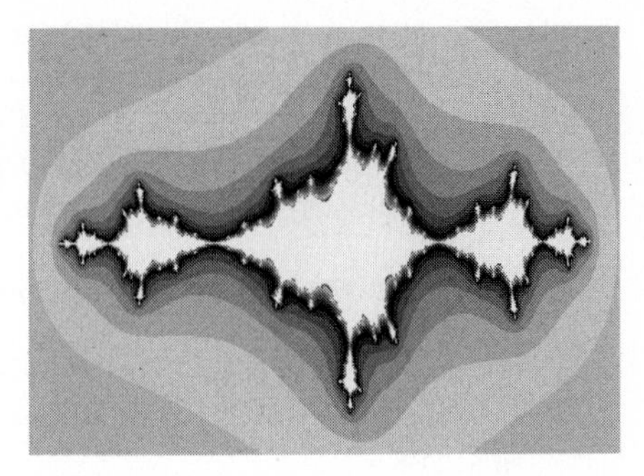

Fig. 11.25. Julia set for a distorted Mandelbrot set. For the parameters, see text.

11.6 Real Experiments and Empirical Evidence

Discrete iterated maps appear quite naturally in nonlinear dynamical systems, as 'first return' mappings, stroboscopic snapshots, or Poincaré maps of phase space sections, for instance. Such phase space maps are most often two-dimensional. In most realistic cases, the mappings are, however, *not* given by simple equations and the quadratic maps studied here can only serve as a qualitative introduction to the basic features of these mappings. A discussion of the role of discrete mappings in experimental studies of chaotic systems can be found in Chap. 9 in the context of one-dimensional maps.

References

[11.1] H. G. Schuster, *Deterministic Chaos* (VCH, Weinheim 1988)

[11.2] R. L. Devaney, *An Introduction to Chaotic Dynamical Systems* (Addison–Wesley, New York 1987)

[11.3] B. B. Mandelbrot, *The fractal geometry of nature* (Freeman, San Francisco 1982)

[11.4] H. O. Peitgen, D. Saupe, and F. v. Haeseler, *Cayley's problem and Julia sets*, Mathem. Intelligencer **6** (1984) 11

[11.5] H.-O. Peitgen and P. H. Richter, *The Beauty of Fractals* (Springer, Berlin 1986)

[11.6] A. K. Dewdney, *Computer Recreations: A computer microscope zooms in for a look at the most complex object in mathematics*, Scientific American August (1985) 8

[11.7] J. Peinke, J. Parisi, B. Röhricht, and O. E. Rössler, *Instability of the Mandelbrot set*, Z. Naturforsch. 42a (1987) 263

[11.8] H.-O. Peitgen, *Fantastic deterministic fractals*, in: H.-O. Peitgen and D. Saupe, editors, *The Science of Fractal Images*, Springer, Berlin 1988

[11.9] J. Peinke, J. Parisi, B. Röhricht, O. E. Rössler, and W. Metzler, *Smooth decomposition of generalized Fatou set explains smooth structure in generalized Mandelbrot set*, Z. Naturforsch. 43a (1988) 14

[11.10] K.-H. Becker and M. Dörfler, *Dynamical Systems and Fractals – Computergraphics Experiments in Pascal* (Cambridge University Press, Cambridge 1989)

[11.11] G. Julia, *Mèmoire sur l'itération des fonctions rationnelles*, J. Math. Pures et Appl. **4** (1918) 47

[11.12] P. Fatou, *Sur les équations fonctionnelles*, Soc. Math. France **47** (1919) 161

[11.13] P. Fatou, *Sur les équations fonctionnelles*, Bull. Soc. Math. Fr. **48** (1920) 33 and 208

[11.14] P. J. Myrberg, *Sur l'itération des polynomes réeles quadratiques*, J. Math. Pures et Appl. **ser. 9.41** (1962) 339

[11.15] A. Douady and J. H. Hubbart, *Iteration des polynomes quadratiques complexes*, CRAS Paris **294** (1982) 123

[11.16] M. Hénon, *A two dimensional map with a strange attractor*, Commun. Math. Phys. **50** (1976) 69 (reprinted in: B.-L. Hao, *Chaos* (World Scientific, Singapore 1984) and P. Cvitanović, *Universality in Chaos* (Adam Hilger, Bristol 1984)

[11.17] J. Guckenheimer and P. Holmes, *Nonlinear Oscillations, Dynamical Systems, and Bifurcations of Vector Fields* (Springer, New York 1983)

[11.18] R. Devaney and Z. Nitecki, *Shift automorphisms in the Hénon mapping*, Commun. Math. Phys. **67** (1979) 137 (reprinted in: R. S. MacKay and J. D. Meiss, *Hamiltonian Dynamical Systems*, (Adam Hilger, Bristol 1987)

[11.19] D. Whitley, *Discrete dynamical systems in dimensions one and two*, Bull. London Math. Soc. **15** (1983) 177

[11.20] N. MacDonald and R. R. Whitehead, *Introducing students to nonlinearity: Computer experiments with Burgers mappings*, Eur. J. Phys. **6** (1985) 143

[11.21] H. Koçak, *Differential and Difference Equations through Computer Experiments* (Springer, New York 1986)

[11.22] J. M. T. Thompson and H. B. Stewart, *Nonlinear Dynamics and Chaos* (John Wiley, Chichester 1986)

[11.23] J. Maynard Smith, *Mathematical Ideas in Biology* (Cambridge University Press, London 1968)

[11.24] J. Moser, *Lectures on Hamiltonian systems*, Mem. Am. Math. Soc. **81** (1968) 1 (reprinted in: R. S. MacKay and J. D. Meiss, *Hamiltonian Dynamical Systems* (Adam Hilger, Bristol 1987))

[11.25] M. Hénon, *Numerical study of quadratic area–preserving mappings*, Quart. Appl. Math. **27** (1969) 291

[11.26] M. Hénon, *Numerical exploration of Hamiltonian systems*, in: G. Iooss, H. G. Helleman, and R. Stora, editors, *Les–Houches Summer School 1981 on Chaotic Behaviour of Deterministic Systems*, page 53, (North–Holland, Amsterdam 1983)

[11.27] S. Eubank, W. Miner, T. Tajima, and J. Wiley, *Interactive computer simulation and analysis of Newtonian dynamics*, Am. J. Phys. **57** (1989) 457

12. Ordinary Differential Equations

Dynamical systems are often expressed in terms of ordinary differential equations. An example are the canonical equations of motion in Hamiltonian systems

$$\dot{p}_i = -\frac{\partial H}{\partial q_i}\,, \qquad \dot{q}_i = \frac{\partial H}{\partial p_i}\,, \tag{12.1}$$

where the time derivatives of the canonical coordinates and momenta are given by the partial derivatives of the Hamiltonian. Typically, the right hand side of these equations is a nonlinear function of the variables p_i and q_i, i.e. (12.1) is a nonlinear dynamical system.

There are many other cases where the dynamics is determined by differential equations. They sometimes appear as higher order equations. However, such differential equations can easily be rewritten as a system of first order equations by introducing the time derivatives, i.e. the velocities, as additional variables. An example is the harmonic oscillator equation

$$\frac{\mathrm{d}^2 x}{\mathrm{d}t^2} + r\frac{\mathrm{d}x}{\mathrm{d}t} + \omega_0^2\, x = 0\,. \tag{12.2}$$

With $v = \mathrm{d}x/\mathrm{d}t$, this is converted to the first order system

$$\frac{\mathrm{d}v}{\mathrm{d}t} = -\omega_0^2\, x - r\, v \tag{12.3}$$

$$\frac{\mathrm{d}x}{\mathrm{d}t} = v\,. \tag{12.4}$$

Another example is worked out in Chap. 10 for the third order differential equation of the *Chaos Generator* (see (10.15)).

Quite generally, one can therefore study a system

$$\frac{\mathrm{d}x_i}{\mathrm{d}t} = f_i(t, x_1, \ldots, x_N)\,, \qquad i = 1, \ldots, N\,, \tag{12.5}$$

with initial conditions $x_i(t_0) = x_{i,0}\,,\ i = 1, \ldots, N$. The variable t is often denoted as a 'time'. The dynamical system (12.5) generates a flow in an N-dimensional solution space, often denoted as 'phase space'. For an explicitly time-dependent system, N must be at least two in order to generate chaos. For a time-independent (*'autonomous'*) system, N must be at least three.

The program provides a numerical solution of the system (12.5) of first order ordinary differential equations (ODE) and allows various graphical presentations of the solutions. Up to $N = 9$ dependent variables can be used. The solutions are computed as a sequence of solution vectors $\mathbf{x} = (t, x_1, \ldots, x_N)$ and can be stored, loaded, graphically displayed, plotted, and printed.

12.1 Numerical Techniques

A fifth order Runge-Kutta-algorithm with adaptive stepsize control is used in the program to integrate the differential equations. The computation follows the program ODEINT recommended in the book by Press et al. [12.1], where a detailed description can be found.

Starting from an initial solution vector, the integration routine propagates the solution in discrete steps h of the independent variable t, the 'time'. The numerically estimated error Δ_i of the dependent variables x_i for this interval scaled to a unit time interval, i.e. Δ_i/h, is compared with the tolerated error per unit time interval. The stepsize is increased if the error is within the tolerated limit for all N variables, and reduced if it misses the accuracy limit in at least one case. The accuracy test for a variable can be suppressed by assigning zero error tolerance to this variable.

The program also permits the computation of a Poincaré section, i.e. the intersection of the computed trajectory with a hyperplane defined by fixing the projection onto a specified direction $\mathbf{n}$ in solution space. When the projection $\mathbf{x} \cdot \mathbf{n}$ computed as a function of time crosses a specified value a, i.e. $\mathbf{x} \cdot \mathbf{n} - a$ changes sign, the program interpolates the solution vector at the crossing point $\mathbf{x} \cdot \mathbf{n} = a$. Only the sequence of solution vectors at these crossing points is stored.

12.2 Interacting with the Program

In contrast to the preceding programs, ODE is controlled by a single command screen. To distinguish the control of two basic modes of the program — solving the differential equations and displaying the results — the screen mask is horizontally divided into three parts, which appear in different colors. The central part allows specification of the differential equations, the initial conditions, and the parameters for numerical integration and storage of the solution. In the lower part, the graphical representation of the solution can be controlled.

- On-line help is available by pressing ⟨F1⟩.
- Pressing ⟨F10⟩ calls the DOS operating system, while the program ODE remains resident in the memory and is re-entered using the command *exit*.

The *menu line* in the upper part of the screen is used to select various facilities:

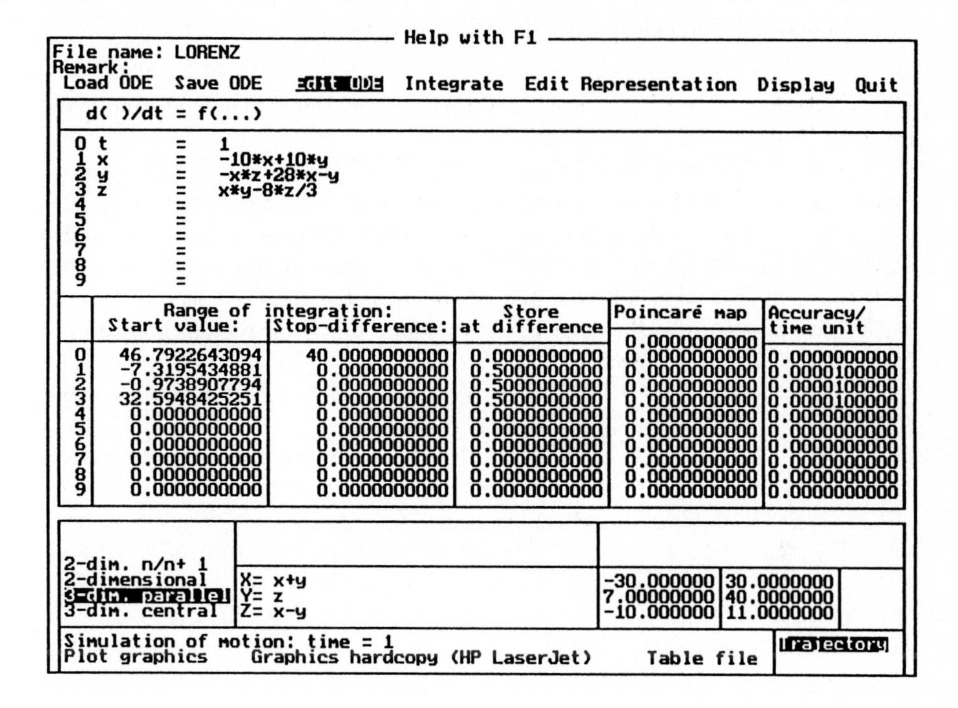

Fig. 12.1. Command screen of program ODE.

- *File name:* — Name of loaded file; a suffix *.ODE is anticipated, files with a different suffix are ignored.
- *Remark:* — Line for informative text with no influence on the program.
- *Load ODE:* — Loads a file *.ODE containing differential equations, parameters, and previous numerical solution vectors.
- *Save ODE:* — Creates a file *.ODE containing differential equations, parameters, and previous numerical solution vectors.
- *Edit ODE:* — Input of differential equations and parameter values (see below).
- *Integrate:* — Starts numerical integration (see below).
- *Edit Representation:* — Input of parameters for graphical output (see below).
- *Display:* — Starts graphical representation of the solution or output of the numerical data on a *.TBL file.

In the following, some entries are described in detail:

• **Edit ODE:** — Activates the central part of the screen. Quit edit mode by pressing ⟨ESC⟩.

Differential equations are written as $\mathrm{d}x_i/\mathrm{d}t = f_i(t, x_1, \ldots, x_9)$, where the x_i denote the variables. These equations must be specified in the column d( )/dt = $f(\ldots)$. The index $i = 0, \ldots, 9$ is listed on the left. On line i, the name of the variable x_i as well as the function f_i must be entered (for a listing of the mathematical operations allowed and pre-defined constants see Sect. B.3).

To give an example, the equations (12.3) for a harmonic oscillator with unit frequency and friction constant

$$\frac{\mathrm{d}v}{\mathrm{d}t} = -x - v \ , \quad \frac{\mathrm{d}x}{\mathrm{d}t} = v$$

require an input of

	d(	)/dt	=	$f(\ldots)$
0	t		=	1
1	v		=	$-x - v$
2	x		=	v

These lines must be written in sequential order without empty lines. Line 0 is pre-defined as $\mathrm{d}t/\mathrm{d}t = 1$ and cannot be modified.

N.B.: if discontinuous functions are used, numerical integration across a point of discontinuity is possible only if the accuracy parameter for this variable is set to zero!

- *Range of integration:*
  - *Start value:* — Input of initial values of variables in same order as in column d( )/dt.
  - *Stop-difference:* — If the absolute value of the difference between the current and initial value exceeds the number given here, the integration is automatically stopped (0 stands for no limit of integration).
- *Store at difference:* — If the absolute value, $\Delta x_{\mathrm{store}}$, of the difference between the current and last stored value of a variable x exceeds the number given here, the current solution vector is stored. If the input is '0' this value is of no significance as far as storing the solution is concerned. Unless a *Poincaré map* is required, at least one of the values given here must be different from zero.
- *Poincaré map:* — The data of the continuous trajectory can be reduced by storing only the intersections with a N-dimensional (hyper)plane in the $(N+1)$-dimensional solution space. This plane is given in Hesse form by

$$\mathbf{x} \cdot \mathbf{n} = a\,, \tag{12.6}$$

where $\mathbf{x}$ is the solution vector, $\mathbf{n}$ a vector orthogonal to the plane, and a a constant. The first number in the column *'Poincaré map'* of the input menu is the value of a, followed by the components of the vector $\mathbf{n}$. The minimum distance of the plane from the origin is $a/|\mathbf{n}|$. Only the solution points where the trajectory passes through this plane are stored.

Examples of the construction of Poincaré sections are worked out in the computer experiments below.

If a Poincaré section is required, *all* entries under *'Store at difference'* must be zero, and $\mathbf{n}$ must be different from the zero-vector.

- *Accuracy/time unit:* — Here, the upper limit of error tolerance can be given, with respect to a unit time interval. Variables with an assigned tolerance of '0' are not checked. The first line for variable t cannot be entered; at least one of the other entries must be different from zero.

• **Integrate:** — Starts the integration of the differential equations. The program generates a temporary file ODE.TMP. The directory for this file can be configured in the DOS environment by the DOS command:

```
SET ODETMP = <path>
```

If no environment variable ODETMP exists, the variable TMP is used, if existent, and otherwise the variable TEMP. If none of these environment variables is set, the temporary file is created in the current directory. The SET command can also be entered in the AUTOEXEC.BAT file.

The numerical integration can be interrupted by pressing any key. The previous integration can be continued if desired. In such a case, one can choose between the following items:

- *Restart integration with start values:* — Integration starts again at the given initial values.

- *Restart integration with current values:* — The initial values are overwritten by the computed instantaneous values. The integration is started again, the old data being discarded.

- *Continue previous integration:* — The computed solution is continued and all computed data remain stored.

- *Start new integration, hold old values:* — Integration starts again using modified initial values, parameters, or differential equations. Previously computed solutions remain stored.

It is therefore possible to suppress transient processes, store sequential integration runs using different initial conditions or parameters, construct synoptic (Poincaré) diagrams, etc.

The computation can be interrupted by pressing any key. During the integration, the data of new calculated solution values are numerically displayed, in order to allow a crude control of the progress of computation. These values appear in the column used for an input of the initial conditions.

• **Edit Representation:** — Activates the lower part of the screen. Quit edit mode by pressing ⟨Esc⟩. In the following, the items in ⟨ ⟩–brackets are highlighted when active.

- ⟨*Representation mode*⟩: Here, one chooses the mode of projection of the solution vector onto the screen. The computed solution variables, e.g. x, y, ... , are stored as vectors, where $x(n)$, $y(n), \ldots$ denotes the nth stored variables x, $y, \ldots$. The graphically displayed variables X, Y, and Z are functions of the variables $x, y, \ldots$ defined in '*Representation of the solution*'.

  - 2-*dim.* $n/(n+k)$: — Two-dimensional return map. X-axis: $X(x(n))$; Y-axis: $X(x(n+k))$. The value of k can be altered ($0 \leq k \leq 32$); $k = 1$ is pre-set.
  - 2-*dimensional*: — Two-dimensional projection. X-axis: $X(x(n))$; Y-axis: $Y(x(n))$
  - 3-*dim. parallel*: — Three-dimensional projection shown in parallel projection. X-axis: $X(x(n))$; Y-axis: $Y(x(n))$; Z-axis: $Z(x(n))$
  - 3-*dim. central*: — Three-dimensional projection shown in central projection. X-axis: $X(x(n))$; Y-axis: $Y(x(n))$; Z-axis: $Z(x(n))$

  In the three-dimensional representation on the screen, the third coordinate is additionally coded by coloring: from white (in front) via yellow and red to dark grey (in the rear).

- *Representation of solution*: Definition of the functions X, Y, Z. Here, the graphical or printer output of a given sequence of solution vectors can be controlled. The solution is graphically displayed in a two- or three-dimensional Cartesian coordinate system, (X, Y) or (X, Y, Z), which is projected onto the screen. The functions X, Y, Z can be specified (see Sect. B.3).

- *Range of representation*: Choice of the intervals displayed on the screen. The option ⟨*Auto.*⟩ activates automatic scaling, where intervals containing the complete computed solution are chosen.

- ⟨*Simulation of motion*⟩: When this mode is active, a sphere is shown moving along the trajectory. The velocity of this motion results from interpreting the function given under the entry '*time=*' as real time in seconds. Possible entries are, e.g., `0.1*t`, `ln(t)`, or any function of the solution vector. One should check that this 'time' never runs backward along the solution curve.

- ⟨*Trajectory*⟩ : Option for connecting the computed solution points by a curve. A spline algorithm is used.
- ⟨*Plot graphics*⟩ : Option for plotting the solution on a plotter. This is not possible when the mode *'Simulation of motion'* is active. Epson HI-80 and HP GL-Plotter, as well as output on a Postscript printer, are supported.
- ⟨*Graphics Hardcopy*⟩ : Option for printer output of the solution. This is not possible when the mode *'Simulation of motion'* is active.
- ⟨*Table file*⟩ : When this item is active, the output is not displayed on the screen. Instead all solution vectors are written as lines on a text file with suffix *.TAB , allowing subsequent processing by other programs.

12.3 Computer Experiments

The program ODE is a general-purpose program for solving ordinary differential equations. It can, therefore, be used to recompute some of the systems studied above. This multipurpose flexibility is achieved at the expense of less direct access to the numerical results, a more complicated method of parameter modification, and — last but not least — reduced graphical and interactive features. Furthermore, the program is controlled by a single menu, which displays all features of the program on the screen. In our experience, efficient use of the program ODE requires practice. We therefore begin our series of numerical experiments using ODE by looking first at some simple systems with regular motion and then at chaotic ones, starting with cases already discussed in preceding chapters.

12.3.1 The Pendulum

Let us first study a well-known example in order to become familiar with the operation of the program. We study a pendulum of mass m with a fixed length l in the gravitational field. The equations of motion are

$$\frac{\mathrm{d}x}{\mathrm{d}t} = v \ , \quad \frac{\mathrm{d}v}{\mathrm{d}t} = -\frac{g}{l}\sin x \, , \tag{12.7}$$

where x denotes the angle coordinate and v the angular velocity. For small amplitude x, the motion is harmonic with angular frequency $\omega = \sqrt{g/l}$. With increasing amplitude, nonlinear effects must be taken into account. The dynamics of the system is, of course, always regular, because the energy

$$E = \frac{1}{2} m l^2 v^2 - m g l \cos x \tag{12.8}$$

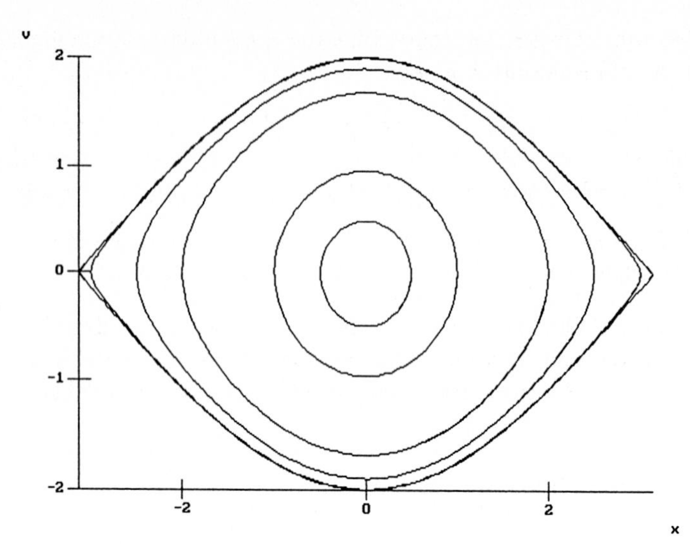

Fig. 12.2. Phase space trajectories for the simple pendulum (12.7) for various initial angles x and an initial velocity $v = 0$. This figure is stored on the disk.

is conserved. The existence of a single integral of motion guarantees the integrability of the system.

For an energy of $E = mgl$ the pendulum can reach the maximum value of the potential energy at $x = \pi$, i.e. an upright position. A trajectory with this energy will ultimately converge to the point $(x, v) = (\pi, 0)$ in infinite time.

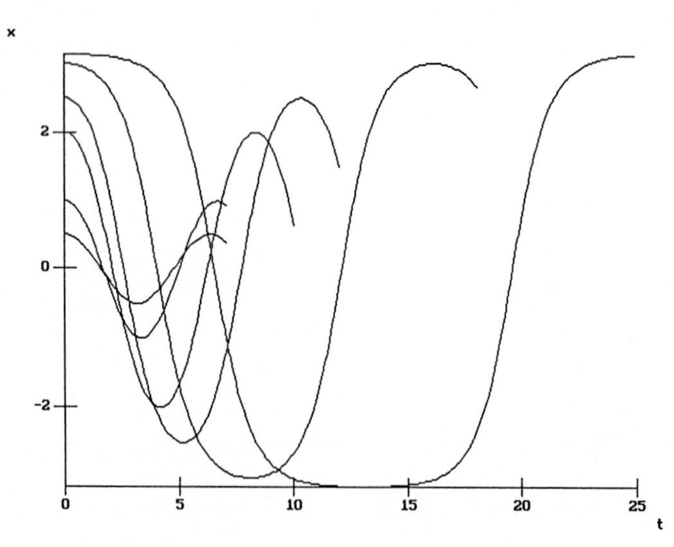

Fig. 12.3. As in Fig. 12.2, but plotted as $x(t)$.

Such a trajectory separates the librational motion for lower energies from the rotational one at higher energies. This *'separatrix'* is the stable manifold of the unstable fixed point at $(x,p) = (\pi, 0)$.

Figure 12.2 shows phase space trajectories for a pendulum ($g/l = 1$) started with various initial angles x and zero angular velocity. It might be illustrative to display the motion along the orbits in real time, using the option *'Simulation of motion'*. This clearly shows the slowing down in the vicinity of the turning points and the increasing period for higher energies. A quantitative determination of the period can be obtained by plotting the time dependence of the angle as shown in Fig. 12.3. For high excitation, where the trajectory comes close to the separatrix, the period is long and the pendulum spends longer and longer times in the vicinity of the turning points, i.e. close to the upward position $x = \pm\pi$.

12.3.2 A Simple Hopf Bifurcation

As a second example, we look at the Hopf bifurcation, discussed in Sect. 2.4.6, which can be observed in many dynamical systems (see, e.g. the electronic circuit in Sect. 10.4.1). A differential equation modeling a simple Hopf bifurcation is given in polar coordinates in (2.115). In Cartesian coordinates, the equations read

$$\begin{aligned}\frac{\mathrm{d}x}{\mathrm{d}t} &= -(g + x^2 + y^2)x - \omega y \\ \frac{\mathrm{d}y}{\mathrm{d}t} &= -(g + x^2 + y^2)y + \omega x\,.\end{aligned} \tag{12.9}$$

The reader will recall that these equations support a point attractor at $x = y = 0$ for $g > 0$, which bifurcates into a limit circle of radius $\sqrt{-g}$ for negative values of g.

Despite the fact that the equations can be solved in closed form, it is interesting to monitor the dynamical behavior on the screen. As an introductory experiment, we edit the differential equations (12.9) for fixed values of the parameters $g = -0.5$ and $\omega = 4$ in the program mode *'Edit Ode'*. In this and subsequent computations the accuracy parameter is chosen as 10^{-3}. Integrating with an initial condition $x = 0$, $y = 0.01$ over a time interval $0 \le t \le 20$ and storing the results at differences $\Delta t_{\text{store}} = \Delta x_{\text{store}} = \Delta y_{\text{store}} = 0.05$, we obtain a solution vector which can then be displayed on the screen in various ways. Fig. 12.4 shows the trajectory in two-dimensional (x,y)-space. Here, the displayed region is chosen as $-1.2 \le x, y \le 1.2$, but the automatic scaling (activate option *'Auto.'*) of all displayed variables in the first instance can be recommended. The command *'Trajectory'* was activated to connect the computed data points by smooth lines. The trajectory spirals outward, approaching a limit cycle $x^2 + y^2 = -g = 0.5$. This relaxation towards the attractor can also be displayed in the time-dependence of the coordinates by plotting x as a function of t in a two-dimensional diagram, or (t,y,x) in a three-dimensional

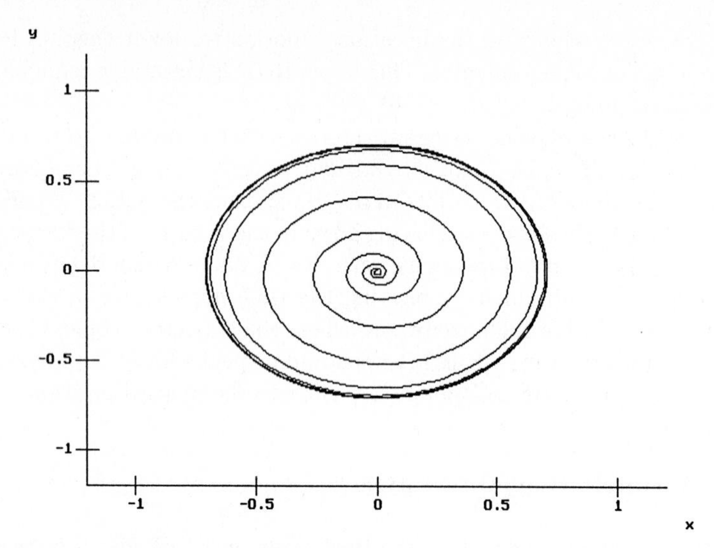

Fig. 12.4. Trajectory for the simple Hopf equations (12.9) for $g = -0.5$ approaching a limit cycle. This figure is stored on the disk.

projection. When the computation is repeated for $g = 0.5$, the limit cycle changes into a point attractor at the origin, as shown in Fig. 12.5.

It is also possible to construct combined diagrams such as Fig. 12.6, which was obtained by inserting a 'g' into the differential equations in place of the fixed

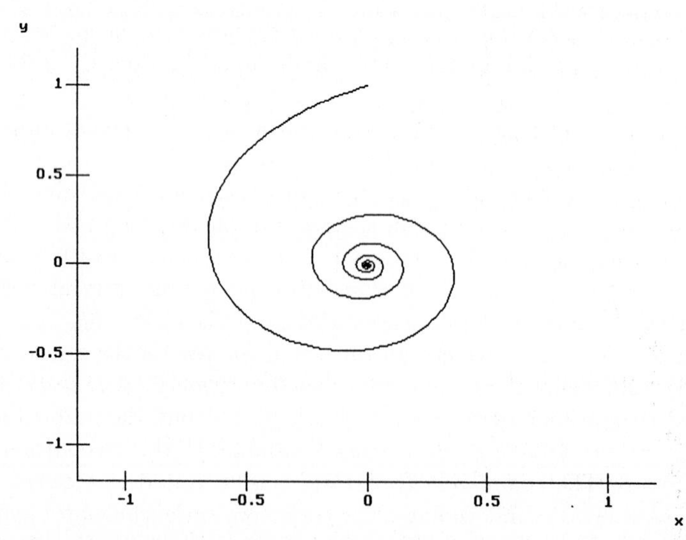

Fig. 12.5. Trajectory for the simple Hopf equations (12.9) for $g = +0.5$ approaching a point attractor.

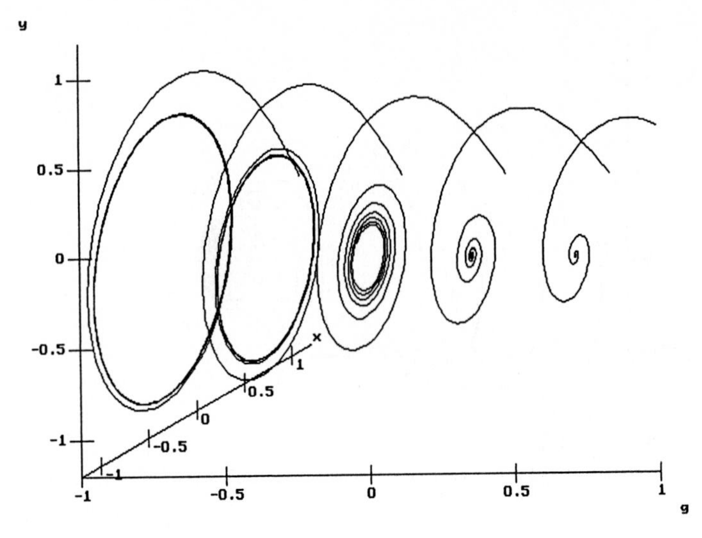

Fig. 12.6. Simultaneous plot of trajectories solving (12.9) for five values of the parameter g, all starting at $x = 2,\ y = 0$.

numerical values used previously, and adding an auxiliary differential equation

$$\dot{g} = 0 \tag{12.10}$$

to the equations of motion (12.9). Solving with an initial value of $g = -0.5$ or $+0.5$, we can reproduce the previous results. We can now add several subsequent runs of the program for different values of g by means of *'Start new integration, hold old values'* and display the results in a three-dimensional (g, y, x) projection, as in Fig. 12.6 for $g = -1.0, -0.5, 0.0, 0.5, 1.0$. In all these plots, the initial condition is $x = 2,\ y = 0$ and the time-interval is chosen as $0 \le t \le 10$. One can observe the characteristic slowing down in the relaxation towards the attractor in the vicinity of the bifurcation point.

It is also possible to change the parameters adiabatically, monitoring the transition in the dynamics induced by this variation. Let us alter the auxiliary equation (12.10) into

$$\dot{g} = 0.005 \tag{12.11}$$

with initial condition $g = -1.0$, and integrate over a g-interval of length 1.4. Starting now a trajectory at $(x, y) = (0., 1)$, i.e. on the limit cycle for the parameter $g = -1$, the trajectory follows almost adiabatically the time-evolving limit cycle, showing the Hopf transition to a point attractor at $g = 0$. The 'velocity' $\dot{g}$ in (12.11) should be small compared to the relaxation to the attractor. This condition is violated in the neighborhood of the bifurcation point and the displayed surface deviates from the paraboloid, which is to be expected in the adiabatic limit. This deviation can be reduced by using a smaller velocity,

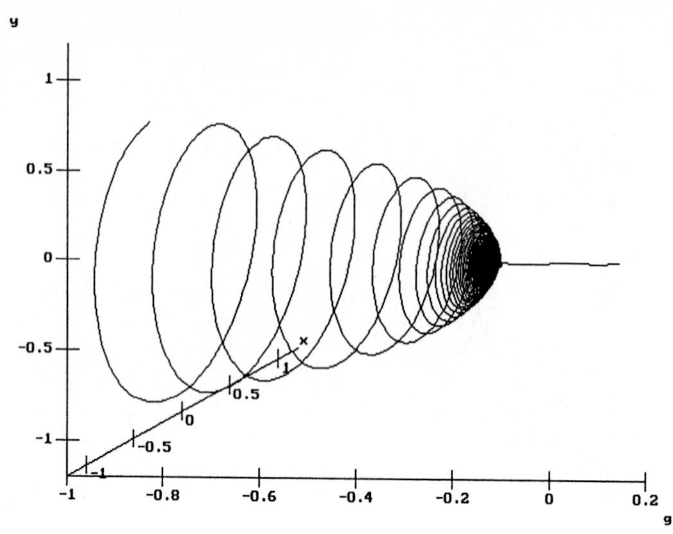

Fig. 12.7. Trajectory initially on the limit cycle under adiabatic variation of the control parameter g (see text).

e.g. $\dot{g} = 0.0001$, with the consequence that the computation time is much longer. In a more refined version, one can prescribe a special functional form of the parameter variation, emphasizing critical regions as, for instance,

$$\dot{g} = 0.0001 + 0.1 \tanh(10\, g^2)\,, \tag{12.12}$$

which was used in the computation of Fig. 12.7 ($\Delta t_{\text{store}} = 0$, $\Delta x_{\text{store}} = \Delta y_{\text{store}} = 0.05$).

12.3.3 The Duffing Oscillator Revisited

The Duffing oscillator is discussed in Chap. 8. Using the pre-set parameter values of the program DUFFING, the second order differential equation (8.1) reads rewritten as first order equations

$$\begin{aligned} \dot{x} &= v \\ \dot{v} &= -0.2v + x - 0.1x^3 + \cos t\,, \end{aligned} \tag{12.13}$$

which can easily be edited in the program mode *'Edit Ode'*. Starting the integration with the initial conditions $t = x = v = 0$ with an accuracy parameter 10^{-4} for x and v and storing the results at time differences $\Delta t_{\text{store}} = 0.1$ up to time $t = 400$, a solution vector is generated. This solution can then be displayed on the screen in various ways. In most cases, it can be recommended that the results be first plotted using automatic scaling of all variables.

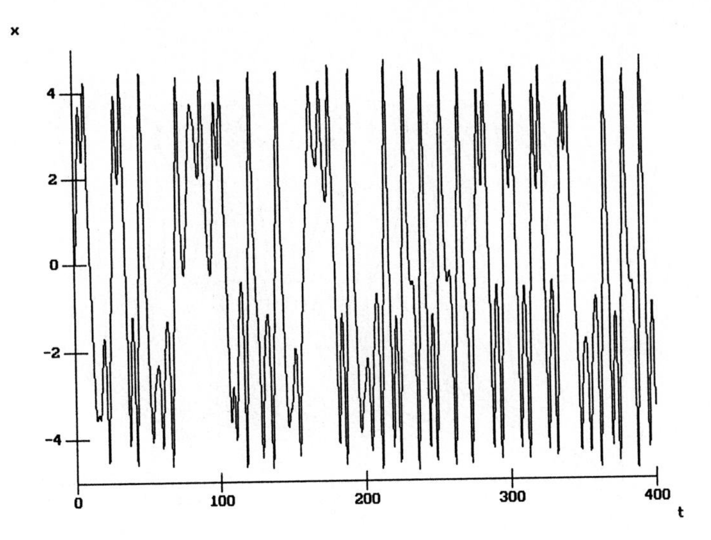

Fig. 12.8. Amplitude x of the Duffing oscillator (12.13) as a function of time t.

Figure 12.8 shows the amplitude x as a function of time t. This plot is obtained by defining the coordinate axes as $X = t$, $Y = x$ and activating the command *'Trajectory'*.

The same data can be displayed in phase space (x, v) by defining the coordinate axes $X = x$ and $Y = v$, as shown in Fig. 12.9. Comparing with Fig. 8.3 — computed for the same parameters — we observe similarities, but also diffe-

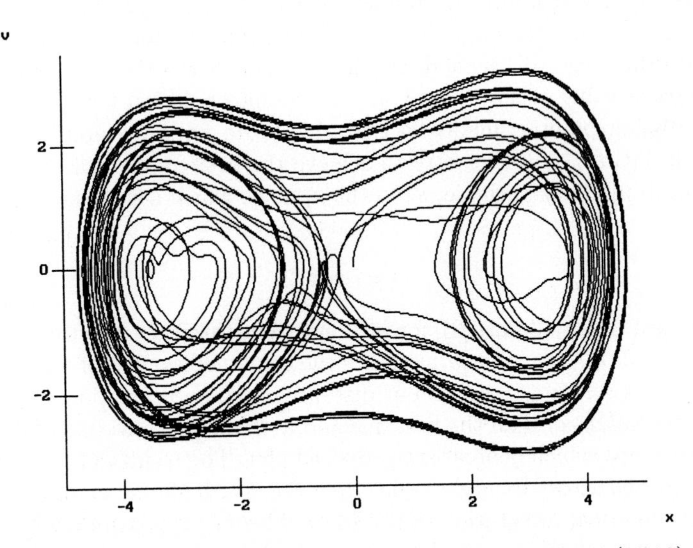

Fig. 12.9. Phase space trajectory of the Duffing oscillator (12.13).

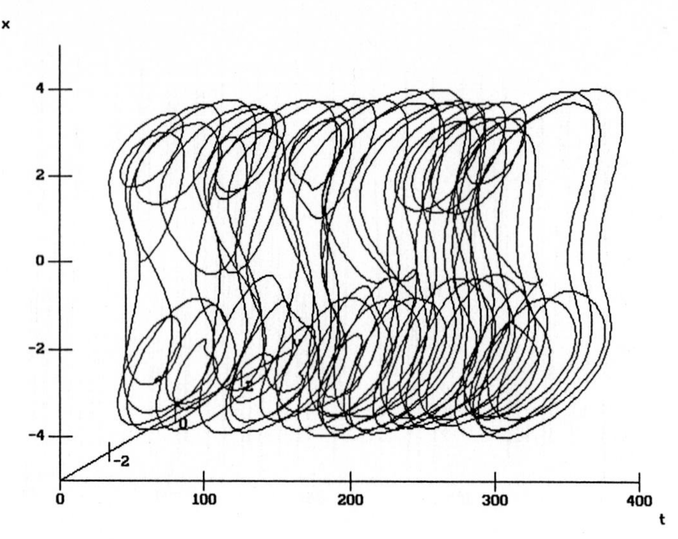

Fig. 12.10. Same trajectory in (t, x, v) space.

rences at longer times, which are due to the extreme sensitivity of the irregular trajectories to small deviations. It should be noted that Figs. 12.9 and 8.3 are computed using different algorithms and different accuracy. Finally, Fig. 12.10 shows the same data again in three-dimensional parallel projection with axes $X = t$, $Y = x$, and $Z = v$.

It is also possible to generate certain maps which are not directly supported by the program. If one wishes, for example, to compute a 'stroboscopic map', — a phase space diagram at equidistant times $t = T, 2T, \ldots$ — one cannot simply enter a value of, e.g., $T = 2\pi$ on the input line for the time variable in 'Store at difference'. If this is done, the program stores the data points *after* t has crossed a value of $2\pi n$; $n = 1, 2, \ldots$. No interpolation for a determination of the crossing point is made, so that this brute force method may result in serious deviations. A better way to achieve the desired stroboscopic map is to add a third auxiliary variable, s, to the system, which satisfies the differential equation

$$\dot{s} = \cos(t/2) \tag{12.14}$$

with integral $s(t) = 2\sin(t/2)$ for $s(0) = 0$. The auxiliary function $s(t)$ is zero at times $t = 2\pi n$. We can now integrate the combined system of three equations and construct a Poincaré section at distance $a = 0$ and direction s by inserting a '0' for the values of a on the first line and a '1' on the third line of the column *'Poincaré map'*, which generates the desired plot. The result in Fig. 12.11 shows the strange attractor already familiar from Fig. 8.4. It should be noted, however, that the numerical integration of (12.14) yields only approximately the desired $\sin(t/2)$-solution and, hence, the numerical zeros will tend to deviate from $t = 2\pi n$ at long times.

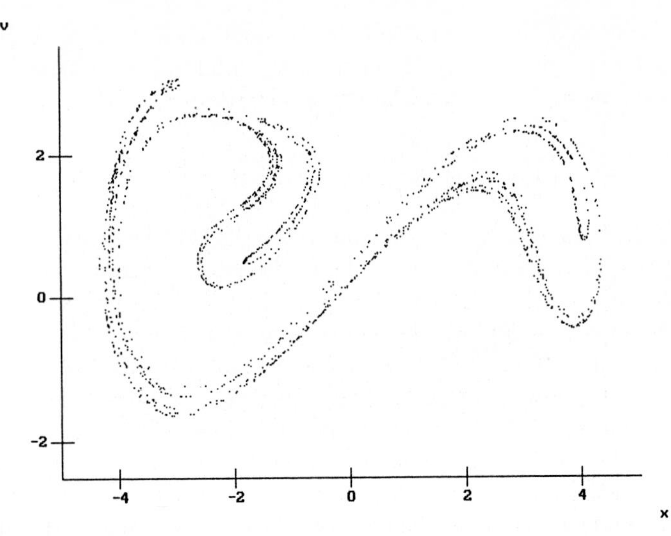

Fig. 12.11. Stroboscopic Poincaré section for the Duffing oscillator. This figure is stored on the disk.

12.3.4 Hill's Equation

For the Duffing oscillator studied in the preceding section, the excitation appears in the inhomogeneity of the differential equation. There are, however, many examples in Physics and Engineering where the excitation arises from time-varying coefficients (the 'parameters') of the differential equations [12.2, Chap. 5]. Such systems are said to be 'parametrically excited'.

In this section, we study the dynamics of the parametrically excited harmonic oscillator

$$\frac{\mathrm{d}^2x}{\mathrm{d}t^2} + \omega^2(t)x = 0 \tag{12.15}$$

with a time-periodic frequency

$$\omega(t+T) = \omega(t)\,. \tag{12.16}$$

This equation is also known as *Hill's equation* [12.2, 12.3]. Here, the real function $\omega^2(t)$ is only a convenient notation, and does not imply that this function is positive. A widely known special case is the *Mathieu equation* with

$$\omega^2(t) = \omega_0^2(1 + \lambda\cos\omega t\,)\,, \tag{12.17}$$

which can be rescaled to the standard form (see [12.4, Chap. 10] and [12.2, Chap. 5.5])

$$\frac{\mathrm{d}^2x}{\mathrm{d}t^2} + (\delta + 2\varepsilon\cos 2t)x = 0\,. \tag{12.18}$$

The reader may notice that differential equations equivalent to (12.15) appear in different context, e.g. in elementary solid state physics as the one-dimensional time-independent Schrödinger equation

$$\frac{\mathrm{d}^2\psi}{\mathrm{d}x^2} + \frac{2m}{\hbar^2}\{E - V(x)\}\psi = 0 \tag{12.19}$$

(see, e.g., Ref. [12.5]). In comparison with (12.15) the wavefunction $\psi(x)$ takes the role of the variable $x(t)$, the coordinate x replaces the time t, and $k^2(x) = 2m\{E - V(x)\}/\hbar^2$ appears instead of $\omega^2(t)$. The potential $V(x)$ must be periodic, of course, and for the special case $V(x) = V_0 \cos(ax)$ this leads to the Mathieu equation (12.18). The parameters δ and 2ε are the (scaled) energy and the (scaled) potential amplitude V_0, respectively. We recall that, in parameter space (δ, ε), we have stable and unstable regions (see Fig. 12.12) where the solutions are bounded or unbounded, respectively. In the context of solid state physics, one usually shows a horizontal cut through this stability diagram (δ is the energy) and speaks about 'bands' (the stability intervals) and 'gaps'. Here, we prefer to discuss a time-dependent system, but all results can immediately be translated to other situations.

It should be stressed, however, that the time-dependent harmonic oscillator (12.15) is *not* a typical time-periodic system, because it is *integrable* and therefore, of course, *not* chaotic. The integrability can be directly proved by the construction of an — explicitly time-dependent — integral of motion, namely the so-called *Lewis invariant* [12.6]

$$I(t) = \frac{1}{2}\left[(r(t)\dot{x} - \dot{r}(t)x)^2 + (x/r(t))^2\right], \tag{12.20}$$

where the function $r(t)$ is an *arbitrary* solution of an auxiliary equation

$$\ddot{r}(t) + \omega^2(t)r(t) = \frac{1}{r^3(t)}, \tag{12.21}$$

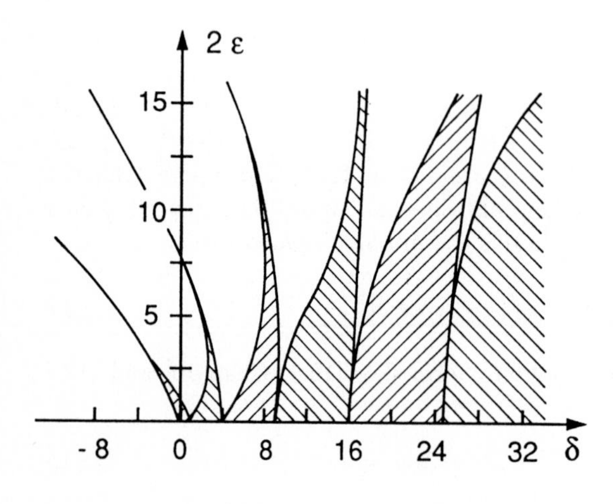

Fig. 12.12. Stable (shaded) and unstable regions for the Mathieu equation.

the so-called Milne equation [12.7]. The numerical value of I is constant along the trajectory, which can be directly verified by differentiation: $dI/dt = 0$.

Depending on the function $\omega^2(t)$, the solutions can be stable (bounded) or unstable (unbounded), as discussed above for the case of the Mathieu equation. In the stable case, a real valued *T-periodic* solution $r(t)$ of the Milne equation (12.21) exists. This periodic solution can be constructed, for example, from the fundamental solutions $x_1(t)$ and $x_2(t)$ of Hill's equation, which satisfy the initial conditions $x_1(0) = \dot{x}_2(0) = 1$ and $\dot{x}_1(0) = x_2(0) = 0$. Writing the general solution of (12.21) as

$$r(t) = \sqrt{Ax_1^2(t) + Bx_2^2(t) + 2Cx_1(t)x_2(t)} \quad \text{with} \quad AB - C^2 = 1 \tag{12.22}$$

[12.8] and demanding periodicity of $r(t)$, we find the initial conditions for r. We only state here the result for a symmetric function $\omega^2(-t) = \omega^2(t)$:

$$r(0) = \sqrt{\frac{x_2(T)}{\sqrt{1 - \dot{x}_2^2(T)}}} \quad \text{and} \quad \dot{r}(0) = 0 \tag{12.23}$$

(for more details see Refs. [12.8] and [12.9]). In the following, we shall assume that $r(t)$ is such a periodic solution. In this case, the invariant is a T-periodic function in phase space: $I(\dot{x}, x, t + T) = I(\dot{x}, x, t)$. The surfaces defined by constant values of the Lewis invariant

$$I(\dot{x}, x, t) = \frac{1}{2}\left[(r\dot{x} - \dot{r}x)^2 + (x/r)^2\right] = const. \tag{12.24}$$

form vortex tubes ('flux tubes'), which provide a stratification of the extended phase space $(\dot{x}, x, t)$. Any trajectory remains for all times on the surface of such a flux tube, which has the topology of a cylinder or torus, when the dynamics is taken modulo T. These flux tubes fill the phase space. At a fixed time, (12.24) is a conic section in phase space and a stroboscopic phase space section at time $t = t_0$ modulo T yields an ellipse in the case of stability (a hyperbola for instability). These similar ellipses fill the stroboscopic phase space section. They are rotated by an angle

$$\phi(t) = \frac{1}{2}\arctan\left\{\frac{2r(t)\dot{r}(t)}{r^2(t) - r^{-2}(t) + \dot{r}^2(t)}\right\} \tag{12.25}$$

with respect to the x-axis. The ratio between the smaller, b, and larger axis, a, is given by

$$b/a = \sigma - \sqrt{\sigma^2 - 1}\,, \quad \text{with} \quad \sigma = (r^2(t) + r^{-2}(t) + \dot{r}^2(t))/2 \geq 1\,, \tag{12.26}$$

and the area of the ellipse equals $\pi ab = 2\pi I$, which is constant in time.

There is an even number of times t_k within each period with $\dot{r}(t_k) = 0$. At times t_k, the half axes of the ellipses $I = const.$ are parallel to the $x, \dot{x}$-axes and are given by $\sqrt{2I}\, r(t_k)$ or $\sqrt{2I}\, /r(t_k)$, respectively. The overall motion and

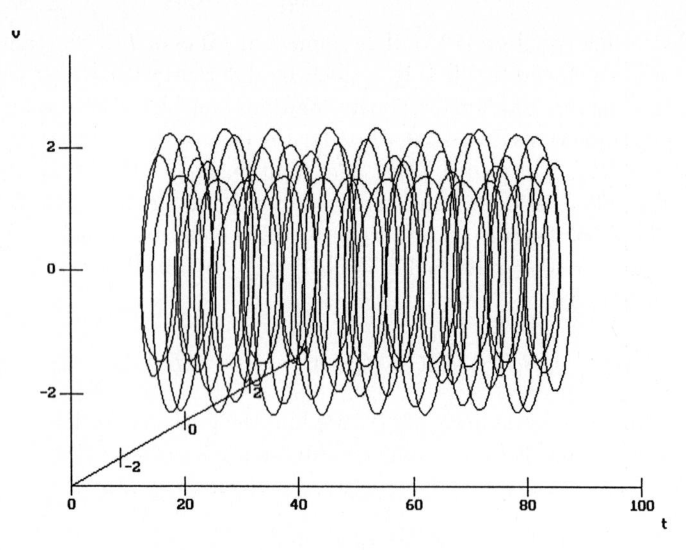

Fig. 12.13. Solution of the Mathieu equation for $\delta = 8$, $2\varepsilon = 5$ and initial conditions $x(0) = 1$, $\dot{x} = 0$ in extended phase space $(x, \dot{x}, t)$ $(0 \leq t \leq 100)$. This figure is stored on the disk.

deformation of the elliptic flux tube is sensitively dependent on the functional form of $\omega^2(t)$. A libration is found for small values of $r^2(t)$ and a rotation, where the whole flux tube is twisted in phase space, may occur for $r^2(t) > 1$ when the denominator in (12.25) can pass through zero.

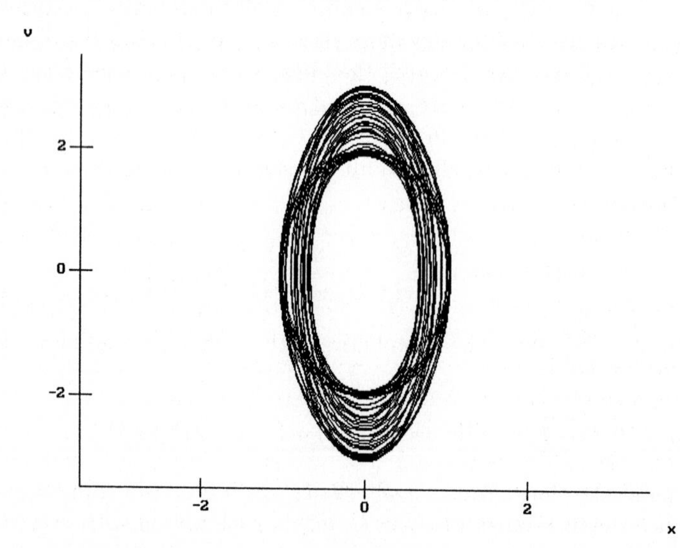

Fig. 12.14. Phase space projection of the Mathieu trajectory shown in Fig. 12.13.

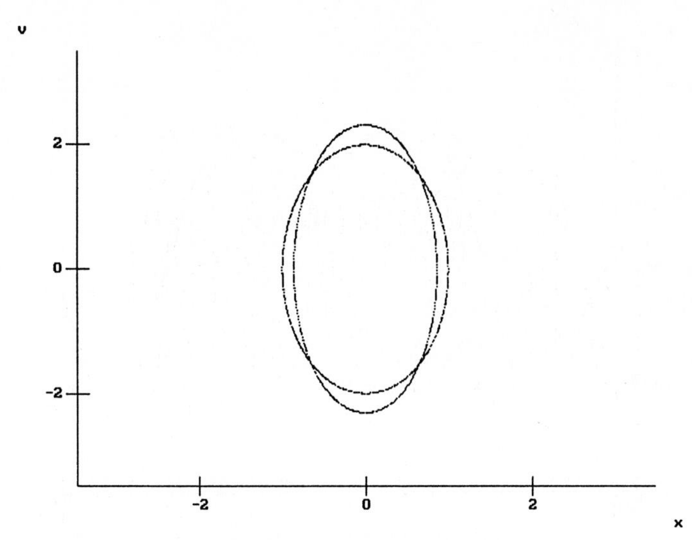

Fig. 12.15. Two stroboscopic Poincaré sections for the Mathieu trajectory shown in Figs. 12.13 and 12.14 at times nT and $(n+1/2)T$ ($0 \leq t \leq 2000$).

As an example, let us now study the Mathieu equation (12.18) numerically. The period of the parametric force is $T = \pi$. We transform the second order differential equation to two first order equations, introducing the velocity $v = \dot{x}$, and solve these equations by means of the program ODE. First, we choose the parameters $\delta = 8$ and $2\varepsilon = 5$. The solutions are stable (compare Fig. 12.12) and Fig. 12.13 shows a computed trajectory with initial conditions $x(0) = 1$ and $\dot{x} = 0$ in extended phase space $(x, \dot{x}, t)$ ($\Delta t_{\text{store}} = \Delta x_{\text{store}} = \Delta v_{\text{store}} = 0.05$, accuracy parameters 10^{-6}). The trajectory oscillates with an amplitude varying in time and there seems to be an underlying regular pattern. The phase space projection in Fig. 12.14 displays some elliptically shaped caustics, but the true regular phase space organization is only visible in the stroboscopic Poincaré plot in Fig. 12.15, which shows two sections at times nT and $(n+1/2)T$. These stroboscopic sections were constructed as described above for the Duffing oscillator. Both sections yield the predicted ellipses (12.24), which are more eccentric at times $(n+1/2)T$.

The reader may also wish to integrate the Milne equation (12.21) and verify numerically the existence of a periodic solution. One can very easily add the nonlinear inverse cubic term to the equation and explore its behavior. The *periodic* solution can be constructed using a brute force 'shooting' method, i.e. by integrating with initial conditions $\dot{r}(0) = 0$ and varying $r(0)$ systematically until the conditions $\dot{r}(T) = 0$ and $r(T) = r(0)$ are satisfied. It is more efficient, however, to use the theoretical considerations outlined above. In this case, one first computes the fundamental solution $x_2(t)$, integrating the Mathieu equation starting at $x(0) = 0$, $\dot{x}(0) = 1$ and generating a stroboscopic Poincaré section at nT. Only the first point is needed here. The data at time T can then be read

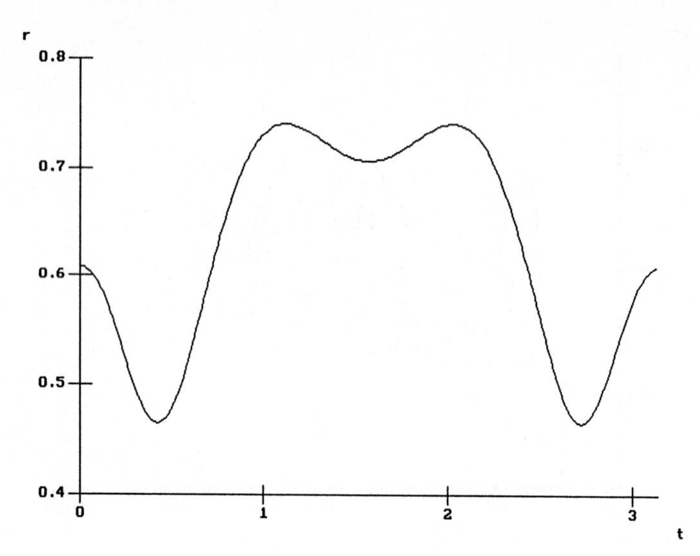

Fig. 12.16. Periodic solution $r(t)$ of the auxiliary Milne equation 12.21.

from the $*$.TBL file. We obtain $x_2(T) = 0.2695$, $\dot{x}_2(T) = -0.6866$ and (12.23) leads to an initial condition of $r(0) = 0.609$. On computing a Milne solution with this value, we indeed find the desired periodic solution, as shown in Figs. 12.16 and 12.17 ($\Delta t_{\text{store}} = 0.01$ $\Delta x_{\text{store}} = \Delta v_{\text{store}} = 0$). We note that $\dot{r}(t)$ vanishes six times per period, i.e. there are six times for which the axis of the ellipse is

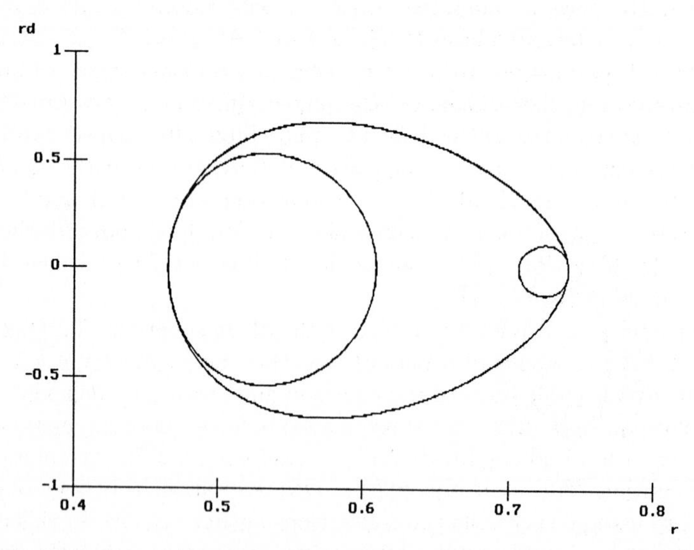

Fig. 12.17. Periodic solution of the Milne equation shown in Fig. 12.16 in $(r, \dot{r})$-phase space ($rd = \dot{r}$).

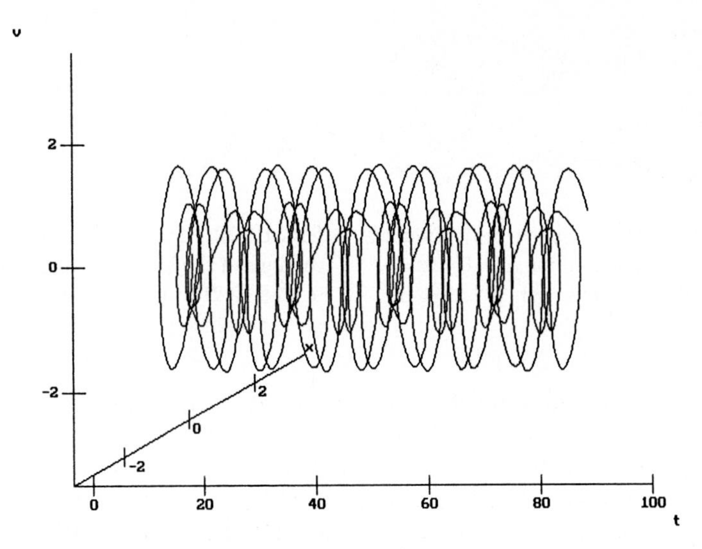

Fig. 12.18. Solution of the Mathieu equation for $\delta = 6$, $2\varepsilon = 5$ and initial conditions $x(0) = 1$, $\dot{x} = 0$ in extended phase space $(x, \dot{x}, t)$ $(0 \leq t \leq 100)$.

horizontal. The interested reader can also compute the time dependence of the orientation angle, the eccentricity, etc. from the $r(t)$-function.

It is interesting to explore the behavior of the Mathieu solutions for other parameter combinations. A much more complicated motion of the trajectories can be detected, but in all cases the underlying dynamics is governed by the quadratic Lewis invariant (12.24). As an example, Figs. 12.18–12.20 show the results for $\delta = 6$, $2\varepsilon = 5$, which is still inside the same stability zone as in the previous case ($\Delta t_{\text{store}} = \Delta x_{\text{store}} = \Delta v_{\text{store}} = 0.05$, accuracy parameters 10^{-6}). Results for the parameters $\delta = 0$, $2\varepsilon = 1$ in the lowest stability zone are finally displayed in Figs. 12.21 and 12.22. In all these cases, the stroboscopic Poincaré

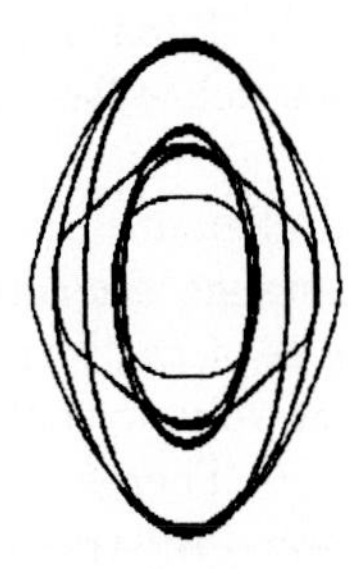

Fig. 12.19. Phase space projection of the Mathieu trajectory shown in Fig. 12.18.

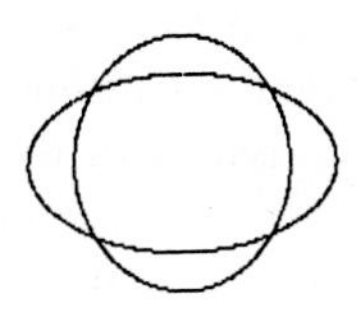

Fig. 12.20. Two stroboscopic Poincaré sections of the Mathieu trajectory shown in Figs. 12.18 and 12.19 at times nT and $(n+1/2)T$ $(0 \leq t \leq 2000)$.

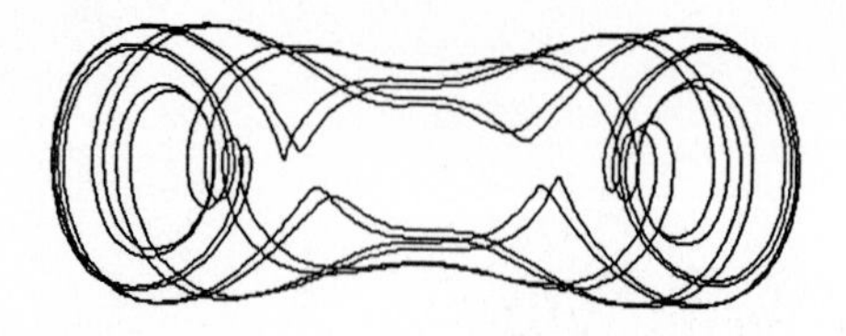

Fig. 12.21. Phase space projection of a Mathieu trajectory for $\delta = 0$, $2\varepsilon = 1$ and initial conditions $x(0) = 1$, $\dot{x} = 0$.

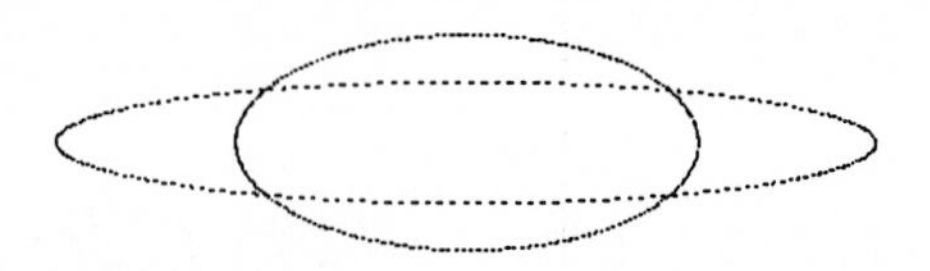

Fig. 12.22. Two stroboscopic Poincaré sections of the Mathieu trajectory shown in Fig. 12.21 at times nT and $(n+1/2)T$ ($0 \le t \le 2000$).

sections are simple ellipses with constant area at all times, but their shape and orientation can vary wildly with time.

The reader is invited to explore the dynamics of the interesting, but still *non-chaotic*, Hill equation in more detail. One can try to find characteristic features for the different stability regions, study the unbounded motion, find the (existing) T- or $2T$-periodic solutions of the Hill equation at the stability boundaries, look at Hill equations whose time-dependence is different from the Mathieu case, etc. Many interesting examples can be found in the literature (see, e.g. [12.3, 12.2] and references given there).

12.3.5 The Lorenz Attractor

The set of three autonomous equations

$$\begin{aligned} \dot{x} &= \sigma(y - x) \\ \dot{y} &= (r - z)x - y \\ \dot{z} &= xy - bz\,, \end{aligned} \tag{12.27}$$

studied by Lorenz [12.10] have been derived in an attempt to explore the transition to turbulence, e.g. in a theoretical analysis of the Bénard experiment, where a fluid layer in a gravitational field is heated from below (see Ref. [12.11, App.A] for details).

The Lorenz equations (12.27) are one of the paradigmatic systems used in nonlinear dynamics to introduce the concept of strange attractors. First, we ensure that the system is dissipative. The system is of the form $d\mathbf{r}/dt = \mathbf{v}(\mathbf{r})$ discussed in Chap. 2.3 and, hence, the phase space volume contracts with constant rate $\operatorname{div}\mathbf{v} = -(\sigma + 1 + b) > 0$ for $\sigma > 0$, $b > 0$. Therefore, any finite phase space volume element will be compressed to zero and the trajectories will approach an *'attractor'* in the long time limit.

Let us now improve our skills and compute the famous *Lorenz attractor* found for the standard set of parameters $\sigma = 10$, $r = 28$, and $b = 8/3$. A trajectory is attracted by a finite region in phase space, where the motion is

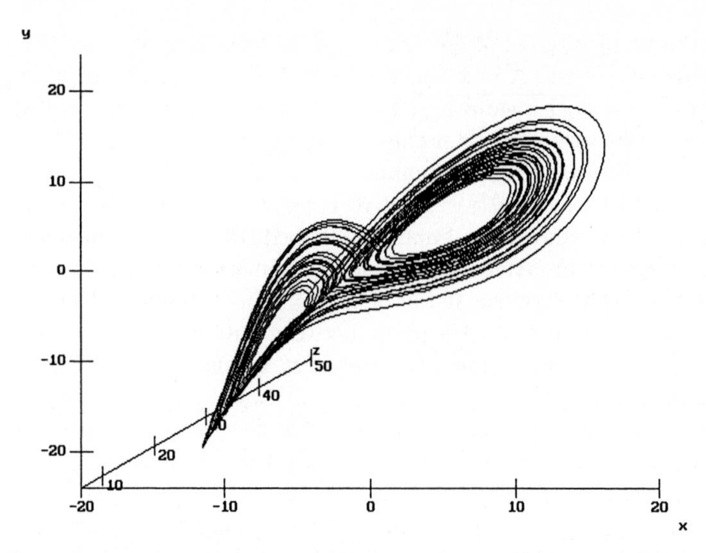

Fig. 12.23. Lorenz attractor with axes chosen as $X = x$, $Y = y$, and $Z = z$ ($\sigma = 10$, $r = 28$, and $b = 8/3$) ($0 \le t \le 40$).

erratic. This so-called *strange attractor* is shown in Fig. 12.23 in three-dimensional X, Y, Z-space with axes $X = x$, $Y = y$, and $Z = z$ ($\Delta t_{\text{store}} = 0$, $\Delta x_{\text{store}} = \Delta y_{\text{store}} = \Delta z_{\text{store}} = 0.5$, accuracy parameters 10^{-5}).

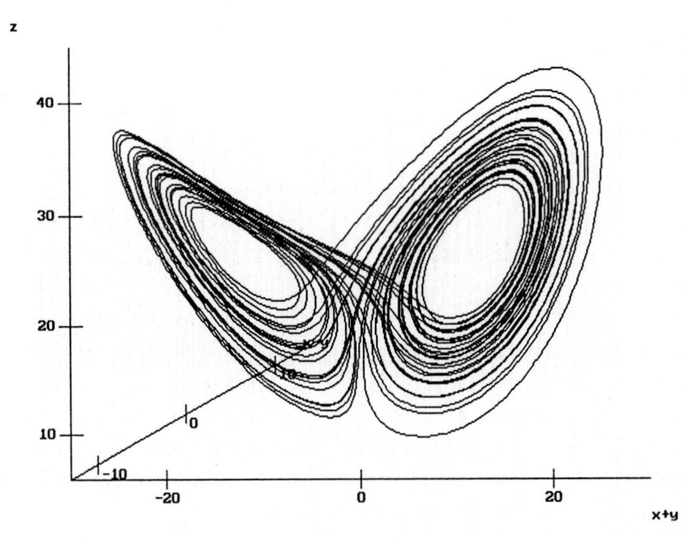

Fig. 12.24. Lorenz attractor shown in Fig. 12.23 with axes chosen as $X = x + y$, $Y = z$, and $Z = x - y$. This figure is stored (file LORENZ.ODE).

The object of interest is not easily visible from this plot, so we rotate in space choosing the axes $X = x+y$, $Y = z$, and $Z = x-y$. The resulting plot in Fig. 12.24 gives a much clearer impression of the attractor, the so-called *'Lorenz mask'*. The trajectory makes a number of loops to the left, followed by loops to the right, in a seemingly erratic manner.

The structure of the attractor is very complicated. It is a fractal set and certainly *not* a two-dimensional manifold embedded in three-dimensional phase space. The reader can explore the structure numerically, e.g., by computing a Poincaré section through the attractor at $z = 0$ in (x, y)-space, and investigate the dynamical mapping on this plane between subsequent points $(x_n, y_n) \rightarrow (x_{n+1}, y_{n+1})$ by plotting x_{n+1} as a function of x_n by means of the *'Representation mode'* '2-*dim.* $n/n+1$'.

In addition, the dynamics at the attractor is sensitively dependent on the initial conditions, which can be visualized by running trajectories with small differences in the initial conditions.

In another numerical experiment we can re-compute a diagram shown by Lorenz [12.10]. We start by integrating the Lorenz equations (12.27) with initial values $x = -8$, $y = -1$, and $z = 33$ at time zero up to $t = 40$. A plot of the z-coordinate versus time, as shown in Fig. 12.25, reveals almost regular oscillations with increasing amplitude, interrupted by erratic amplitude jumps. We can analyze this behavior in more detail. First, the successive maxima z_n, $n = 1, 2, \ldots$, of the z-coordinate are calculated. This can be done, for instance, from the numerical data of the solution obtained by activating the command *'Table file'* in *'Edit Representation'*. A *'Display'* of the results then generates a data file $*$.TBL of the solution vector (t, x, y, z). The data can then be accessed by any editor and modified as desired and/or used as input for appropriate software

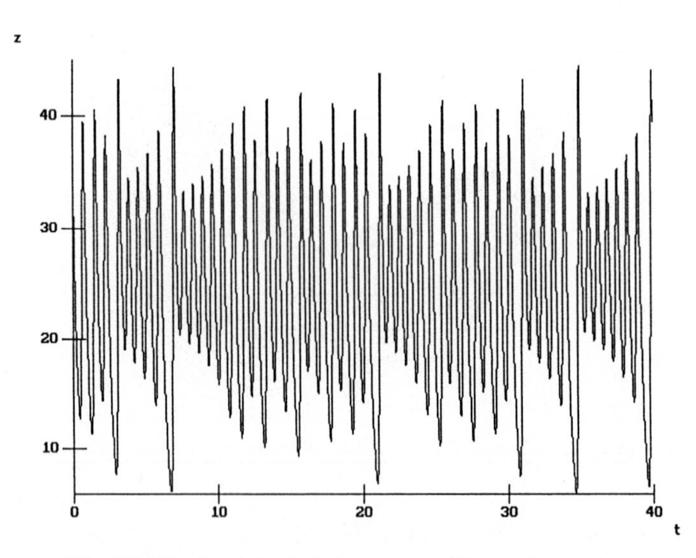

Fig. 12.25. Lorenz attractor: z-coordinate versus time.

Table 12.1. Lorenz attractor: Solution vector (t_n, x_n, y_n, z_n) at the nth maximum of the z-coordinate as read from the *.TBL file.

6.95369112E-0001	-1.41545901E+0001	-8.24604288E+0000	3.97658081E+0001
1.45750662E+0000	1.51542653E+0001	1.18829590E+0001	3.90208310E+0001
2.23123607E+0000	-1.49913181E+0001	-1.16677912E+0001	3.88341776E+0001
3.04435409E+0000	1.43987920E+0001	8.34384815E+0000	4.02006349E+0001
3.78085379E+0000	-1.45422976E+0001	-1.09271821E+0001	3.84529441E+0001
4.65938179E+0000	1.61984785E+0001	1.19914644E+0001	4.12837800E+0001
5.33538606E+0000	-1.31148005E+0001	-1.06496591E+0001	3.55179384E+0001
6.06218364E+0000	-1.36673890E+0001	-9.95361848E+0000	3.73508088E+0001
6.87517878E+0000	-1.52424740E+0001	-1.06081112E+0001	4.02529784E+0001
7.59849114E+0000	1.41885746E+0001	1.09840071E+0001	3.76191312E+0001

tools. As an example, Table 12.1 lists the solution vector (t_n, x_n, y_n, z_n) at the nth maximum of the z-coordinate for $n = 1, \ldots, 10$ (compare Fig. 12.25). The solution vectors at other times have been deleted.

This procedure is, however, somewhat complicated and was only explained in detail in view of possible applications to other problems. In the case considered here we can use a much more convenient method. We wish to find the maximum amplitudes of the z-coordinate. We therefore introduce once more an additional variable $w = \dot{z}$ satisfying the auxiliary differential equation

$$\dot{w} = \sigma(y - x)y + (-xz + rx - y)x - (xy - bz)b \tag{12.28}$$

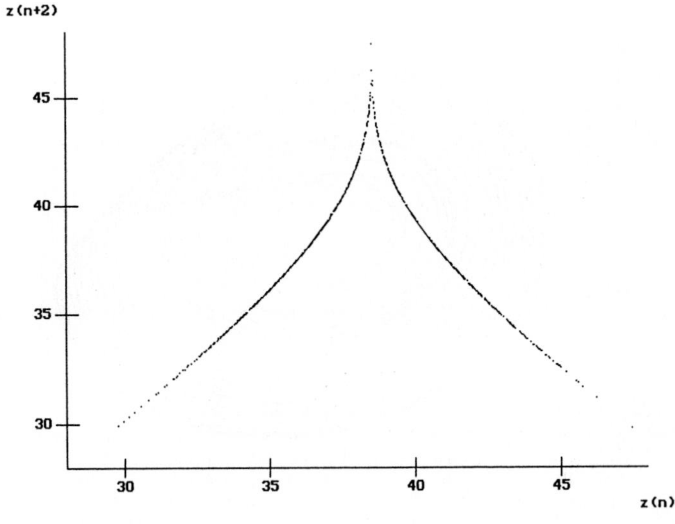

Fig. 12.26. Maximum z_{n+2} as a function of the last but one maximum z_n.

with initial condition $w(0) = x(0)y(0) - bz(0)$. We now compute a Poincaré section whenever w crosses zero, which immediately yields the desired list of successive extrema of $z(t)$.

In Fig. 12.26, this series z_n, $n = 1, \ldots, 2000$, is analyzed by plotting z_{n+2} as a function of the next but one extremum z_n, in order to separate the series of maxima and minima. Only the maxima are shown in Fig. 12.26. This plot is generated by means of the representation mode '2-*dim.* $n/n+2$'.

The discrete mapping $z_n \to z_{n+2}$ is approximately triangular and can be modeled by the tent map studied in Sect. 9.3.4, which generates chaotic sequences.

Further numerical experiments exploring the Lorenz equations are suggested in Sect. 12.3.8 below. For more details the reader is referred to the literature (see, e.g., Refs. [12.11, 12.12] and references given there).

12.3.6 The Rössler Attractor

The flow at the Lorenz attractor oscillates erratically between two 'leaves', which are mutually connected in quite an intricate manner. A much simpler object is the single-leafed attractor in the Rössler system [12.13]

$$\begin{aligned} \dot{x} &= -y - z \\ \dot{y} &= z + ay \\ \dot{z} &= b + z(x - c), \end{aligned} \tag{12.29}$$

with standard parameter values $a = b = 0.2$, $c = 5.7$. Figures 12.27 and 12.28 show a two- and a three-dimensional projection of a trajectory for the system

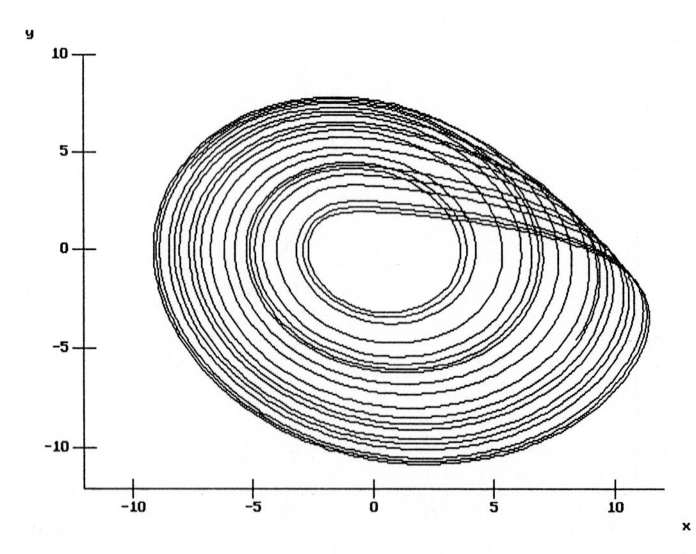

Fig. 12.27. Rössler attractor in a two-dimensional projection onto the (x, y)-plane.

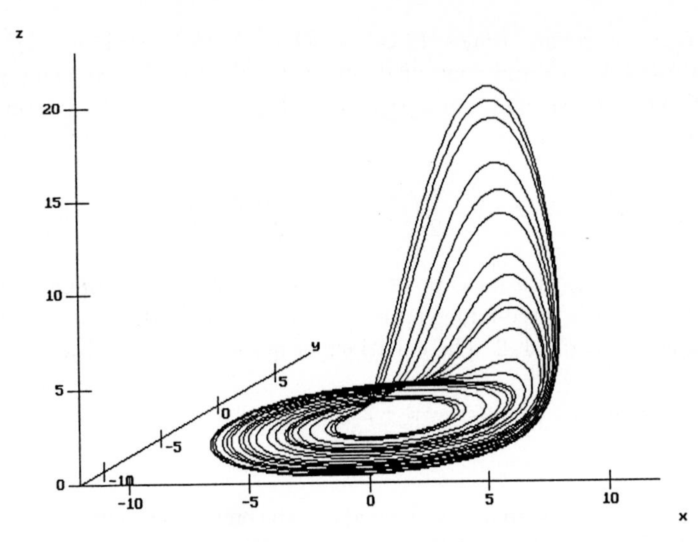

Fig. 12.28. Rössler attractor in three-dimensional x, y, z-space. This figure is stored on the disk.

(12.29) approaching this attractor ($\Delta x_{\text{store}} = \Delta y_{\text{store}} = \Delta z_{\text{store}} = 0.2$, accuracy parameters 10^{-7}). The transient motion at short times has been erased.

For this *'Rössler attractor'* — also known as *'Rössler's band'* — a qualitative understanding of the chaotic flow is much easier to obtain. As a first approximation, one starts with a two-dimensional 'band' which is bent into phase space, cut, twisted, and glued together as in a Möbius strip. This twist moves a trajectory from the outer to the inner part of the band. Secondly, distinct trajectories of a dynamical system cannot join. Therefore, the attractor cannot be a simple surface. What happens is that the 'twist and glue' is accompanied by a stretching and folding of the band, so that one actually has to glue a U-shaped double-sheeted band onto a single-sheeted one, which is impossible in any clean way. In the course of the next loop the band is stretched and folded again and returns as a four-sided double-U band, and so on. The 'band' appears as a thickened multilayered surface with a complicated microstructure. During the process of stretching and folding, the contraction of phase space continues and the limiting object, a strange attractor, is clearly a fractal set.

As an exercise, the reader can compute the sequence of the nth maximum amplitudes of the oscillations on the attractor and construct a $n \to n+1$ or $n \to n+2$ map similar to Fig. 12.26 for the Lorenz attractor.

12.3.7 The Hénon–Heiles System

The Hénon–Heiles system is one of the early examples of Hamiltonian systems where computational studies caused a change in the *qualitative* picture of classical dynamics as acquired by most physicists in Mechanics lecture courses. The original article by Hénon–Heiles [12.14], which dates from 1964, was devoted to studies in astrophysics and contained surprising results of some computer experiments, which developed into a prototype of such studies. The Hénon–Heiles Hamiltonian

$$H = \frac{1}{2}\left(p_x^2 + p_y^2\right) + V(x,y) = \frac{1}{2}\left(p_x^2 + p_y^2\right) + \frac{1}{2}\left(x^2 + y^2\right) + x^2 y - \frac{1}{3}y^3 \quad (12.30)$$

describes the motion of a mass point with unit mass in a two-dimensional potential $V(x,y)$ with three-fold symmetry, a minimum at the center, and three saddle points of height 1/6. The equations of motion are given by

$$\begin{aligned} \dot{x} &= p_x \quad , \quad \dot{p_x} = -x - 2xy \\ \dot{y} &= p_y \quad , \quad \dot{p_y} = -y - x^2 + y^2 \,. \end{aligned} \quad (12.31)$$

The character of the solutions depends on the energy. For energies $E < 1/6$, the internal motion is bounded. At low energies the dynamics is almost regular. With increasing values of E, the characteristic Poincaré scenario develops and more and more phase space is filled by chaotic dynamics with a more or less sharp onset at $E \approx 0.11$ For an energy $E = 0.125$, regular and chaotic motions coexist: figures 12.29 and 12.30 show a regular and a chaotic trajectory in coordinate space, started at $x = 0$, $y = 0.1$, and $p_x = 0.49057$, $p_y = 0$ or at $x = y = 0$, $p_x = p_y = 0.35355$, respectively. ($\Delta x_{\text{store}} = \Delta y_{\text{store}} = 0.015$, $\Delta px_{\text{store}} = \Delta py_{\text{store}} = 0.05$, accuracy parameters for x, y, p_x, p_y chosen as $10^{-7}, 10^{-5}, 10^{-5}, 10^{-5}$).

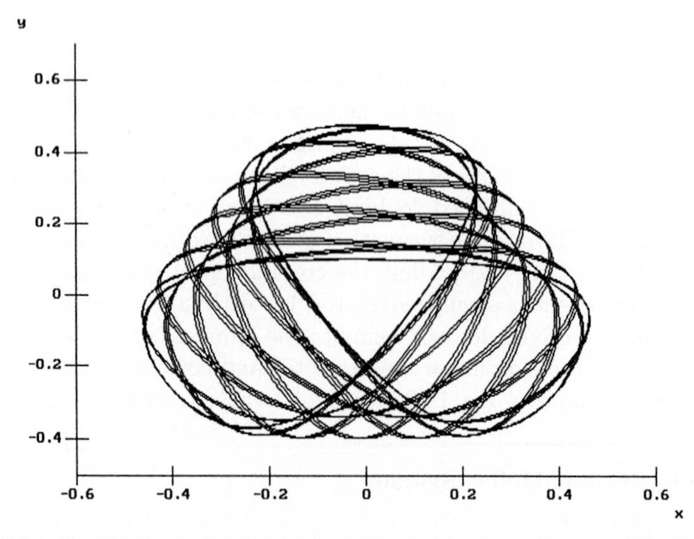

Fig. 12.29. Regular trajectory in coordinate space for the Hénon–Heiles system. Initial conditions $x = 0$, $y = 0.1$ and $p_x = 0.49057$, $p_y = 0$ ($0 \le t \le 200$).

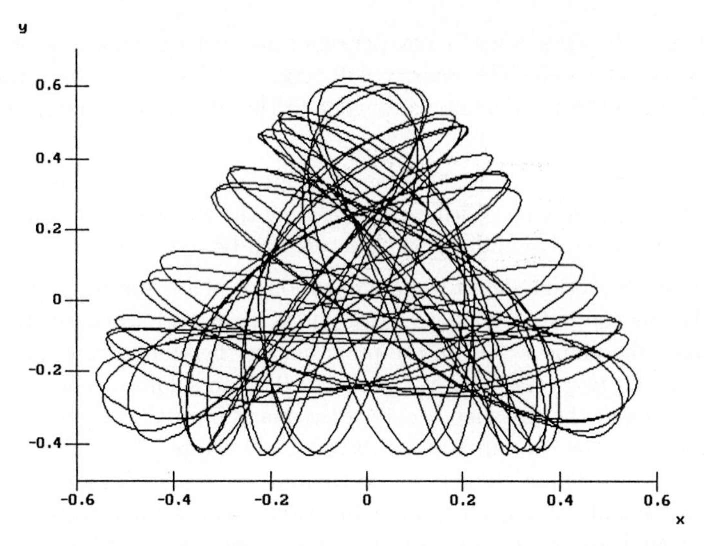

Fig. 12.30. Chaotic trajectory in coordinate space for the Hénon–Heiles system. Initial conditions $x = y = 0$ and $p_x = p_y = 0.35355$ ($0 \le t \le 200$).

The regular trajectory in Fig. 12.29 traces out a regular pattern in coordinate space. In four-dimensional phase space the trajectory is confined to a two-dimensional manifold, which appears in Fig. 12.29 as a projection onto the

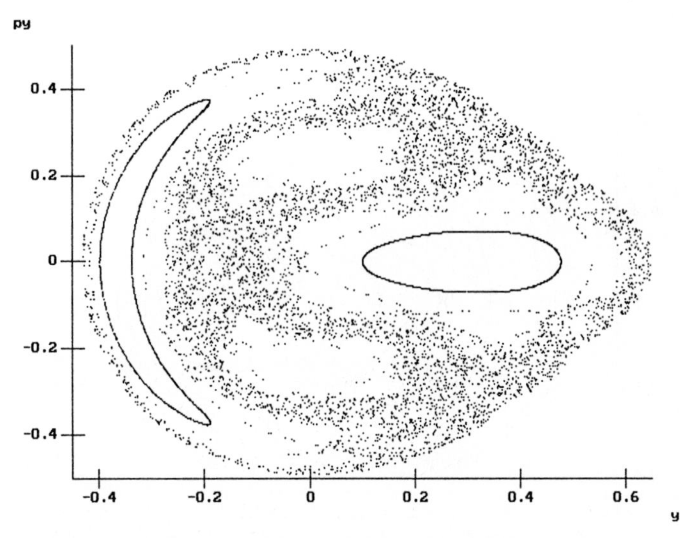

Fig. 12.31. Poincaré section at $x = 0$ for the Hénon–Heiles system. A regular and a chaotic trajectory for $E = 0.125$ and initial conditions of Figs. 12.29 and 12.30. This figure is stored (file HENON.ODE).

(x, y)-plane. The trajectory is not periodic, and with increasing time it will densely cover a 'band'. The density will tend to infinity at points where the projection onto the (x, y)-plane is tangent to the phase space manifold, i.e. at the *caustics*.

The trajectory in Fig. 12.30 appears as an irregular tangle, only confined by conservation of energy. With increasing time the whole energetically allowed region will be filled. No underlying structure can be observed.

A Poincaré section shows the organization of phase space dynamics more clearly. Let us compute the phase space points (y, p_x, p_y) whenever the trajectory crosses the plane $x = 0$. In the program, we therefore define the plane by the unit vector $\mathbf{n} = (t, x, y, p_x, p_y) = (0, 1, 0, 0, 0)$ in solution space orthogonal to this plane and the distance $a = 0$ of the plane from the origin (see (12.6)). These data, e.g. the column $0, 0, 1, 0, 0, 0, \ldots$, are the required input under the headline *'Poincaré map'*. The program computes the solution and, in addition, the intersection of the trajectory with this plane within a certain accuracy limit. Therefore, at least one of the corresponding entries in the column *'Accuracy'* must be different from zero. In the present computation, we require an accuracy of 10^{-7} for the x-coordinate (same as above). Figure 12.31 show the (y, p_y) values of this section. The plot has been generated by selecting a *'2-dim.'* representation mode in the menu *'Edit Representation'* and choosing axis '$X = y$' and '$Y = p_y$'. It should be noted that the third phase space variable, p_x, is determined up to a sign by conservation of energy. Results for the regular and chaotic trajectory of Figs. 12.29 and 12.30 are shown simultaneously. For the regular case, the trajectory is confined to an invariant torus, which is cut twice

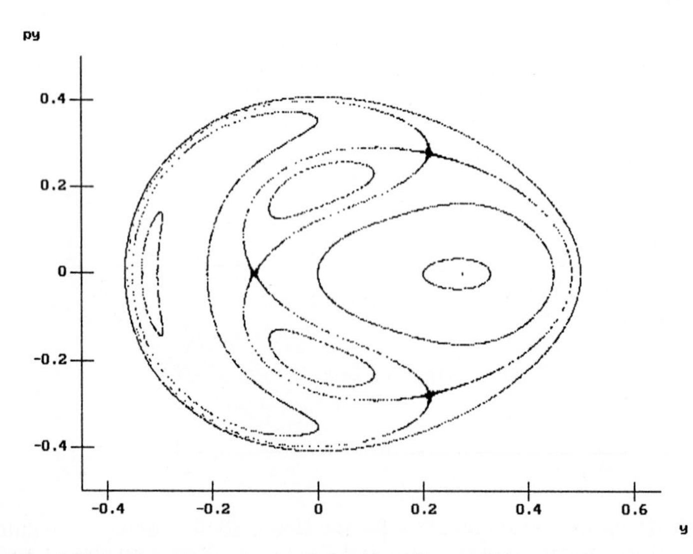

Fig. 12.32. Poincaré section at $x = 0$ for the Hénon–Heiles system. Six trajectories for $E = 0.083$ and different initial conditions are shown.

in the Poincaré section, where the phase space points trace out a smooth curve. The chaotic trajectory fills an area in an irregular manner.

The 'chaotic sea' in Fig. 12.31 arises from the destruction of a separatrix at lower energies. In the Poincaré section at $E = 0.083$ shown in Fig. 12.32 (accuracy parameters for x, y, p_x, p_y, t chosen as $10^{-8}, 10^{-7}, 10^{-7}, 10^{-7}$).

The whole phase space still seems to be filled with invariant curves and the separatrix is clearly visible. Five trajectories with different initial conditions have been integrated. In addition, the confinement due to total energy conservation is displayed by computing a trajectory embedded in the Poincaré section, which can be computed from the initial conditions $x = y = p_x = 0$ and $p_y = \sqrt{2E}$.

Further information on the numerical exploration of the Hénon–Heiles and similar Hamiltonian systems can be found in the literature (see, e.g. Sect. 2.2 and [12.15, 12.16]).

12.3.8 Suggestions for Additional Experiments

Lorenz System: Limit Cycles and Intermittency. The Lorenz equations (12.27) studied in the numerical experiment in Sect. 12.3.5 above for the traditional parameter values $\sigma = 10$, $r = 28$, and $b = 8/3$ display a rich structure of dynamical behavior. Following the numerical experiments in the literature, we vary the control parameter r, which is proportional to the temperature gradient. When we increase r from zero for fixed values of $\sigma = 10$ and $b = 8/3$, the onset of chaotic behavior occurs at $r = 24.74$ and we observe the (stran-

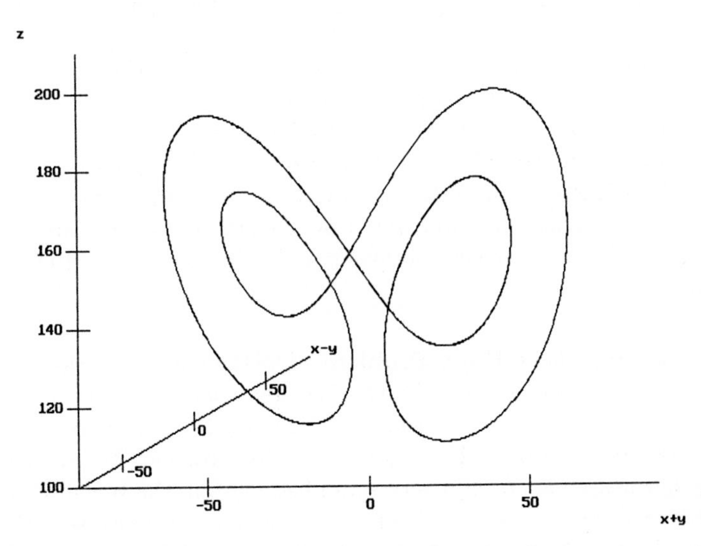

Fig. 12.33. Limit cycle C_1 of the Lorenz system for control parameter $r = 155$ with axes $X = x + y$, $Y = z$, and $Z = x - y$ $(0 \leq t \leq 10)$.

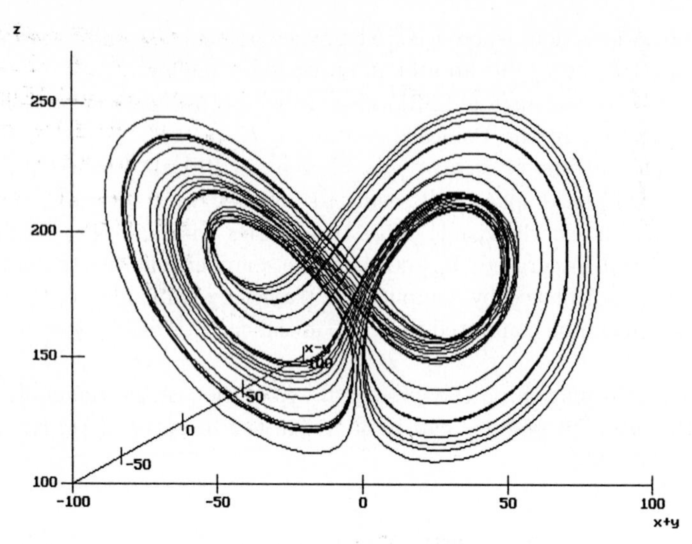

Fig. 12.34. Strange attractor L_2 of the Lorenz system for control parameter $r = 180$ with axes $X = x + y$, $Y = z$, and $Z = x - y$ $(0 \leq t \leq 20)$.

ge) Lorenz attractor L_1. At $r = 145$, a gradual transition to a coexisting limit cycle C_1 begins, which ends at about $r = 148.4$. The limit cycle C_1 can be found up to $r = 166.1$, where a strange attractor L_2 appears with transition to a limit cycle C_2 in the interval $166.1 < r < 233.5$ [12.17, 12.18]. Examples of the limit cycle C_1 and the strange attractor L_2 are shown in Figs. 12.33 ($\Delta x_{\text{store}} = \Delta y_{\text{store}} = \Delta z_{\text{store}} = 0.5$, $\Delta t_{\text{store}} = 0$, accuracy parameters 10^{-5}) and 12.34 ($\Delta x_{\text{store}} = \Delta y_{\text{store}} = \Delta z_{\text{store}} = 2$, $\Delta t_{\text{store}} = 0$, accuracy parameters 10^{-5}). Many more details can be explored by closer inspection. The interested reader should consult Refs. [12.19] and [12.12, Chap. 10].

In addition, the Lorenz system also displays intermittency in the transition region from the limit cycle C_1 to the strange attractor L_2. Choosing, for example, the control parameters $r = 166.0$, $r = 166.1, \ldots, r = 167.0$, one observes long 'laminar' phases erratically interrupted by 'chaotic' ones. The duration of the chaotic phases becomes longer and longer with increasing r (see Ref. [12.18] for more details).

The Restricted Three Body Problem. The three body problem of celestial mechanics, namely the motion of three mass points interacting via gravitational forces, is one of the classic problems in dynamics. Despite enormous effort, it has not been possible to find a 'solution' to this problem. Today, it is known to be a chaotic system. Even a very simplified version, the so-called *'restricted three body problem'*, is chaotic. This model is the limit, where one of the masses is very small, i.e., the motion of a satellite in the field of two suns with masses m_1 and m_2. The motion of the massive bodies m_1 and m_2 with coordinates $\mathbf{r}_1(t)$ and $\mathbf{r}_2(t)$ is independent of the small mass, and the time-dependent Hamiltonian

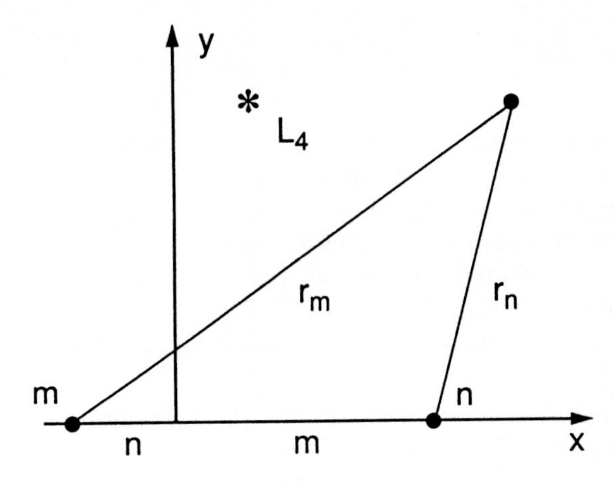

Fig. 12.35. Restricted three body problem. Coordinates and equilateral equilibrium point L_4.

determining the motion of the small mass in the center of mass system is simply

$$H(t) = \frac{1}{2}\mathbf{p}^2 - \kappa\left(\frac{m_1}{|\mathbf{r}_1 - \mathbf{x}|} + \frac{m_2}{|\mathbf{r}_2 - \mathbf{x}|}\right), \tag{12.32}$$

where κ is the gravitational constant. The mass of the light particle may be set equal to unity, as it factors out of the equations of motion. The heavy masses move on Kepler ellipses, which are taken to be circles having a fixed distance R for the two masses. From Kepler's third law, we know the period of rotation to be $T^2/R^3 = 4\pi^2/\kappa(m_1 + m_2)$. To simplify the equations, we choose units where $\kappa(m_1 + m_2) = 1$ and measure distances in units of R, i.e. the frequency of rotation $\omega = 2\pi/T$ is equal to one. We furthermore use relative masses $m = m_1/(m_1 + m_2)$ and $n = m_2/(m_1 + m_2)$ with $m + n = 1$.

We now transform to a coordinate frame rotating with the heavy masses. In this frame, the centers of force are fixed and — restricting our analysis to a planar motion — the Hamiltonian is

$$H = \frac{1}{2}(p^2+q^2) - xq + yp - \frac{m}{[(x+n)^2+y^2]^{1/2}} - \frac{n}{[(x-m)^2+y^2]^{1/2}}, \tag{12.33}$$

where the p and q are the canonical momenta of the coordinates x and y, respectively. The angular momentum term $\omega(-xq + yp)$ introduces Coriolis and centrifugal forces into the equations of motion

$$\begin{aligned} \dot{x} &= p + y \quad , \quad \dot{p} = q - \frac{m(x+n)}{[(x+n)^2 + y^2]^{3/2}} - \frac{n(x-m)}{[(x-m)^2 + y^2]^{3/2}} \\ \dot{y} &= q - x \quad , \quad \dot{q} = -p - \frac{my}{[(x+n)^2 + y^2]^{3/2}} - \frac{ny}{[(x-m)^2 + y^2]^{3/2}} . \end{aligned} \tag{12.34}$$

It is easy to show that there are five equilibrium positions $\dot{x} = \dot{y} = \dot{p} = \dot{q} = 0$, the Lagrange points L_i, $i = 1, \ldots 5$. Three of them, L_1, L_2, L_3, are found on the x-axis at points satisfying

$$m \frac{x+n}{|x+n|^3} + n \frac{x-m}{|x-m|^3} = x \,. \tag{12.35}$$

The other two, L_4, L_5, appear at $(x, y) = (0.5 - n, \pm\sqrt{3}/2)$, where the three masses are placed at the corners of an equilateral triangle (see Fig. 12.35). A stability analysis shows that the collinear equilibrium points are always unstable, whereas stability of the equilateral solutions is possible if $mn \leq 1/27$, i.e. when the ratio of the masses is smaller than 0.04. In the solar system this is satisfied for Jupiter and the sun, whose mass ratio is about 10^{-3}. There is, in fact, a group of asteroids, the Trojans, which move close to an equilateral equilibrium point.

The equations of motion (12.34) can be integrated numerically by means of the program ODE. On loading the pre-computed example in file 3BODY.ODE, the reader will observe that, in order to shorten the formulae, the constants m and n have been introduced by solving the auxiliary differential equations $\mathrm{dm}/\mathrm{d}t = 0$ and $\mathrm{dn}/\mathrm{d}t = 0$ with initial conditions $n(0) = n$, $m(0) = 1 - n$.

Figure 12.36 shows for $n = 0.02$ (and therefore $m = 1 - n = 0.98$) a trajectory staying close to the equilibrium point L_4 at $(\bar{x}, \bar{y}) = (0.48, 0.866)$. The trajectory is computed for initial conditions $(x, y, p, q) = (0.44, \bar{y}, -\bar{y}, \bar{x})$ close to the (stable) equilibrium solution $(x, y, p, q) = (\bar{x}, \bar{y}, -\bar{y}, \bar{x})$ ($\Delta t_{\text{store}} = \Delta x_{\text{store}} = \Delta y_{\text{store}} = 0.05$, $\Delta p_{\text{store}} = \Delta q_{\text{store}} = 0.2$, accuracy parameters 10^{-3}).

By changing the initial conditions and the mass parameter, one can explore the intricate dynamics of the restricted three body problem using the various features of program ODE. Of interest are periodic orbits, their stability proper-

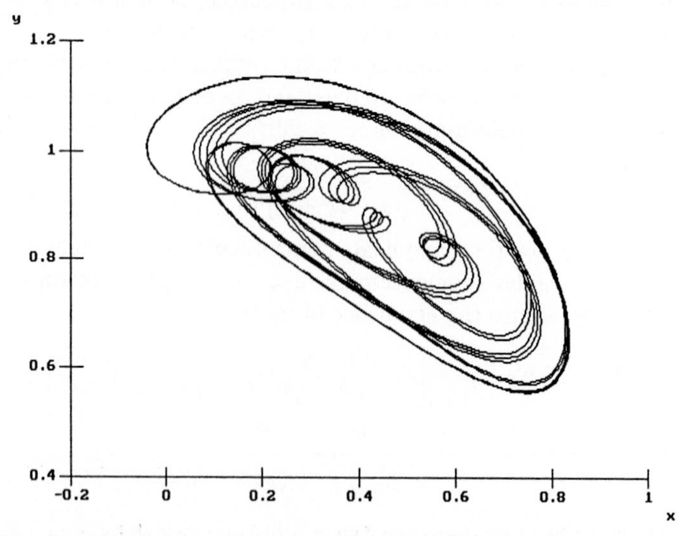

Fig. 12.36. Trajectory in the restricted three body problem staying close to the equilateral point L_4 (masses $m = 0.98$, $n = 0.02$) ($0 \leq t \leq 200$).

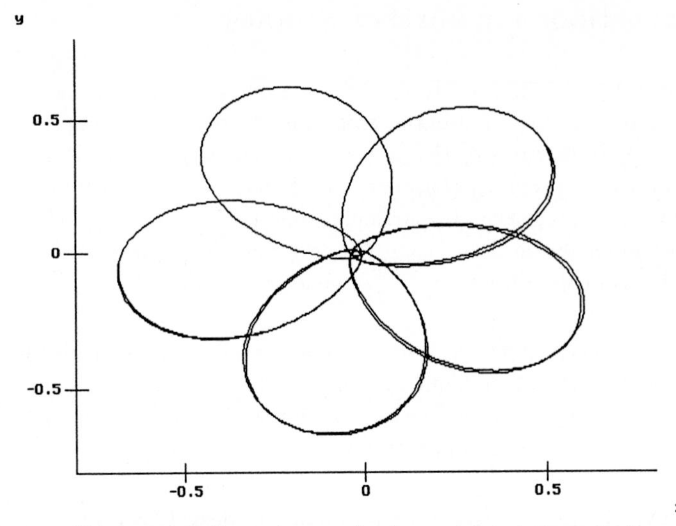

Fig. 12.37. Almost periodic trajectory in the restricted three body problem with mass $n = 0.02$ for initial conditions $(x, y, p, q) = (0.5, 0.4, 0.0, -0.3)$ $(0 \leq t \leq 10)$. This figure is stored on the disk.

ties, the conditions for bounded motion, escape orbits, etc. An example of an almost periodic orbit is shown in Fig. 12.37 (same parameters as above).

The Hamiltonian (12.33) is conserved, i.e. $H = E$, which is known as *Jacobi's constant*. This is the only constant of motion and the restricted three body problem is therefore non-integrable, showing chaotic motion. An integrable special case of some interest is, however, the so-called *two center problem*, where the heavy masses are fixed in space, i.e. $\omega = 0$. The corresponding differential equations can be obtained from (12.34) by erasing the angular momentum terms.

Furthermore, it is helpful for understanding the global restrictions on the dynamics of the restricted three body system to rewrite the Hamiltonian (12.33) as

$$H = \frac{1}{2}\left((p+y)^2 + (q-x)^2\right) + \Omega(x,y) = E \tag{12.36}$$

with

$$\Omega(x,y) = -\frac{1}{2}(x^2+y^2) - \frac{m}{[(x+n)^2+y^2]^{1/2}} - \frac{n}{[(x-m)^2+y^2]^{1/2}} \,. \tag{12.37}$$

The positivity of the kinetic term yields the constraint

$$\Omega(x,y) \leq E \tag{12.38}$$

for the allowed region in coordinate space.

More information on the three body problem of celestial mechanics can be found in the textbooks [12.20]–[12.23].

12.4 Suggestions for Further Studies

The program ODE can be used to study numerous problems in science formulated in terms of ordinary differential equations. In particular, one can use this program to explore features of the systems studied by means of the more specialized programs in previous chapters, which may be of interest, but are not supported by these programs. In particular, one may wish to export numerical data, which can easily be done using the program ODE. Here, we only give a few further hints in the direct context of chaotic dynamics.

Integrable Hamiltonian Systems are rare. For simplicity, we consider here only two-dimensional time-independent systems. Well-known are, of course, spherically symmetric cases, where the second integral of motion is the angular momentum, but there are also non-trivial cases:

(a) The Hénon–Heiles type Hamiltonian (compare Sect. 12.3.7)

$$H = \frac{1}{2}\left(p_x^2 + p_y^2 + Ax^2 + By^2\right) + x^2 y - \frac{\epsilon}{3} y^3 \tag{12.39}$$

is integrable for $A = B$, $\epsilon = 1$, where the equations of motion decouple in $x+y$ and $x-y$ variables, and — less trivial — for $\epsilon = 6$, where an explicit integral of motion is

$$x^4 + 4x^2y^2 + 4(\dot{x}y - \dot{y}x) - 4Ax^2y + (4A - B)(\dot{x}^2 + Ax^2) = const. \tag{12.40}$$

(b) Two coupled quartic oscillators:

$$H = \frac{1}{2}\left(p_x^2 + p_y^2 + Ax^2 + By^2\right) + \frac{x^4}{4} + \frac{\sigma x^4}{4} + \frac{\rho}{2} x^2 y^2 \tag{12.41}$$

for the parameters $A = B$, $\sigma = \rho = 1$, where the integral is the angular momentum

$$x\dot{y} - y\dot{x} = const., \tag{12.42}$$

and for $A = B$, $\sigma = 1$, $\rho = 3$, where the integral is

$$\dot{x}\dot{y} + xy(A + x^2 + y^2) = const. \tag{12.43}$$

These systems can be explored numerically by studying, for example, the sensitivity with respect to small deviations from integrability. More examples of integrable systems can be found in Refs. [12.24, 12.25].

Nonharmonic Forcing Functions: In most of the explicitly time-dependent systems with periodic force or periodically varying parameters, the driving terms are assumed to be harmonic with frequency $\omega = 2\pi/T$, where T is the period of the driving term. For linear systems, e.g., the forced harmonic oscillator, this is sufficient because the behavior for any non-harmonic excitation can be analyzed by Fourier analysis, expanding the driving term as well as the solution in a Fourier series. For nonlinear systems, the superposition principle is no longer valid and new effects may be observed. One can therefore study the dynamics of, e.g., the Duffing oscillator under the action of, say, a triangular force modeling, e.g., the excitation of a string-instrument (for a simple input of a triangular function see Sect. B.3).

Time Series Analysis. One often has only limited information about the true dynamics of the system. In many cases, one has only the time signal of a single physical observable at certain times. It is therefore of interest to reconstruct the full dynamics, if possible, from these data. As an example, we would like to reconstruct an attractor of the system $\mathrm{d}\mathbf{x}/\mathrm{d}t = \mathbf{F}(\mathbf{x})$ in d-dimensional phase space $\mathbf{x} = (x_1, \ldots, x_d)$ from a time series of a single component, say $x_k = x$. It has been shown [12.26] that this can be achieved in terms of the $(2d+1)$-dimensional vectors

$$\mathbf{y}(t) = (\, x(t), x(t+\tau), x(t+2\tau), \ldots, x(t+(2d+1)\tau)\,)\,. \tag{12.44}$$

In this way it is possible to regain information about the geometry of the attractor, to compute the Lyapunov exponent, etc. The time delay τ should be chosen, if possible, with some care. Too small and too large values yield little information. See, e.g. Ref. [12.11, Chap. 5.3] or [12.27] for more details.

Numerically, such an analysis of a time series can be done by studying the data of a dynamical system computed by means of the program ODE and

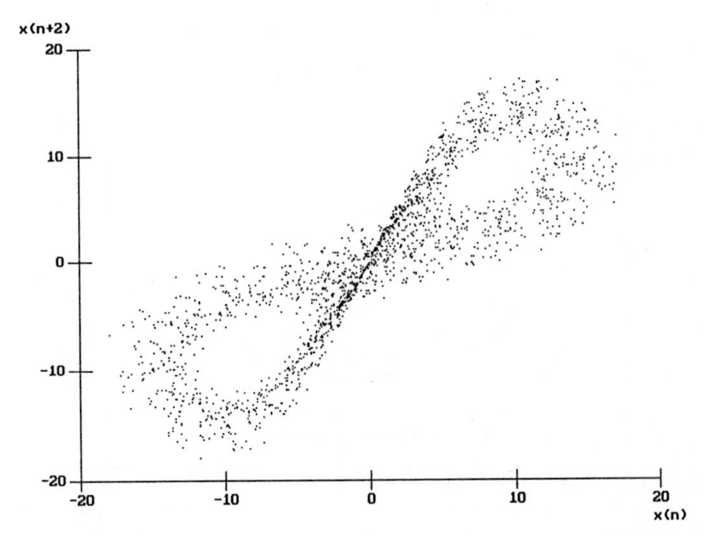

Fig. 12.38. Reconstruction of the Lorenz attractor from the time evolution of the x-component in two-dimensional $(\, x(n), x(n+2)\,) = (\, x(t_n), x(t_n+2\tau)\,)$ space.

stored on a *.TBL file. As a rather crude example of such a reconstruction, Fig. 12.38 shows the x-component of a trajectory on the Lorenz attractor shown in Figs. 12.23 and 12.24 for $0 < t < 100$ stored at time intervals $\Delta t_{\text{store}} = 0.05$. Here, the data are $x(n) = x(t_n) = x(n\Delta t)$ and plotted is the two-dimensional vector

$$(\,x(n), x(n+2)) = (x(t_n), x(t_n+\tau)\,) \tag{12.45}$$

with $\tau = 2\Delta t$. The similarity to the structure of the Lorenz attractor is evident.

van der Pol's Oscillator. The van der Pol equation, originally introduced to study oscillations in a vacuum tube unit [12.28], is given by

$$\ddot{x} + \alpha\phi(x)\dot{x} + x = \beta f(t)\,, \tag{12.46}$$

where $f(t)$ is a T-periodic forcing function, $\phi(x)$ is an even function with $\phi(x) \gtrless 0$ for $x \gtrless 0$, and $\alpha \geq 0$ and $\beta \geq 0$ are control parameters. Here, the nonlinearity enters via the frictional term. Moreover, energy is dissipated at large amplitudes and generated at low ones. The second order differential equation (12.46) can be transformed to first order form

$$\begin{aligned} \dot{x} &= y - \alpha\Phi(x) \\ \dot{y} &= -x + \beta f(t) \end{aligned} \tag{12.47}$$

with $\Phi'(x) = \phi(x)$ and $\Phi(0) = 0$. The most common choices are

$$\phi(x) = x^2 - 1\ , \quad \Phi(x) = x^3/3 - x\ , \quad f(t) = \cos\omega t\,. \tag{12.48}$$

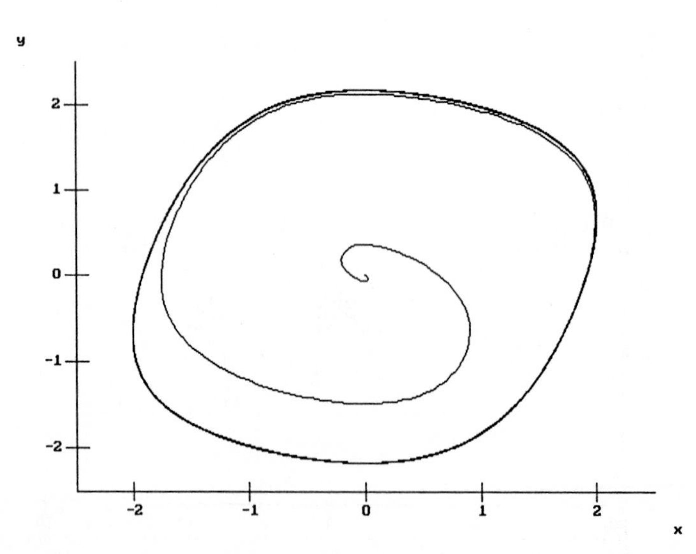

Fig. 12.39. Phase space trajectory for the unforced van der Pol oscillator (12.47) with $\Phi = x^3/3 - x$, and $\alpha = 1$ approaching a limit cycle $(\,0 \leq t \leq 400\,)$.

The van der Pol oscillator constitutes an active system, showing many interesting phenomena. Even without external forcing, the system starts to oscillate by amplifying small fluctuations, i.e., it is self-exciting and approaches a limit cycle oscillating with an internal frequency ω_0. As an example, Fig. 12.39 shows a phase space trajectory for the van der Pol equation (12.47) for the standard choice $\Phi = x^3/3 - x$, $f(t) = 0$ and $\alpha = 1$ ($\Delta t_{\text{store}} = 0.05$, $\Delta x_{\text{store}} = \Delta y_{\text{store}} = 0.1$, accuracy parameters 10^{-4}). The trajectory is started close to the unstable fixed point $x = y = 0$ and approaches a limit circle.

If a force term $f \cos \omega t$ is added, where ω is close to ω_0, the resulting periodic motion is *'entrained'* at the driving frequency ω. On increasing the detuning $\omega - \omega_0$, one observes *'combination oscillations'*, a quasiperiodic superposition of oscillations with frequency ω and ω_0,.

More information can be found in the literature, e.g., in the books on nonlinear oscillations [12.29, 12.2] or chaotic dynamics [12.30, 12.31].

Intermittency in Josephson Junctions. The study of the chaotic behavior of (quantum) Josephson junctions is of much fundamental and even practical interest [12.11]–[12.34]. Written in dimensionless form, the differential equation for the quantum phase difference, ϕ, across the junction is given by

$$\ddot{\phi} + \frac{1}{\sqrt{\beta_c}}\dot{\phi} + \sin \phi = A \sin \Omega t\,, \qquad (12.49)$$

where β_c is the so-called McCumber parameter and Ω is the (normalized) angular frequency of the driving current (see Refs. [12.35, 12.36] for details). With $u = \phi$ and $v = \dot{\phi}$, this can be written as

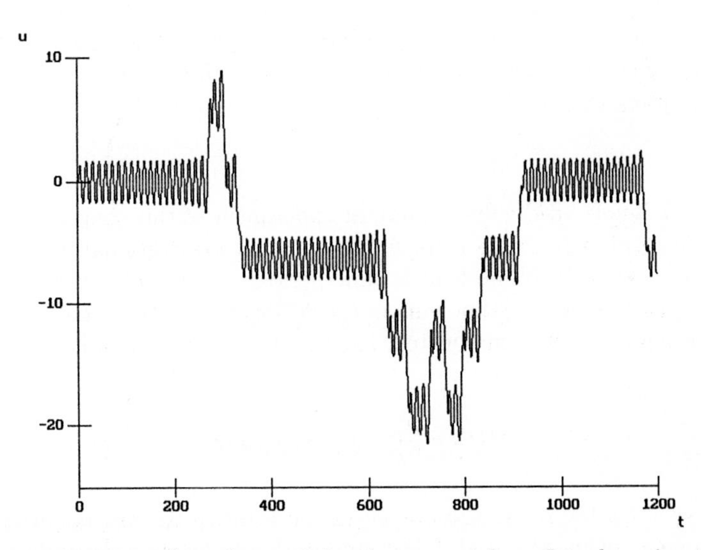

Fig. 12.40. Time dependence of the phase $u = \phi$ for a Josephson junction.

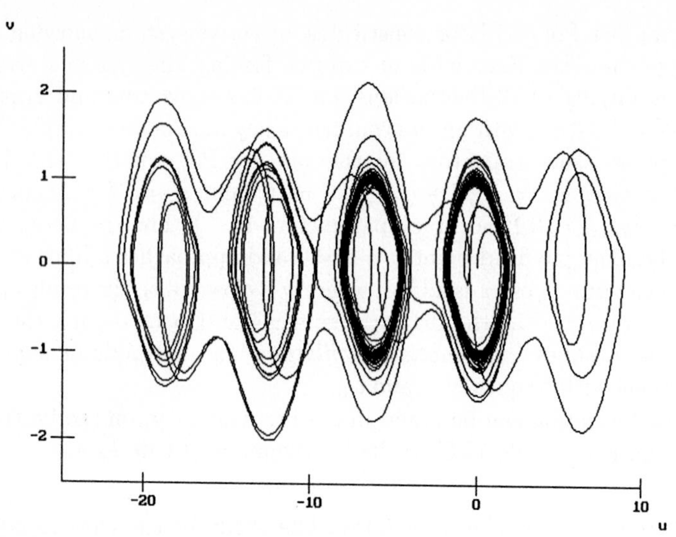

Fig. 12.41. Phase space trajectory for a Josephson junction ($0 \le t \le 1200$).

$$
\begin{aligned}
\dot{u} &= v \\
\dot{v} &= A \sin \Omega t - r\,v - \sin u
\end{aligned}
\qquad (12.50)
$$

which is used in the program ODE. As an example, Figs. 12.40 and 12.41 show the time dependence of $u(t)$ and the phase space trajectory (u, v) for parameter values $A = 0.9045$, $r = 0.5$ and $\Omega = 0.47$ ($\Delta t_{\mathrm{store}} = 0.5$, $\Delta u_{\mathrm{store}} = \Delta v_{\mathrm{store}} = 0.1$, accuracy parameters 10^{-7}). The dynamics is characterized by almost regular oscillations, interrupted by sudden 2π-jumps of the angle, i.e., irregular phase slips, which are related to turbulent voltage bursts across the junction [12.36].

The Driven Rotor. In the first computer experiment of this chapter we studied the simple pendulum, i.e., a rotor in the constant gravitational field, which is an integrable system. If the field is time-periodic, as is for instance the force on an electric or magnetic dipole in an oscillating electric or magnetic field, the dynamics is much more complicated, namely chaotic. The Hamiltonian is given by

$$
H(t) = \frac{p^2}{2I} - f\, \cos \omega t\, \cos x\,, \qquad (12.51)
$$

where x denotes the orientation angle of the rotor, p its angular momentum, I the moment of inertia, and f the amplitude of the time dependent torque $f(t) = f \cos \omega t$. The equations of motion read

$$\begin{aligned}\frac{\mathrm{d}x}{\mathrm{d}t} &= \frac{p}{I} \\ \frac{\mathrm{d}p}{\mathrm{d}t} &= -f\,\cos\omega t\,\sin x\,.\end{aligned} \tag{12.52}$$

This system, sometimes modified by including a dissipative term, models a spinning magnet in an oscillating magnetic field, i.e., a magnetic compass needle (see Ref. [12.31] for more details) or rotational excitations of molecules in microwave or laser fields. In the latter case, the system serves as a model for studying quantum chaos (see, e.g., Refs. [12.37, 12.38] and references given there).

References

[12.1] W. H. Press, B. P. Flannery, S. A. Teukolsky, and W. T. Vetterling, *Numerical Recipes* (Cambridge University Press, Cambridge 1986)

[12.2] A. H. Nayfeh and D. T. Mook, *Nonlinear Oscillations* (John Wiley, New York 1979)

[12.3] W. Magnus and S. Winkler, *Hill's Equation* (Wiley Interscience, New York 1966)

[12.4] M. Abramowitz and I. A. Stegun, *Handbook of Mathematical Functions* (Dover Publications, New York 1970)

[12.5] N. W. Ashcroft and N. D. Mermin, *Solid State Physics* (Saunders College, Philadelphia 1976)

[12.6] H. R. Lewis, *Class of exact invariants for classical and quantum time-dependent harmonic oscillators*, J. Math. Phys. **9** (1968) 1976

[12.7] E. W. Milne, *The numerical determination of characteristic numbers*, Phys. Rev. **35** (1930) 863

[12.8] J. A. Nuñez, F. Bensch, and H. J. Korsch, *On the solution of Hill's equation using Milne's method*, J. Phys. A **24** (1991) 2069

[12.9] F. Bensch, *Thesis*, (Univ. Kaiserslautern, 1993)

[12.10] E. N. Lorenz, *Deterministic nonperiodic flow*, J. Atmos. Sci. **20** (1963) 130 (reprinted in: P. Cvitanović, *Universality in Chaos* (Ada) Hilger, Bristol 1984).

[12.11] H. G. Schuster, *Deterministic Chaos* (VCH, Weinheim 1988)

[12.12] J. Frøyland, *Introduction to Chaos and Coherence* (IOP Publishing, Bristol 1992)

[12.13] O. E. Rössler, *An equation for continuous chaos*, Phys. Lett. A **57** (1976) 397

[12.14] M. Hénon and C. Heiles, *The applicability of the third integral of motion: some numerical experiments*, Astron. J. **69** (1964) 73

[12.15] M. V. Berry, *Regular and irregular motion*, in: S. Jorna, editor, *Topics in Nonlinear Dynamics*, page 16. Am. Inst. Phys. Conf. Proc. Vol, 46 1978 (reprinted in: R. S. MacKay and J. D. Meiss, *Hamiltonian Dynamical Systems* (Adam Hilger, Bristol 1987).

[12.16] M. Hénon, *Numerical exploration of Hamiltonian systems*, in: G. Iooss, H. G. Helleman, and R. Stora, editors, *Les–Houches Summer School 1981 on*

Chaotic Behaviour of Deterministic Systems, page 53, North–Holland, Amsterdam 1983

[12.17] R. May, *Simple mathematical models with very complicated dynamics*, Nature **261** (1976) 459 (reprinted in: B.-L. Hao , *Chaos* (World Scientific, Singapore) 1984) and P. Cvitanović, *Universality in Chaos* (Adam Hilger, Bristol, 1984).

[12.18] P. Manneville and Y. Pomeau, *Intermittency and the Lorenz model*, Phys. Lett. A **75** (1979) 1

[12.19] J. Frøyland and K. H. Alfsen, *Lyapunov-exponent spectra for the Lorenz model*, Phys. Rev. A **29** (1984) 2928

[12.20] C. L. Siegel and J. K. Moser, *Lectures on Celestial Mechanics* (Springer, Berlin-Heidelberg-New York 1971)

[12.21] R. Abraham and J. E. Marsden, *Foundations of Mechanics* (Benjamin, Reading 1978)

[12.22] W. E. Thirring, *Classical Dynamical Systems* (Springer, New York 1973)

[12.23] J. M. A. Danby, *Celestial Mechanics* (Willman-Bell, Richmond 1989)

[12.24] M. Tabor, *Chaos and Integrability in Nonlinear Dynamics* (John Wiley, New York 1989)

[12.25] J. Hietarinta, *Direct methods for the search of the second invariant*, Phys. Rep. **147** (1987) 87

[12.26] F. Takens, *Lecture Notes in Math. Vol.* **898** (Springer, Heidelberg-New York 1991)

[12.27] A. Wolf, J. B. Swift, H. L. Swinney, and J. A. Vastano, *Determining Lyapunov exponents from a time series*, Physica D **16** (1985) 285

[12.28] B. van der Pol and J. van der Mark, *Frequency demultiplication*, Nature **120** (1927) 363

[12.29] J. J. Stoker, *Nonlinear Vibrations* (Interscience, New York 1950)

[12.30] J. Guckenheimer and P. Holmes, *Nonlinear Oscillations, Dynamical Systems, and Bifurcations of Vector Fields. Springer, New York 1983*,

[12.31] F. C. Moon, *Chaotic Vibrations* (John Wiley, New York 1987)

[12.32] R. Graham, *Chaos in lasers*, in: E. Frehland, editor, *Synergetics – From Microscopic to Macroscopic Order* (Springer, Berlin-Heidelberg-New York 1984)

[12.33] R. Graham and J. Keymer, *Level repulsion in power spectra of chaotic Josephson junctions*, Phys. Rev. A **44** (1991) 6281

[12.34] R. Graham, M. Schlautmann, and J. Keymer, *Dynamical localization in Josephson junctions*, Phys. Rev. Lett. **67** (1991) 255

[12.35] M. Cirillo and N. F. Pedersen, *On bifurcation and transition to chaos in a Josephson junction*, Phys. Lett. A **90** (1982) 150

[12.36] W. J. Yeh and Y. H. Kao, *Intermittency in Josephson junctions*, Appl. Phys. Lett. **42** (1983) 299

[12.37] A. R. Kolovsky, *Steady-state regime for the rotational dynamics of a molecule at the condition of quantum chaos*, Phys. Rev. A **48** (1993) 3072

[12.38] N. Moiseyev, H. J. Korsch, and B. Mirbach, *Classical and quantum chaos in molecular rotational excitation by AC electric fields*, Z. Phys. D **29** (1994) 125

A. System Requirements and Program Installation

A.1 System Requirements

The programs are designed to run with following system requirements:

- CPU: 80286 processor or higher
- Operating systems: MS-DOS 3.0; Win 3.11; Win95; WinNT 4.0; OS/2
- System memory: 640K (extended memory is not supported)
- Disk space: 2.0 MBytes for executable program files and 1.0 MBytes for essential examples. Additional examples are provided in subdirectory `examples` on the disk (1.4 MBytes).
- All programs run under standard VGA resolution (640x480). The old CGA format is also supported everywhere, some other formats partly.
- Laser printers (HP-Laserjet compatible), PostScript printers, and dot matrix printers (Epson compatible) are supported for printing graphics.
- Some programs also allow output onto a plotter. Supported plotter types are Epson HI-80, HP GL, and Postscript.
- Not required, but supported by most of the programs, is a mouse (serial or bus mouse).

A.2 Installing the Programs

The programs in this collection were originally written to run within the operating system DOS. They have been tested to run under Windows 95 and NT as well. Problems still may occur; thus we do not guarantee that they will run under Windows 95 or Windows NT under all possible circumstances.

A.2.1 Installing the programs on the hard disk

The program packages in this collection are stored in a compressed archive on the installation disk. The programs must be installed on the hard disk by using the program INSTALL, which can be found on the CD.

The destination directory `<PATH>` (drive and path) must be specified, as for example `C:\CHAOS`. If the subdirectory (e.g. `\Chaos`) does not exist, it is created.

After opening a DOS window and changing to the CD ROM drive, the command

```
INSTALL <PATH>
```

will start the installation. If, for example, E: is the CD ROM drive, the command

```
E:
```

changes to the CD ROM drive and

```
INSTALL C:\CHAOS
```

installs the programs on the hard disk (drive C:) in subdirectory `\CHAOS`.

In the course of the decompression the names of all copied files are displayed.

A.2.2 Starting the programs

Executing

```
<PATH>MENU.BAT
```

provides a convenient menu for calling all programs comfortably. The programs require the full screen modus, which can be activated, e.g., under Windows 95 or Windows NT by pressing the keys ⟨ALT⟩ and ⟨ENTER⟩ simultaneously.

A.3 Programs and Files

The program collection contains the programs

- **Billiard** — *'Point mass on a billiard table'*
- **Wedge** — *'A particle jumping in a wedge under gravitational force'*
- **Dpend** — *'The double pendulum'*
- **3Disk** — *'Scattering off three disks'*
- **Fermi** — *'The Fermi-acceleration'*
- **Duffing** — *'The Duffing-oscillator'*
- **Feigbaum** — *'One-dimensional iterative maps of an interval'*

- **Chaosgen** — *'An electronic chaos-generator'*
- **Mandelbr** — *'Mandelbrot and Julia sets'*
- **ODE** — *'Ordinary differential equations'*

which are briefly described in Sect. 1.2. The essential files for running a program are listed in Table A.1. In most cases, pre-computed examples can be loaded when running a program. The names of these program files are also listed in Table A.1.

Additional examples are provided in the subdirectory `examples` on the disk. These files contain some of the examples shown in the figures in Chapters 3 – 12. A listing of these figures and program files is given in Table A.2 (note that the filename agrees with the number of the figure).

Table A.1. Program files and pre-computed examples

Program Name	Essential Program Files Suffix	Pre-computed Examples Files
BILLIARD	EXE, MSK, POY, INF	ELLIPSE.PIC, –.DAT STADIUM.PIC, –.DAT OVAL.PIC, –.DAT
WEDGE	EXE, MEN	20.PIC, –.PAR, 30.PIC, –.PAR 40.PIC, –.PAR, 46.PIC, –.PAR
DPEND	EXE, MSK, POY, INF	PEND10.POI, –.POP E100.POI, –.POP
3DISK	EXE, MSK, POY	
FERMI	EXE, MSK, POY INF	CUBIC10.FMI, SAW10.FMI, SINE10.FMI
DUFFING	EXE, INF	RES1.RES, RES2.RES, RES3.RES, FEIGB.FGB
FEIGBAUM	EXE	LOGISTIC.NLD, –.DAT
CHAOSGEN	EXE, MSK, POY, INF	BIFURC.PIC, –.DAT
MANDELBR	EXE, MSK, POY, MDB, DAT	JULIA.MDB, –.DAT, MANDLSON.MDB, –.DAT, MDLBRVGA.MDB, –.DAT, VARIANT.MDM, –.DAT
ODE	EXE, MSK, POY, INF	HENON.ODE, LORENZ.ODE

Table A.2. Pre-computed examples in subdirectory `examples` on the disk

Program	Figures	Files (Prefix)	Suffix
BILLIARD	3.9a, 3.17a	03_09A, 03_17A	–.PIC, –.DAT
WEDGE	4.17a,b,c, 4.21, 4.22, 4.23	04_17A, 04_17B, 04_17C, 04_21, 04_22, 04_23	–.PIC, –.PAR
DPEND	5.13, 5.22 5.23, 5.24	05_13, 05_22, 05_23, 05_24	–.POI, –.POP
FERMI	7.8, 7.9, 7.10 7.11, 7.18	07_08, 07_09, 07_10 07_11, 07_18	–.FMI
FEIGBAUM	9.9, 9.10, 9.15 9.20, 9.21, 9.22 9.23	09_09, 09_10, 09_15 09_20, 09_21, 09_22 09_23	–.NLD, –.DAT
MANDELBR	11.11, 11.12b,f 11.13, 11.16	11_11, 11_12B, 11_12F, 11_13, 11_16,	–.MDB, –.DAT
ODE	12.2, 12.4, 12.11 12.18, 12.28, 12.37	12_02, 12_04, 12_11 12_18, 12_28, 12_37	–.ODE

B. General Remarks on Using the Programs

B.1 Mask Menus

Many programs use masks which provide a convenient way of changing parameters and starting procedures. A mask contains options which can be selected by means of the cursor keys. There are the following types of options:

- **Menu items:** Pressing ⟨ENTER⟩ starts the selected action.
- **Status fields:** The ⟨ENTER⟩ key may
  - switch from *'on'* (normal representation) to *'off'* (inverted representation),
  - change the inscription of the field,
  - switch the current field on and others off.
- **Editable fields:** After the cursor has been placed on such a field, the whole content can be deleted by pressing ⟨ENTER⟩ or ⟨ESC⟩. It can be restored, if necessary, immediately after the deletion with ⟨→⟩ and ⟨ENTER⟩. The text can be edited by using the ⟨←⟩, ⟨→⟩, ⟨DEL⟩, and ⟨BACKSPACE⟩ keys (the last two delete characters at the cursor position or left of it, respectively). ⟨INS⟩ switches between the insert and overwrite modes. ⟨HOME⟩ and ⟨END⟩ move the cursor to the beginning or end of the text, respectively. The field is left by pressing ⟨ENTER⟩, ⟨↑⟩, or ⟨↓⟩, and the cursor jumps to the next field. Furthermore, pressing ⟨←⟩ or ⟨→⟩ at the beginning or end of the field, respectively, leaves the edit mode.

  In editable fields used for the input of numerical data, the permitted range may be restricted according to the context. Such fields can only be left after an acceptable value has been entered.

One can quit a mask menu by pressing ⟨ESC⟩ or by activating an appropriate menu item.
While a mask appears on the screen, many programs provide on-line help via the ⟨F1⟩ key. It is possible to browse through the help text using ⟨↑⟩, ⟨↓⟩, ⟨PGUP⟩, ⟨PGDN⟩, ⟨HOME⟩ or ⟨END⟩. ⟨ESC⟩ quits the on-line help function and returns to the menu.

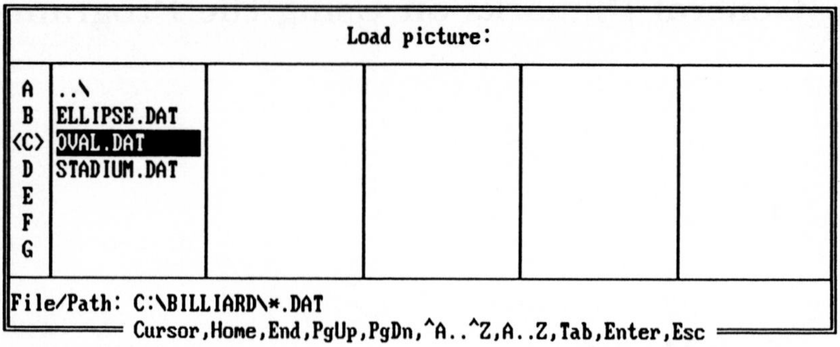

Load picture:

A ..\
B ELLIPSE.DAT
<C> OVAL.DAT
D STADIUM.DAT
E
F
G

File/Path: C:\BILLIARD*.DAT
Cursor,Home,End,PgUp,PgDn,^A..^Z,A..Z,Tab,Enter,Esc

Fig. B.1. The file-select box.

B.2 The File-Select Box

The file-select box allows one to set a file name including drive and path specification for loading and storing operations. When the box appears, it displays a list of files in the current directory, with one entry highlighted. An example is shown in Fig. B.1. It is possible to move the cursor to other entries using the cursor keys. ⟨PgUp⟩ and ⟨PgDn⟩ allow one to browse one page at a time. ⟨Home⟩ or ⟨End⟩ set the cursor to the first or last entry in the display, respectively, whereas ⟨Ctrl-PgUp⟩ or ⟨Ctrl-PgDn⟩ go to the first or last entry of the current directory.

If a letter is entered, the cursor goes to the next file name starting with this letter. If the ⟨Shift⟩ key is pressed simultaneously, the cursor is set to the corresponding sub-directory. The period key, ⟨.⟩, selects the entry ..\. If a sub-directory is selected with ⟨Enter⟩, the select box displays its contents. The ⟨\⟩ key sets the current directory to the root directory. The current drive can be changed by pressing the ⟨Ctrl⟩ key together with the name of the new drive (⟨A⟩–⟨Z⟩). ⟨Enter⟩ accepts the file selection, whereas ⟨Esc⟩ cancels.

When ⟨Tab⟩ is used, the file-select box displays an edit cursor in the bottom line. A new file name or a new file mask can be entered, but the directory path cannot be changed. The cursor keys ⟨←⟩ and ⟨→⟩ may be used to edit. ⟨Home⟩ or ⟨End⟩ set the cursor to the beginning or the end of the line, respectively. ⟨Ins⟩ switches between the insert and overwrite modes, ⟨Del⟩ deletes the character at the cursor position, and ⟨Backspace⟩ deletes the character left of the cursor. ⟨Esc⟩ or ⟨Tab⟩ reactivate the file-selection cursor, and if ⟨Enter⟩ is used the input is accepted and the file-select box terminated. If a new file mask has been entered (i.e. a file name containing the wildcards '*' and '?'), the new mask is updated by means of ⟨Enter⟩ or ⟨Tab⟩ and the file-select box returns to the file selection.

Table B.1. Mathematical operations:

( )	brackets
+ - * /	add, subtract, multiply, divide
^	exponentiation
sqr	square
sqrt	square root
abs	absolute value
sgn	sign function
ln	natural logarithm
lg	logarithm of base 10
exp	exponential function
sin, cos, tan	trigonometric functions
asn, acs, atn	inverse trigonometric functions
sinh, cosh, tanh	hyperbolic functions
asinh, acosh, atanh	inverse hyperbolic functions
Hv	Heaviside step function
fac	faculty
round	rounding to integer
trunc	truncation to integer

B.3 Input of Mathematical Expressions

Several programs permit the input of user specified functions from the keyboard. A list of the mathematical operations allowed, as well as the pre-defined mathematical functions, is given in Table B.1 . Other functions of interest can be generated by combinations of these functions, e.g. the function $f(x) = x$ modulo π can be written as

```
x+pi*(Hv(-x)-trunc(x/pi)) .
```

A rectangular pulse $f(x) = 1$ for $|x| < 1$ and zero for $|x| > 1$ is given by

```
Hv(x+1)-Hv(x-1) ,
```

and a triangular periodic oscillation of period 2π and unit amplitude can be generated by

```
2*asn(sin(x))/pi .
```

In addition, pre-defined fundamental constants, which are listed in Table B.2, can be used. An example is 'pi', which appears in the functions listed above. For the physical constants, the CODATA recommended values are chosen (E. R. Cohen and B. N. Taylor, *The 1986 CODATA recommended values of the fundamental physical constants*, J. Phys. Chem. **17** (1988) 1795).

Table B.2. Pre-defined constants:

pi	=	3.1415926535897932		π
eu	=	2.7182818284590452		Euler's number e
c	=	$2.99792458 \cdot 10^{8}$	$\mathrm{m\,s^{-1}}$	velocity of light
e	=	$1.60217733 \cdot 10^{-19}$	C	elementary charge
h	=	$6.6260755 \cdot 10^{-34}$	Js	Planck constant
hq	=	$1.05457266 \cdot 10^{-34}$	Js	$\hbar$
me	=	$9.1093897 \cdot 10^{-31}$	kg	electron rest mass
mp	=	$1.6726231 \cdot 10^{-27}$	kg	proton rest mass
mn	=	$1.6749286 \cdot 10^{-27}$	kg	neutron rest mass
a0	=	$5.29177249 \cdot 10^{-11}$	m	Bohr radius
Na	=	$6.0221367 \cdot 10^{23}$	$\mathrm{mol^{-1}}$	Avogadro constant
k	=	$1.380658 \cdot 10^{-23}$	$\mathrm{J\,K^{-1}}$	Boltzmann constant
R	=	8.314510	$\mathrm{J\,K^{-1}\,mol^{-1}}$	gas constant
E0	=	$8.854187817 \cdot 10^{-12}$	$\mathrm{C^2N^{-1}m^2}$	vacuum permittivity
u0	=	$1.2566370614 \cdot 10^{-6}$	$\mathrm{Vs\,Am^{-1}}$	vacuum permeability
f	=	$6.67259 \cdot 10^{-11}$	$\mathrm{Nm^2\,kg^{-2}}$	gravitational constant
g	=	9.80665	$\mathrm{m\,s^{-2}}$	standard value of gravity
Ry	=	$1.0973731534 \cdot 10^{7}$	$\mathrm{m^{-1}}$	Rydberg constant
ub	=	$9.2740154 \cdot 10^{-24}$	$\mathrm{J\,T^{-1}}$	Bohr magneton
LCe	=	$2.42631058 \cdot 10^{-12}$	m	Compton wavelength of electron
LCp	=	$1.32141002 \cdot 10^{-15}$	m	Compton wavelength of proton

B.4 Selection of Points or Areas in Graphics

Some programs allow a certain point on a diagram to be selected by means of the cursor keys ⟨↑⟩, ⟨↓⟩, ⟨←⟩, and ⟨→⟩. Using ⟨TAB⟩, one can control the step size of the cursor movement. ⟨ESC⟩ cancels the operation, whereas ⟨ENTER⟩ accepts the selection.

Selecting an area of a diagram ('zooming') is possible by means of a movable frame, which can be changed in size. In addition to the keys for selecting points, the active edge of the frame (i.e. the moveable edge) can be toggled with the ⟨SPACE BAR⟩ from the upper left to the lower right.

Glossary

General Terms in Chaotic and Nonlinear Dynamics

Edited by Bruno Mirbach

Cross references are *italicized.*

Almost periodic (also quasi-periodic): Motion governed by a number of discrete incommensurate frequencies. Trajectories return after a fixed time-period almost to the initial point. They are neither stable nor unstable, but neutrally stable. Almost periodic motion occurs typically in non-chaotic systems and is confined to *toroidal* surfaces in *phase space.*

Arnold tongues: In a parameter dependent, *dissipative* system the motion may become *mode-locked* at certain rational values r/s (r, s integers) of the *winding number*. The term *Arnold tongue* (named after the mathematical physicist V. I. Arnold) refers to such a locked region in parameter space. The non-locked regions typically show *fractal* structures at critical parameter values.

Attractor: A geometrical object in *phase space* towards which trajectories converge in the long-time limit. Attractors can be of various dimensionality. The simplest case of an attractor is an *equilibrium* point, also called a *fixed point* in the context of maps. *Limit cycles* are attractors of dimension one and represent a periodic motion. *Strange attractors* occur in *chaotic* systems and have *fractal dimension.*

Basin of attraction: The set of all *initial conditions* in *phase space* for which the trajectory approaches a particular *attractor*. The boundaries between basins of different attractors can be *fractal.*

Bifurcation: Rapid change in the type of dynamics when parameters in the system are varied. *Period doubling* is a typical bifurcation phenomenon.

Cantor set: A set of points on the unit interval obtained by dividing the interval into three equal segments from which the middle one is removed. If this process is repeated successively on the remaining interval segments, the set converges to an uncountable number of points forming a *fractal* of dimension smaller than one. This special set is called the 'middle-thirds Cantor set'. More generally, a Cantor set is a compact, uncountable set, which is nowhere dense. One can distinguish between Cantor sets of measure zero (such as the middle-thirds Cantor set) or of finite measure ('fat' Cantor sets). The name refers to the mathematician G. Cantor.

Chaos: Irregular (non-periodic) motion due to the *nonlinear* nature of a system. Although chaotic motion appears to be erratic or random, it is governed by *deterministic* equations of motion, in contrast to *stochastic* motion which is influenced by noise. In particular, chaos does not require high dimensionality of the system.

Chaotic: Refers to the property of a system displaying *chaos* in its time-evolution. Characteristic of chaotic motion is the sensitivity to *initial conditions*: initially close *phase space* trajectories separate exponentially in time which is measured in terms of the *Lyapunov exponent*. This limits the predictability of chaotic motion to short time intervals. Chaotic motion is characterized by a positive *Lyapunov exponent*.

Degrees of freedom: The number of independent coordinates of a system. The dimension of the *phase space* in which the dynamics of the system is described is twice the number of degrees of freedom.

Deterministic: Refers to a dynamical system whose equations of motion do not contain any random or *stochastic* parameters. The time evolution is uniquely determined by the *initial conditions*. Nevertheless, motion in such a system may appear completely random.

Difference equation: Equation that relates the value of variables at one time to the value at a previous time. Typically, a difference equation is given as $x_{n+1} = f(x_n)$, where x_n is the nth iterate of the *phase space* variable x. They appear in iterated maps and play a role similar to that of *differential equations* in systems with continuous time evolution.

Differential equation: Equation determining the dynamics of a system by relating variables of the system to their derivatives. Ordinary differential equations contain only derivatives with respect to time, whereas partial differential equations contain, in addition, derivatives with respect to the variables. If a differential equation contains derivatives up to the kth order, it is called a differential equation of kth order.

Dissipative: A dynamical system, for which the volume elements in *phase space* shrink under time evolution, is said to be dissipative. In contrast, the phase volume is conserved in *Hamiltonian systems*. Dissipation is a prerequisite for the existence of *attractors* in the system.

Duffing's equation: A second-order *differential equation* with a harmonic driving force and cubic nonlinearity. Named after the German engineer G. Duffing, who explored this system as early as 1918.

Dynamical system: A system described by time-dependent variables. The time evolution of the variables is given by a set of *differential equations* and *initial conditions*. Iterated maps described by *difference equations* can be considered as dynamical systems with discrete time variable. Dynamical systems can be controlled by parameters. One distinguishes between *deterministic* and *stochastic* systems, whereby the former ones can be *Hamiltonian* or *dissipative*.

Equilibrium: A state of a system staying constant in time. In dynamical systems this usually means a *phase space* point with zero velocity. Equilibria can be either stable or unstable. In the latter case, the slightest perturbation causes the system to move far away from the equilibrium state, whereas it stays forever in the vicinity of the equilibrium in the case of stability.

Ergodic: Motion in a dynamical system is ergodic if almost every trajectory gets arbitrarily close to any point of the *phase space* in the course of time. This definition can also be restricted to a part of the *phase space*, e.g. the energy surface. Chaotic motion is always ergodic, but the opposite does not generally hold.

Feigenbaum constant: The ratio of successive differences between *bifurcation* parameters of a *period doubling* sequence approaches a limit, the Feigenbaum constant 4.669201609... . This number is a *universal* constant for a large class of dynamical systems. The name refers to the the pioneering work by M. J. Feigenbaum.

Feigenbaum scenario: An infinite *period doubling* sequence which is a route to *chaos*. Feigenbaum proved that this transition towards a *chaotic* system is *universal* and can be characterized by the *Feigenbaum constant*.

Fibonacci sequence: An infinite sequence of integer numbers in which each term is the sum of the two previous ones, i.e. the elements x_n, $n = 0, 1, 2, \ldots$ of the sequence fulfill the recursion relation $x_{n+1} = x_n + x_{n-1}$, where the *initial conditions* are in most cases $x_0 = 0$, $x_1 = 1$. The ratio x_n/x_{n+1} of two adjacent Fibonacci numbers approaches the *golden mean* for $n \to \infty$. The properties of these numbers were first studied by the medieval mathematician Leonardo ('Fibonacci') Pisano.

Fixed point (of a mapping): A point which is mapped onto itself after one or several iterations of a map. Besides this number of iterations, the period of the fixed point, the stability is an important characteristic of a fixed point. In dynamical systems with continuous time evolution, the intersection points of a periodic orbit with a surface are fixed points of the corresponding *Poincaré map*. The *bifurcation* properties of fixed points are a widely studied phenomenon, which leads to an understanding of the transition to *chaos* (see *Feigenbaum scenario*).

Fourier transform: Transforms a time-series of some variable into a frequency spectrum. In the case of *(quasi-)periodic* motion, the Fourier transform of a time-evolved variable consists of a finite number of peaks at certain frequencies. For *chaotic* motion, the transform of any time-dependent variable shows a continuous frequency spectrum.

Fractal: Geometrical objects whose dimensionality is non-integer (see *Fractal dimension*). Characteristic for fractals is their *self-similarity*. *Strange attractors* are fractals by definition; other *invariant sets* and *basin* boundaries can also be fractal sets.

Fractal dimension: The generalization of the mathematical term 'dimension' (also 'Hausdorff dimension') to non-integer values. The fractal dimension characterizes *strange* sets (*fractals*) and can be defined by the scaling properties of the set.

Golden mean: Dividing an interval into two segments such that the ratio of the shorter segment to the longer one equals the ratio of the longer one to the total interval length yields the golden mean ratio. Its value is $(\sqrt{5}-1)/2 = 0.61803\ldots$. This 'golden number' can also be obtained as the limit of the ratio of subsequent *Fibonacci numbers*, or as the most irrational number in the unit interval, i.e. the one, which can be least well approximated by rational ones. The last property is the most important in the study of dynamical systems.

Hamiltonian system: Dynamical systems in which the equations of motion can be derived from a scalar function $H(p_1, \ldots, p_n, q_1, \ldots, q_n, t)$, the Hamiltonian function, by differentiation: $\dot{p}_i = -\partial H/\partial q_i$, $\dot{q}_i = \partial H/\partial p_i$. In contrast to *dissipative* systems, Hamiltonian systems are area-preserving. A *phase space* segment keeps its size during time evolution. Nevertheless, *chaos* typically occurs in Hamiltonian systems due to a stretching and folding of volume elements in *phase space*. The name refers to the mathematician and astronomer W. R. Hamilton.

Hénon–Heiles system : This system of two coupled *nonlinear* oscillators is one of the most intensively studied *Hamiltonian systems* exhibiting *chaotic* motion. It was first studied in 1968 by M. Hénon and C. Heiles in the context of problems in astrophysics.

Hopf bifurcation: Transition from an *equilibrium* state of a system, a point *attractor*, to a *limit cycle*. Named after the mathematician E. Hopf who studied this *bifurcation* for dynamical systems with more than two dimensions in 1942.

Initial condition: The values of the variables describing a dynamical system at a certain instant in time. In *deterministic* systems, *phase space* trajectories are uniquely determined by their initial conditions.

Instability: The characteristic of a state to change dramatically under perturbation. Instabilities appear in all kinds of dynamical systems. Whereas in regular systems only isolated states are unstable, e.g. an inverted pendulum, in *chaotic* systems motion is generally unstable.

Intermittency: A type of *chaotic* motion, where time sequences of almost periodic motion are interrupted by short bursts of random motion. The length of the regular sequences are not predictable, but also random.

Invariant set: A set which is mapped onto itself under a mapping or a continuous time evolution. An *attractor* is an invariant set by definition. In non-chaotic *Hamiltonian systems* invariant sets are toroidal surfaces in *phase space*.

Julia set: The complex map $z \to z^2 + c$, where c is a complex constant, has an *attractor* at infinity. The boundary of its *basin of attraction* is called the Julia set. It is sensitively dependent on the value of c (see *Mandelbrot set*) and shows a *fractal* structure. The French mathematician G. Julia recognized the complicated structure of this set as early as 1918.

KAM theorem: A very important theorem for *Hamiltonian systems* named after the scientists A. N. Kolmogorov, V. I. Arnold, and J. K. Moser. This theorem states that an *invariant torus* in *phase space* persists under perturbations if the *rotation number* on this *torus* is 'sufficiently irrational', otherwise it is destroyed. The *torus* withstanding the strongest perturbation is the *golden mean torus*, i.e. the *torus* with the *golden mean* value of the winding number.

Limit cycle: A frequent realization of an *attractor*. A limit cycle corresponds to a closed curve in *phase space* and represents a periodic motion of the system. An elementary well-known limit cycle is the long-time motion of a damped and harmonically forced *linear* oscillator.

Linear: A dynamical system is linear if its response to the change in a variable is proportional to the value of the variable. *Chaos*, for which an exponential amplification of changes in the system is characteristic, cannot occur in linear systems.

Logistic map: The map $x \to x(1-x)$ was originally introduced to describe the population dynamics of some species in a closed area. Now it has become a standard example of an iterated map showing *universal chaotic* properties.

Lorenz equation: A set of three first-order *differential equations* introduced by E. N. Lorenz in 1963 as a model for atmospheric convection, which turned out to exhibit *chaotic* motion.

Lyapunov exponent: Measures the exponential divergence (or attraction) in time of trajectories which stem from slightly different *initial conditions*. A positive Lyapunov exponent indicates *chaos*. Named after the dynamicist A. M. Lyapunov (also spelled Liapunov).

Mandelbrot set: The most famous and perhaps most beautiful *fractal* induced by the complex map $z \to z^2+c$, where c is a complex constant. The Mandelbrot set is the set of all parameters c for which the iteration of the origin $z = 0$ does not grow without limit. The Mandelbrot set can be considered as a one-page catalogue of all *Julia sets*. Named after the mathematician B. B. Mandelbrot.

Mode-locking: Refers to a phenomenon in *dissipative*, parameter dependent systems, where motion may approach periodic orbits with certain rational values r/s (r, s integers) of the winding number. The system is then said to be mode-locked. In parameter space, the mode-locked regions are often characteristically shaped (see *Arnold tongues*).

Nonlinear: Dynamical systems in which the response to the change in some variable is not *linear*. Nonlinearity is an inherent characteristic of most dynamical systems. Changes in the variables, e.g. due to perturbation, may be exponentially amplified in time. *Chaotic* systems are always nonlinear.

Period Doubling: One characteristic type of *bifurcation* occurring typically in *nonlinear dynamical systems* which are parameter dependent. As a parameter is varied, a *limit cycle* of a system suddenly changes into a cycle of twice the period. In a period doubling sequence this process is repeated infinitely often, where the parameter intervals between two period doublings decline exponentially (compare *Feigenbaum scenario*). Beyond a critical parameter value, the motion is *chaotic* and the bifurcating *attractor* has become a *strange attractor*.

Phase space: The space of generalized coordinates and momenta of a dynamical system. The dimension of phase space is twice the number of coordinates of the system. In phase space the time-evolution of a system can be described by first order *differential equations*. A point in phase space uniquely determines the future of a system. Therefore, different trajectories in phase space can not intersect.

Poincaré map: Mapping of an intersection point of a trajectory with a *surface of section* onto the subsequent intersection point. In this way continuous time evolution in *phase space* is reduced to a map of a lower dimensional plane onto itself. In explicitly time-dependent systems, which are T-periodic, the natural choice for a Poincaré map is the 'stroboscopic map' mapping the variables at time t onto those at $t+T$. The name refers to the mathematician and mathematical physicist H. Poincaré.

Power spectrum: The absolute amplitude square of the *Fourier transform* of a time-series of some variable. The power spectrum measures the relative importance of different frequencies.

Quasi-periodic: See 'almost periodic'.

Rayleigh–Bénard convection: A thin fluid layer heated from below in a gravitational field: for small temperature differences ΔT, heat transport occurs via uniform heat conduction. At a critical value of ΔT, a regular pattern of fluid rolls develops, showing a transition to *chaotic* motion at a second threshold.

Rotation number (also winding number): The frequency ratio ω_1/ω_2 for the regular motion on a two-dimensional *torus*. Irrational rotational numbers correspond to almost periodic motion, rational ones to periodic motion. According to the *KAM theorem*, the robustness of a *torus* under perturbation is determined by the sequence of best rational approximations to the rotation number.

Self-similarity: Property of a set whose geometrical structure is scale invariant. This means that magnification of the set by a certain scale factor reproduces the original structure of the set. The geometrical structure of such a set is contained in any of its details. *Fractals* are often self-similar objects.

Stochastic: Originally used to characterize a type of motion which is governed by random processes. Typical of stochastic motion is the independence of the future of the system from its history ('loss of memory'). Nowadays the term *stochastic* also refers to *chaotic* motion in *deterministic* systems, which show the same characteristics as random processes despite their determinism.

Strange attractor: A complex geometric object in *phase space* towards which *chaotic* trajectories move in the course of time. In contrast to other *attractors* such as equilibria or *limit cycles*, a strange attractor is a *fractal* set. The motion on a strange attractor is itself *chaotic*.

Surface of section: A (hyper)plane in d-dimensional *phase space* introduced to visualize the dynamics in a $(d-1)$-dimensional space. Only the sequence of points, where the trajectory cuts the surface of section in a specified direction are recorded (see *Poincaré map*).

Symbolic dynamics: Refers to a type of dynamics described by an infinite sequence of a finite set of symbols. Flipping a coin, for example, produces a sequence of two symbols. If every symbol can appear at any position in the sequence independently of the previous symbol, the motion is equivalent to a *stochastic* motion. This type of *deterministic* motion is referred to as *'hard chaos'*.

Torus: A closed surface in the shape of a doughnut generated by a closed loop rotated in space around an axis which does not cut the loop is called a torus. Alternatively, it can be characterized as the topological space obtained by identifying the opposite sides of a rectangle. Tori in *phase space* are the characteristic *invariant sets* in non-chaotic *Hamiltonian systems*. A trajectory following a d-dimensional torus is characterized by d fixed frequencies. Depending on their ratios, the motion on the torus is either periodic or *quasi-periodic*.

Universal property: A property of a class of dynamical systems which does not depend on the particular system. The route to *chaos* via an infinite sequence of *period doubling bifurcations*, for example, has the universal property that the bifurcation parameters are related to each other by an *universal* constant, the *Feigenbaum constant*.

van der Pol equation: A second-order *differential equation* with a *linear* restoring and a harmonic driving force, but *nonlinear* damping. The van der Pol oscillator shows self-excitation, *limit cycle attractors* and combination oscillations. Named after B. van der Pol who studied oscillations in vacuum tube circuits (1927).

Winding number: See *rotation number*.

Index

Printing: Mercedesdruck, Berlin
Binding: Buchbinderei Lüderitz & Bauer, Berlin